THE
BIOMECHANICS
OF TRAUMA

THE
BIOMECHANICS
OF TRAUMA

Edited by

Alan M. Nahum, M.D.
Professor of Surgery
Chief, Division of Head and Neck Surgery
University of California School of Medicine
San Diego, California

John Melvin, Ph.D.
Head, Biosciences Division
Transportation Research Institute
Associate Professor of Mechanical Engineering
 and Applied Mechanics
University of Michigan
Ann Arbor, Michigan

APPLETON-CENTURY-CROFTS/Norwalk, Connecticut

0-8385-0660-7

85 86 87 88 89 90 / 10 9 8 7 6 5 4 3 2 1

Prentice-Hall International, Inc., London
Prentice-Hall of Australia, Pty. Ltd., Sydney
Prentice-Hall Canada, Inc.
Prentice-Hall of India Private Limited, New Delhi
Prentice-Hall of Japan, Inc., Tokyo
Prentice-Hall of Southeast Asia (Pte.) Ltd., Singapore
Whitehall Books Ltd., Wellington, New Zealand
Editora Prentice-Hall do Brasil Ltda., Rio de Janeiro
Prentice-Hall Hispanoamerica, S.A., Mexico

Library of Congress Cataloging in Publication Data
Main entry under title:

The biomechanics of trauma.

Includes index.
1. Traumatology.   2. Human mechanics.   I. Nahum,
Alan M., 1931–   . II. Melvin, John. [DNLM:
1. Biomechanics.   2. Wounds and Injuries. WO 700 B615]
RD93.B45   1985   617′.107   84-18618
ISBN 0-8385-0660-7

PRINTED IN THE UNITED STATES OF AMERICA

For Pat, Julie, David, and Bob,
whose unqualified love and support
make all things possible

Success in . . . is like success in anything,
not available to those who don't know what's possible.

Wrinkles, by Charles Simmons

Contributors

Wayne H. Akeson, M.D.
Professor of Surgery
Chief, Division of Orthopaedic Surgery and Rehabilitation
University of California School of Medicine, San Diego
Division of Orthopaedic Surgery
Veterans Administration Medical Center
San Diego, California

David Amiel, M.D., Dit. Ing.
Director, Orthopaedic Biochemistry
Assistant Adjunct Professor of Surgery
University of California School of Medicine
San Diego, California

Robert Craig Bone, M.D., F.A.C.S.
Associate Adjunct Professor of Surgery
Division of Head and Neck Surgery
University of California School of Medicine, San Diego
Chief, Division of Head and Neck Surgery
Scripps Clinic and Research Foundation
La Jolla, California

Dennis R. Carter, Ph.D.
Director, Orthopaedic Biomechanics Program
Rehabilitation Research and Development Center
Veterans Administration Medical Center
Palo Alto, California
Associate Professor
Department of Mechanical Engineering
Stanford University
Stanford, California

Richard F. Chandler
Chief, Protection and Survival Lab
FAA-Civil Aeromedical Institute
Oklahoma City, Oklahoma

Rolf H. Eppinger, Ph.D.
Head, Biomechanics Group
Research and Development Office of Vehicle Research
National Highway Traffic Safety Administration
United States Department of Transportation
Washington, D.C.

F. Gaynor Evans, Ph.D.
Professor Emeritus
Department of Anatomy
University of Michigan Medical School
Ann Arbor, Michigan

Cyril B. Frank, M.D.
Alberta Heritage Research Fellow
Department of Surgery
Division of Orthopaedic Surgery
University of California School of Medicine, San Diego
Division of Orthopaedic Surgery
Veterans Administration Medical Center
San Diego, California

Y. C. Fung, Ph.D.
Professor of Applied Mechanics and Bioengineering
Department of Applied Mechanics and Engineering
  Sciences/Bioengineering
University of California School of Medicine
San Diego, California

Steven R. Garfin, M.D.
Assistant Professor of Surgery
Division of Orthopaedics and Rehabilitation
University of California School of Medicine
San Diego, California

David H. Gershuni, M.D., F.R.C.S. (Eng. and Edin.)
Associate Professor of Orthopaedic Surgery
University of California, San Diego
Co-Chief, Fracture Service
University of California Medical Center, San Diego

Chief, Orthopaedic Surgery
Veterans Administration Medical Center
San Diego, California

Mark A. Gomez, M.S.
Department of Surgery
University of California School of Medicine, San Diego
Veterans Administration Medical Center
San Diego, California

Jose Guerra, Jr., M.D.
Assistant Professor of Radiology
University of California School of Medicine
San Diego, California

Michael M. Katz, M.D.
Clinical Fellow
Division of Orthopaedics and Rehabilitation
University of California Medical Center
San Diego, California

Albert I. King, Ph.D.
Professor and Director, Bioengineering Center
Wayne State University
College of Engineering
Detroit, Michigan

Lawrence F. Marshall, M.D.
Associate Professor of Surgery
Division of Neurosurgery
University of California School of Medicine
San Diego, California

John W. Melvin, Ph.D.
Director of Biosciences Division
Transportation Research Institute
Project Director of Rehabilitation Engineering Center
Associate Professor of Applied Mechanics and Engineering
University of Michigan
Ann Arbor, Michigan

Harold J. Mertz, Ph.D.
Safety and Crashworthiness Systems
Current Product Engineering
General Motors Corporation
Warren, Michigan

Gopal Nagendra
Project Engineer
MacNeal Schwendler Corporation
Los Angeles, California

Ayub K. Ommaya, M.D., F.R.C.S., F.A.C.S.
Clinical Professor of Neurosurgery
George Washington University
Chief Medical Advisor
National Highway Traffic Safety Administration
United States Department of Transportation
Washington, D.C.

Brian O'Neill
Senior Vice President
Insurance Institute for Highway Safety
Washington, D.C.

Richard M. Peters, M.D.
Professor of Surgery and Bioengineering
University of California School of Medicine
San Diego, California

Frank O. Raasch, Jr., M.D.
Associate Clinical Professor
Department of Pathology
University of California, San Diego
Forensic Pathologist for the San Diego County Coroner
San Diego, California

Dennis C. Schneider, Ph.D.
Assistant Adjunct Professor of Surgery
University of California School of Medicine
San Diego, California

John M. Seelig, M.D.
Assistant Professor
Department of Surgery
Division of Neurosurgery
University of California School of Medicine
San Diego, California

Carley C. Ward, Ph.D.
Biodynamics/Engineering, Inc.
Pacific Palisades, California

Assistant Research Bioengineer
University of California School of Medicine
San Diego, California

Savio L-Y Woo, Ph.D.
Professor of Surgery and Bioengineering
Department of Surgery
University of California School of Medicine, San Diego
Veterans Administration Medical Center
San Diego, California

Contents

Preface

Improved methods of transportation, especially of the high-speed variety, have created a serious societal health problem, the deaths and injuries associated with malfunctions of the transportation system. This book is devoted to a special area of biomechanics, the mechanisms, treatment, and prevention of accidental trauma. Situated between the classic disciplines of engineering and medicine, biomechanics provides a means of applying engineering mechanics theory to the problem of accidental injury.

The text has been divided into two parts. The first part deals with basic issues in biomechanics such as restraint systems, structural properties of human tissues, and mathematical simulation of human structures. The second part considers the clinical and experimental aspects of injury to specific anatomic areas. The clinical topics have been prepared by physicians and the experimental aspects by engineers.

The book is intended to serve as an introduction to trauma biomechanics for both physicians and engineers. With an understanding of the physical properties of the human body and how traumatic injuries are produced, it should be possible to develop countermeasures that will prevent exposure or prevent or reduce injury when exposure is unavoidable.

Finally, improved methods of treatment can be developed if the injury mechanisms and human responses are understood. Only a multifaceted approach to accident prevention, injury reduction, and trauma treatment can produce an urgently needed decrease in the current deaths and injuries.

We wish to thank Sherida Bush, editorial consultant, and Mary Bartoo, planning coordinator, for their help in preparing the manuscript for publication.

THE
BIOMECHANICS
OF TRAUMA

CHAPTER 1

The Application of Biomechanics to the Understanding and Analysis of Trauma

Y. C. Fung

INTRODUCTION

Trauma is a Greek word for wound. According to *Webster's Dictionary,* it means an injury to a living body caused by the application of external force or violence. Trauma is the third greatest killer in the United States, especially of the young and vigorous. The minimization of trauma is on everybody's mind. To the medical community, the problem is critical care and management. To engineers, the problem is design for safety and protection. To police and paramedics, the question is quick transportation. To politicians, the problem is legislation. To mothers and fathers, the problem is safety education. Everybody has a role to play. Everybody has a stake. From the point of view of reducing the possibility of getting into a traumatic situation in the first place, it is a problem of culture, of lifestyle, of war and peace, of law and order. When it reaches the stage where bioengineers, physicians, surgeons, and nurses can do something about it, it is already toward the end of the line.

Biomechanics in involved in many of these stages. Biomechanics is nothing but the study of the principles of the action of forces and their effects. It is involved in processes causing traumas, in trauma management, in physiology, in pathology, in recovery, in physical therapy, in rehabilitation, in the design of vehicles, and in personal protection. A clear understanding of biomechanics in all these stages will be helpful to minimize trauma as a national and personal problem.

This chapter offers a grand tour of the whole field emphasizing the principles of mechanics, and briefly touching on topics detailed in other chapters of this book.

LOADINGS THAT CAUSE TRAUMA

Penetration by bullets and bomb fragments are well-known causes of trauma. Strong electromagnetic and heat radiations from nuclear weapons are, of course, traumatic in the extreme, and we hope that such weapons will never be used. Control of these entities requires intelligence of man beyond his technology.

Blunt impact occurs daily and is more controllable. It is our main concern in the following discussion

The Response of an Elastic System to an External Load Depends on How Fast the Load is Applied

The damage that blunt impact can do to a dynamic system depends on how fast the load is applied. To understand this statement, consider the experiment illustrated in Figures 1, 2, and 3. There is a long coil spring with a mass at its end. If a load is put on the spring slowly, the spring extends slowly; the load-deflection relationship follows a straight line defining the *static elastic behavior* of the spring, as shown in Figure 1. If a periodically varying load is put on the spring, the spring–mass system will vibrate. The amplitude of vibration will depend on the frequency of the loading. If the frequency of the load is equal to the natural frequency of free vibration of the system, the amplitude of oscillation will theoretically be infinite if there is no damping in the system. Damping usually exists, and it controls the resonance curve. Since the maximum stress and strain in the spring are proportional to the amplitude of oscillation, it can be seen that in the case of a periodically applied load the maximum stress and strain depend on the

SLOWLY APPLIED LOAD

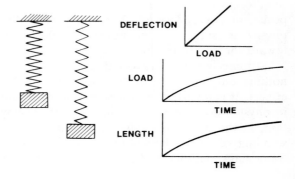

Figure 1. The load-deflection relationship of a spring when the load is applied slowly. How slow the loading must be to qualify for such a static response is discussed in the text.

frequency of the loading, and the severest case occurs at the condition of resonance.

Figure 2 shows a case in which the mass attached to the spring is subjected to an *impulsive load,* which has a time history like a half-sine curve. In this case the response of the spring depends on the ratio of the rise time of the pulse to the half-period of natural free vibration of the system. If this ratio is equal to 1, a condition analogous to resonance is obtained. Although there is insufficient repetition of the loading to achieve resonance, the response is large nevertheless. The response in this case can be presented in terms of a *response spectrum,* as shown in Figure 2, the ordinate of which is the *amplification factor,* defined as the maximum stress in the spring divided by the static elastic stress in the spring if the peak load were applied to the spring statically. The abscissa is the ratio of the rise time of the pulse to the half-period of natural vibration, or $2ft_m$ where f is the frequency (Hz) of natural vibration, t_m is the rise time of the pulse. The amplification factor is sensitive to damping. When the damping is small, the amplification factor is the largest when $2ft_m$ approaches 1. When $2ft_m$ is small ($\ll 1$), i.e., when the pulse is short compared with the period of free vibration, the amplification factor depends solely on the *total impulse* (the area under the pulse–time curve or the integral of the pulse with respect to time). When $2ft_m$ is large ($\gg 1$), the amplification factor tends to 1, i.e., the system responds as if the peak load of the pulse is applied statically.

In Figure 3 we show a case in which an impulsive load is suddenly applied to the mass attached to the spring. The suddenly arrived load causes the mass to move at a finite velocity. An elastic wave is induced in the spring. A point in the spring will not know that a load has been applied to

Airplane Landing Loads
(Drop test, force per wing, References 16 and 4)

The force shock amplification spectrum is the same as the ground acceleration spectrum for $\ddot{s}(t) = F(t)/m$. The t_m for the pulses are the time at which the peak impact force is reached in each impact.

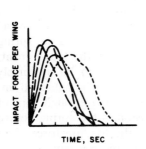 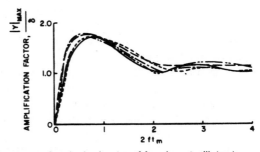

Figure 2. The *amplification spectrum* of a single degree of freedom oscillator to a set of impact loads recorded from airplane landings by a variety of airplanes and airports.

RESPONSE TO IMPULSIVE LOAD

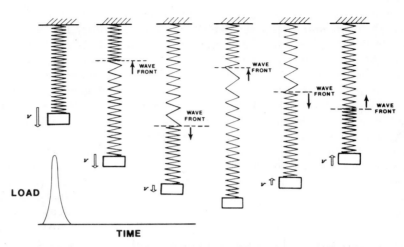

Figure 3. Elastic waves in a spring induced by a load suddenly applied to a mass attached to the end of the spring.

the mass until the elastic wave reaches it. The tension wave moves up the spring until it reaches the fixed end at the top, where the wave is reflected and the tension is doubled. The reflected wave, with double strength, moves down the spring, until it is reflected again by the mass. Since the mass is finite and is moving, the reflection this time reduces the tension and slows down the mass. (If the mass at the free end is effectively zero, a tension wave will be reflected as a compression wave of the same amplitude.) The reflected waves moving up the spring will be reflected again at the clamped end, and so forth. For a relatively small mass, the largest tensile stress in the spring usually occurs at the second reflection at the top.[1] Further reverberations will have decreased tension. Eventually, a static condition prevails.

If a succession of loads is applied on the mass shown in Figure 3, a complex system of waves will be set up in the spring.

Comparing the suddenly applied loading case shown in Figure 3 with the single-pulse case shown in Figure 2, we see that the difference lies in the fact that in Figure 3 the energy of the impulse is propagated into the spring in the form of an elastic wave, with stress localized at the wave front, whereas in Figure 2 the strain in the spring is assumed to be uniformly distributed. The hypothetical condition of Figure 2 cannot be achieved unless the loading pulse can be considered to be so slow that enough trips of the elastic waves can take place to equalize the tension in the spring before any significant change occurs in the magnitude of the loading force. The situations shown in Figure 2 and hence in Figure 1 are approximations of Figure 3 under the hypothesis that the applied pulse loading has a sufficiently slow time history.

Waves in the Body: Shock and Sound Waves

The examples given in the preceding section show that there are shocks and shocks. The response of a given elastic body to one shock can be very different to another. To a layman any sudden change in motion is a shock. We who are trying to understand trauma must classify shocks into various categories. The basis for the classification is the characteristic time of the shock as compared with the characteristic times of the body. The characteristic time of the shock is its rise time. The characteristic times of the body are its periods of natural vibrations and the time it takes for an elastic wave to traverse from one end of the body to another. Shocks are classified according to the relative magnitude of these characteristic times.

It is necessary for us to know something about elastic waves.

Sound waves are elastic waves in which incremental stress and strain caused by the wave motion are linearly related. Waves with greater amplitude may exceed the linear range of the stress–strain relationship. If the stress–strain relationship is *stiffening* (elastic modulus increases with increasing stress), *shock waves* will result. A shock wave will move at a speed faster than a sound wave of the same type. Across a shock wave the stress and strain can have a finite jump. If the stress is so high that the material will behave plastically, plastic flow results as the wave passes.

Penetration wounds suffered from high-speed bullets are likely to be caused by shock waves. For the analysis of shock waves, see the classic study by Courant and Friedrichs.[2]

Trauma due to blunt impact is usually associated with sound waves (i.e., weak shock waves). In either shock or sound wave, the energy can be carried at the wave front and concentrated in a small space, where large local stress can occur.

A well-known result of the wave theory is that in the propagation of plane waves in a linear elastic body the stress is equal to the product of three factors: the velocity of the material particles, the velocity of sound in the material, and the mass density of the material.[3] This result is true for any wave, be it earthquake, arterial pulse wave, or impact of a long bone. Since density and velocity of sound are material properties of the organ and are unaffected by the loading, we see that in an impact condition, the stress level in the tissue is controlled only by the velocity of the material particles of the tissue and is directly proportional to it. Hence, if damage is correlated with the maximum stress, the controlling factor is the particle velocity induced by the shock loading.

In this regard one may recall the famous experiment by Hopkinson,[4] who measured the strength of steel wires by impacting them with a falling weight (Fig. 4). A ball-shaped weight pierced by a hole was threaded on a wire and was dropped from a known height so that it struck a stopper attached to the bottom of the wire. For a given weight we expect a critical height beyond which the falling weight would break the wire. Using different weights dropped from different heights, however, Hopkinson obtained

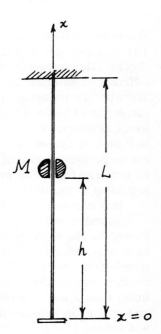

Figure 4. Hopkinson's experiment on a weight falling on a stopper supported by a metal wire. The experiment is designed to test the strength of the metal.

the remarkable result that the minimum height from which a weight had to be dropped to break the wire was nearly *independent* of the size of the weight!

For large and small weights falling from a given height, the velocity with which they strike the ground is the same. Hence, Hopkinson's result tells us that at failure the ultimate stress in the wire is proportional to the velocity at the end of the wire induced by the falling weight. If the shock loading is such that it can impart a finite velocity to the boundary of a body before the interior of the body is disturbed by elastic waves, the shock is carried by the waves into the interior. In this case the critical condition may be brought about by the elastic waves.

As an example, consider the effect on man of a detonation of explosives in air (Fig. 5). The air shock speed is faster than the velocity of sound in air (300 m/sec). The shock wave may have a speed of 400 to 600 m/sec and last a few milliseconds. When it hits a cortical bone, in which the speed of the tension wave is of the order of 3500 m/sec and that of compression wave is of the order of 2000 m/sec, the bone would see the air shock as rather slow. The stress induced in the bone can be computed as if the air shock is applied quasi-statically. On the other hand, the air shock hits the chest and imparts a certain finite velocity to the chest wall. In the chest the speed of elastic waves in the lung is of the order of 30 to 45 m/sec,[5] or roughly 10 times slower than the speed of the blast wave. Thus the lung sees the air blast as

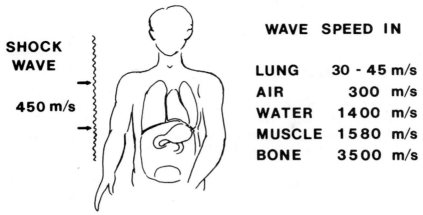

Figure 5. Speeds of shock waves in air and sound waves in lung, air, water, muscles, and bone.

very fast, and the features of elastic waves must be considered if we want to analyze the stress concentration or trauma in the lung under blast loading.

If a man is exposed to strong blast waves, his lung injury may cause edema, hemorrhage, or death. A large literature exists on this subject (see Clemedson,[6] Richmond et al.,[7] and Jönsson et al.[8]). Experience has shown that the velocity of the chest wall caused by the blast is a major parameter for defining the level of tolerance of or injury to the lung. Under this concept, Jönsson has investigated the problem of lung injury in blast wave by impacting an animal in such a way that the velocity of the chest wall is simulated.

Next, consider shocks encountered in various crash conditions of aircraft, automobiles, motorcycles, ejecting seats of supersonic planes, and in sports and stunts. For each organ we have to decide first whether we must consider elastic waves or not. If the answer is no, we must decide whether vibration will be induced. If the answer is again no, the stress distribution can be computed as if the load is applied statically.

A motor vehicle, a man, and an organ have many vibration modes and many natural vibration frequencies. The procedure to decide whether vibration is important or not can be based on the amplification spectrum in Figure 3 for each of these modes. Compare the rise time of the impact with the period of free vibration of the system in that mode. If the rise time to vibration period ratio is much smaller than 1, the stress is proportional to the total impulse (integral of force over time) of the shock. If the rise time to vibration period ratio is much larger than 1, the stress can be calculated as if the shock were applied statically. In between these two extreme conditions, there is dynamic amplification of the response. The amplification is the largest if the rise time of the load is about one quarter of a natural vibration period of the system.

FAILURE MODES OF MATERIALS

It is important to realize that the damage that can be done to an organ by a force depends not only on that force but also on the general stress condition the organ is in. Consider the following simple experiments (Fig. 6):

1. Twine is to be cut by a pair of dull scissors. I have difficulty cutting it when the twine is relaxed, but if I pull it tight and then cut it, it breaks very easily. Why?
2. I have a stalk of fresh celery and another old, dehydrated one. One breaks very easily in bending, the other does not. Practice on carrots also!
3. A balloon is inflated. Another is not inflated but is stretched to a great length. Prick them with a needle. One explodes. The other does not. Why?
4. A metal tube is filled with a liquid. Strike it on one side. It fails on the other side. Why?
5. A ball of Silly Putty bounces like rubber when you strike it hard but flows like a viscous liquid when you leave it alone. It is a typical viscoelastic material. All biologic tissues are viscoelastic.

You can easily think of the biologic analogs of these experiments. The twine is similar to a blood vessel, a tendon, a muscle. The celery is similar to an erectile organ. The balloon experiments shows the difference between the behavior of a material under uniaxial tension and one subjected to biaxial tension. Many organs of our body are subjected to biaxial tension: pericardium, pulmonary pleura, interalveolar septa of the lung, a taut skin, a diaphragm, a filled bladder, and others. Hence, it is instructive to understand what is going on in these examples.

The first example, the twine, can be understood if we postulate that the fibers in the twine break when the maximum principal tensile stress exceeds the ultimate stress. When shear is applied to the twine when it is slack, the principal tensile stress in the twine is numerically equal to the shear stress imposed. See the Mohr's circle on the lower left part of Figure 6, which is a graphic method for determining principal stresses. On the other hand, if the twine is pulled taut and then the shear is applied, the principal tensile stress is numerically equal to the initial tensile stress plus the shear stress and is therefore larger than that in the slack case (again, see the other Mohr's circle in Figure 6)—thus, the ease of cutting when the twine is taut.

In the second example, the specimen fails by bending. In the process, the fibers in the specimen fail by tension. The fibers in the fresh and plump specimen are more likely to be taut; those in the dehydrated specimen are likely to be slack. The difference in failure characteristics can be explained in a way similar to the first example. This is an interesting example. The contrast between a fresh celery and a dehydrated one is not so different from that of some tissues in vivo and in vitro, with blood perfusion and

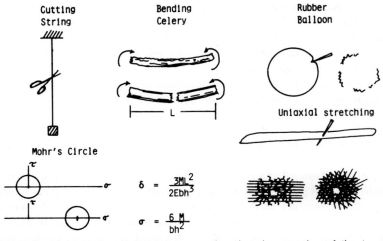

Figure 6. Several experiments demonstrating that the meaning of the term "strength" depends on the condition the specimen is in: on the initial stress or residual stress, internal fluid pressure, uniaxial versus biaxial or triaxial loading condition, and the focusing of elastic waves. See text discussion.

without, edematous or normal. The difference in mechanical property is worth remembering.

The third example is also an interesting one to remember. It shows that the material behaves ductilely under uniaxial loading but becomes brittle under biaxial or triaxial tension. The reason for the difference is illustrated in the sketches in the lower right corner of Figure 6. Rubber is a high polymer that is composed of long-chain molecules. These molecules are bent and twisted in a complex and random fashion. When the rubber membrane is stretched uniaxially, some molecules in the direction of stretching become straightened and take up the load. If a hole is now made in the membrane (as by the needle), those taut molecules in the direction of stretching will be broken, but the molecules in other directions remain bent and twisted, the hole remains a hole, and nothing dramatic happens. Consider, on the other hand, the situation of the inflated balloon. In this case the membrane is stretched in every direction. The long-chain molecules in every direction are stretched straight and taut. If a hole is made in the middle, the chains in every taut molecule intersecting the hole are broken, and an explosion results!

Biologic soft tissues are composed of collagen and elastin fibers and other long-chain molecules embedded in ground substances. The fibers and chains can be stretched when the tissue is under strain. The relevance of the example is evident.

The fourth example shows what focusing of stress waves can do. The compression wave in the fluid initiated by the impact moves to the right. The flexural wave of the metal shell also moves to the right but along the surface of the tube wall. If the flexural wave and the compression wave arrive at the other side simultaneously, a concentration of stress occurs that may exceed the ultimate stress of the materials.

The last example calls attention to viscoelasticity, which is a general property of all biologic tissues.

In examining the questions of strength and tolerance of man to impact loading, it may well pay to remember these simple examples.

STRENGTH AND TOLERANCE

A major direction of trauma research has been the determination of the strength and tolerance of living organs with respect to impact loading. It aims at understanding how trauma is produced. The principal users of this information are engineers who have to design the vehicles for transportation and government regulatory agencies who must set rules to protect the public. For these users, the most suitable form in which the results can be presented is a list of *tolerance specifications,* telling the designer the limit of acceleration or force each organ can tolerate. Physical, medical, economical, and political considerations go into the making of these specifications.

For the scientific community, a knowledge of the *tolerance level,* which is defined as the magnitude of the loading that produces a specific injury level, is more useful. To determine a tolerance level, one has to first define the injury level quantitatively and then correlate the injury level with the level of the impact force.

A study of tolerance level is very difficult. Obviously, we need a coordinated study of the clinical and experimental aspects, together with the records of the impact load and observations on recovery and rehabilitation, on the same patient. In practice, a complete set of data on a trauma of man is very difficult to obtain: it is virtually impossible. How can you reconstruct an acceleration record for a patient you receive at the trauma center? A verbal description of the accident and a look at the wreckage will not give you the acceleration record that you need. You can get an accurate acceleration record on a cadaver, but then you have no opportunity to make clinical observations. To extend the meager data that can be obtained either clinically or experimentally, research workers have constructed mathematical models. Mathematical reasoning must always be complete. Through a mathematical model you hope to connect pieces of information on anatomy, histology, physiology, pathology, medicine, surgery, and clinical observations with the vital statistics of the patient and the data on the accident, the car, the airplane, or other objects involved. The more extensive and more accurate your data are, the more specific the mathematical model can be made to match your patient or your cadaver in the accident. If the mathematical model is well formulated and validated, it will supply the missing information you may wish to have and predict the outcome for specific inputs. It will then become a tool for engineering design of vehicles. Only then can it aid the law and government in legislative, litigative, and administrative processes.

A mathematical model is a concrete embodiment of logical reasoning. It is a tool for extrapolation and interpolation. In trauma research, the need for mathematical modeling is evident. When properly simplified, the mathematical model can become a useful tool in the clinic.

There is a large literature on tolerance levels and tolerance specifications, and later chapters in this book review this literature and bring it up to date. Earlier research on trauma tended to formulate simplified statements of tolerance specifications. Recent trend tends to focus on the validation of mathematical models. Both approaches are useful and needed. The task is not done, and a coordinated national or international program of research is needed.

BIOMECHANICS IN INITIAL EVALUATION AND MANAGEMENT OF THE TRAUMA PATIENT

It is widely recognized that the handling of trauma patients needs a systems approach. Hospitals with trauma centers must provide total preparedness for any life-threatening injury. When a patient is received by the admitting team, he or she will be evaluated according to a list of priorities and treated accordingly. Cardiovascular, respiratory, and neurologic status must receive primary attention. Then all the organs must be examined. In some aspects, biomechanics works only in the background, it does not have to make its presence known. For example, basic researches on the biomechanics of circulation, respiration, and urology have led to a series of procedures and instruments, but you do not have to think of the first principles every time. Only when we think of how to improve the present methods and tools shall we think of biomechanics. On the other hand, surgeons who have to work on the patient must approach every new situation as a new problem, weigh the alternatives, and decide upon a course of action. In their weighing of the pros and cons of every alternative surgical approach, they must hold the principles of biomechanics clearly in view. This is evident in the chapters on clinical aspects of injuries to the head, face and facial bones, vertebral column, joints, chest and abdomen, and extremities.

New tools are important to trauma evaluation and management. There are many new researches going on in this field. To name a few examples:

1. High-frequency small-amplitude ventilation. With stroke volume smaller than the dead space of the lung, mixing and diffusion are so improved by high frequency that blood can be oxygenated and CO_2 removed from the lung. The prospect for clinical application is exciting.

2. Hemodilution. Artificial fluid to make up hypovolemia in traumatized patients is prepared basically on rheologic considerations.
3. The role of white blood cells in reestablishment of perfusion after stoppage of flow in muscles. Difficulties in reestablishment of blood perfusion after stoppage are thought to be caused by sticking of the white blood cells to the endothelium of the arterioles, venules, and capillaries. If this view is established, new pharmacologic treatment may follow.
4. Dissolution and aspiration of thrombus. Small-amplitude vibration is an effective way to dissolve thrombus, and dissolved thrombus can be aspirated away. Ultrasound instruments carried on catheter tips may be an instrument of the future.
5. Medical treatment of neglected trauma.
6. Surgical techniques to minimize tension across a suture.
7. Microsurgery for connecting severed limbs.
8. Transplantation of tendons or other tissues and organs.

The list can be very long. The point I want to make is that these items are part of trauma research.

RECOVERY AND REHABILITATION

When the initial crisis of a severely injured patient is over, a long-term problem of recovery and rehabilitation begins. Now we are concerned with repair, growth, and change of tissues. Since growth and change are modulated by stress and strain, it is clearly necessary to understand (or manipulate) the state of stress and strain in the patient in this period. Thus, biomechanics plays an important role.

From the point of view of thermodynamics, growth is an accumulation of matter (resorption is reduction of matter) and thus is a transport phenomenon in the nonequilibrium state. In a biologic system, matter and other forms of energy, such as heat, strain energy, electric charges, move together. The conjugate generalized forces are the chemical potential, temperature, stresses, electric potential, and so on. Simultaneous action of several irreversible processes may cause interference, typified by such phenomena as the *Soret effect* (the appearance of a concentration gradient of matter due to a temperature gradient), the *Dufour effect* (the appearance of a temperature difference when a concentration gradient exists), the *Peltier effect* (the absorption of heat at junctions of metals due to an electric current), the *Seebeck effect* (the generation of electromotive force due to the contact of different materials), the *Thompson effect* (heat absorption due to electric current and temperature gradient), and others. We expect these and other phenomena to occur in biologic systems. Growth, repair, or resorption means exchange of matter between blood and the cells or interstitial space of an organ. Involved in this process are:

1. Changes of concentrations of materials in different compartments.
2. The transport properties of various membranes, including diffusion, filtration, and active pumping characteristics.
3. The phase changes of matter, such as the polymerization of procollagen into microfibrils of collagen, microfibrils into fibers, cartilage into bone, and so on.

These processes are affected by interference of all the participating irreversible processes. The methodology of irreversible thermodynamics includes two approaches: either use statistical mechanics or use the phenomenologic approach. In the latter approach, one identifies the fluxes J_k and the conjugate forces X_k $(k = 1, 2, \ldots m)$ in such a way that the *entropy production* per unit time may be written as:

$$\sum_{k=1}^{m} J_k X_k$$

J_k may represent mass transport flux, strain rate, heat flow, electric current, chemical reaction rate, or other factors. X_k may represent the concentration gradient, stress, temperature gradient, electric potential, chemical affinity, and others. A linear *phenomenologic law* can be expressed in the form:

$$J_k = \sum_{\ell=1}^{n} L_{k\ell} X_\ell \qquad (k\ell = 1, 2, \ldots n)$$

where $L_{k\ell}$ are constants called the phenomenologic coefficients. A general theory by Onsager states that the matrix of the coefficients $L_{k\ell}$ is symmetric, i.e.:

$$L_{k\ell} = L_{\ell k} \qquad (k\ell = 1, 2, \ldots n).$$

See Fung,[3] Chapter 13, for more details. I believe, however, that for the dependence of tissue growth on stress level, the phenomenological relation is not linear, but quadratic, with a maximum at a certain optimal stress level.

The biomedical community has known for a long time that growth and resorption are affected by stresses. Wolff's law was stated in 1869.[9] Roux formulated his functional adaptation theory in 1880.[10] Orthopedists, such as Pauwel, St Krompecher, and others, used, debated, and modified Wolff's and Roux's ideas. This line of research is continued today; see, for example, Hayes and Snyder[11] and Woo.[12,13] It is undoubtedly one of the most useful lines of thought in orthopedics.

Bone healing assisted by electric field is an illustration of the interaction of mass transport and piezoelectricity. Bassett et al.,[14] Yasuda,[15] and others have promoted this method. Since all soft tissues are piezoelectric,[16] one wonders if healing of soft tissues can be promoted in a similar way.

Bone resorption by lack of exercise or immobilization is well known. A net loss of calcium was reported by Mack et al.[17] for astronauts subjected to weightlessness. Astronauts in the space lab are required to exercise against springs to keep fit.

Cowan and Crystal[18] surgically removed the left lung of a rabbit and measured the rate of growth of the right lung following the surgery. The right lung had to expand to fill the thoracic cavity and thus was subjected to a large tensile stress. These investigators showed that the collagen in the right lung at once assumed a rate of growth approaching that of a neonatal lung of the rabbit. In 2 weeks the rate of growth reached the maximum; then it slowed down to a steady state by the end of the month. This is a clear example of growth induced by tensile stress.

Bevan[19] induced hypertension in dogs by tying off a renal artery and showed that mitotic activity in the auricular artery begins soon afterwards, reaching a maximum in about 2 weeks, and subsides after a month. Such mitotic activity thus seems to be stress related.

These examples entice me to believe that if we understand fully the relationship between stress and growth, we shall obtain one more key to apply to surgery with regard to healing; to medicine with regard to diagnosis, treatment, and prognosis; to rehabilitation with regard to physical therapy, reduction of pain or deformity, and acquiring new physical skills; to physical education to obtain better health; and to sports to obtain better records.

TRENDS OF FUTURE RESEARCH

Trauma is a world problem. Trauma research aims at minimizing the ill effects of trauma. Central to the trauma research is the study of the tolerance level of various organs to impact. The best way to determine the tolerance level is to correlate clinical and pathologic observations with the maximum stresses and strains in the organs that are subjected to impact load. The determination of stresses and strains in organs in response to a given loading requires mathematical modeling. The determination of the loading acting on any organ when a man is subjected to an impact (such as in a car crash) requires mathematical modeling, too. Hence, the development and validation of mathematical modeling is an obvious trend for the future.

The use of biomechanics in the initial evaluation of an injured patient and in the management of trauma, as well as in the recovery process and in rehabilitation, often takes the form of basic research in respiration, circulation, neurology, surgery, anesthesiology, and orthopedics. Especially important and far reaching are the biomechanics of the wound healing process and of growth, resorption, and change in the tissues. Basic research on these subjects must be another trend in biomechanics.

Trauma research is expensive and difficult. Since the beneficiary of such research is all mankind, I would hope that there will be international cooperation to sponsor well-planned programs to get the needed information efficiently, economically, quickly, and with a minimum of duplication. The initial tasks to be accomplished should include at least the following:

1. Collection of anthropomorphic data of man: dimensions, mass, shape, and structure of various organs.
2. Collection of data on mechanical properties of various organs in normal conditions.
3. Data on failure modes of the tissues of these organs.
4. Correlation of clinical observations with pathologic lesions and the maximum normal and shear stresses and strains in the tissues.
5. Validation of computing programs for man and vehicle in impact, with an objective to obtain stress and strain history at any point in any organ.
6. Improved regulation for the design of motor vehicles and aircraft or other vehicles of transportation, with the safety of man as a central consideration.
7. Further research into clinical treatment of trauma patients.
8. Advancement of the art of rehabilitation of the severely injured.

Are these biomechanics? Yes! Biomechanics is in every aspect of these tasks. Good biomechanics needs a solid data base. There is no substitute for hard work in a systematic collection of the needed data. Good biomechanics leads to good computing programs, good management, good methods of treatment, and a good program of rehabilitation.

REFERENCES

1. Taylor GI: The testing of materials at high rates of loading. *J Inst Civil Eng* 1946; 26:486–519.
2. Courant R, Friedrichs KO: *Supersonic Flow and Shock Waves.* New York, Interscience, 1948.
3. Fung YC: *Foundations of Solid Mechanics.* Englewood Cliffs, NJ, Prentice-Hall, 1965.
4. Hopkinson J: *Collected Scientific Papers,* 1872, Vol II, p 316. From Taylor, Ref. 1.
5. Yen, MRT, Fung YC, Ho HH, et al.: Elastic wave speed in the lung. AMES/Bioengineering Report No. 83-2, San Diego, University of California, 1983.
6. Clemedson CJ: Blast injury, *Physiol Rev* 1956; 36:336–354.
7. Richmond DR, Damon EG, Fletcher ER, et al.: The relationship between selected blast-wave parameters and the response of mammals exposed to air blast. *Ann NY Acad Sci* 1968; 152:103–121.
8. Jönsson A, Clemedson CJ, Sundqvist AB, et al.: Dynamic factors influencing the production of lung injury in rabbits subjected to blunt chest wall impact. *Aviat Space Environ Med* 1979; 50:325–337.

9. Wolff J: Uber die innere Architektur der Knochen und ihre Bedeutung fur die Frage vom Knochenwachstum. *Arch Pathol Anat Physiol Klin Med* (*Virchows Archiv*) 1870; 50:389–453.
10. Roux W: *Gesammelte Abhandlungen über die Entwicklungs Mechanik der Organismen.* Leipzig, W. Engelmann, 1895.
11. Hayes, WC, Snyder, B: Toward a quantitative formulation of Wolff's law in trabecular bone, in Cowin SC (ed): *Mechanical Properties of Bone.* New York, American Society of Mechanical Engineers, 1981, AME Vol 45, pp 43–68.
12. Woo, SLY: The relationships of changes in stress levels on long bone remodeling, in Cowin SC (ed) *Mechanical Properties of Bone.* New York, American Society of Mechanical Engineers, 1981, AME Vol 45, pp 107–129.
13. Woo, SLY, Gomez, MA, Woo YK, et al.: Mechanical properties of tendons and ligaments: II. The relationships of immobilization and exercise on tissue remodeling. *Biorheology* 1982; 19(3):397–408.
14. Bassett AL, Pawluk RJ, Pilla, AA: Acceleration of fracture repair by electromagnetic fields. A surgically noninvasive method. *Ann NY Acad Sci* 1974; 238:242–262.
15. Yasuda I: Mechanical and electrical callus. *Ann NY Acad Sci* 1974; 238:457–464.
16. Fukada E: Piezoelectric properties of biological macromolecules. *Adv Biophys* 1974; 6:121–155.
17. Mack PB, LaChange PA, Vost GP, et al.: Bone demineralization of foot and hand of Gemini–Titan IV, V and VII astronauts during orbital flight. *Am J Roentgenol* 1967; 100:503–511.
18. Cowan MJ, Crystal RG: Lung growth after unilateral pneumonectomy: Quantitation of collagen synthesis and content. *Am Rev Respir Dis* 1975; 3:267–276.
19. Bevan RD: An autoradiographic and pathological study of cellular proliferation in rabbit arteries correlated with an increase in arterial pressure. *Blood Vessels* 1976; 13:100–218.

CHAPTER 2

The Statistics of Trauma

Brian O'Neill

INTRODUCTION

Trauma—or injury—is a major public health problem worldwide. In the United States, injuries are the fourth leading cause of death, surpassed only by heart disease, cancer, and stroke.[1] The treatment of nonfatal injuries is one of the major burdens on the health care system.

Despite the magnitude of the injury problem, only limited public attention has been directed at preventive efforts. To some extent this is because of the pervasive but erroneous belief that accidents, which produce most injuries, can be prevented only by changing human behavior. The connotations of chance, luck, and unpredictability that are now firmly associated with the word accident* continue to foster this viewpoint. In fact, injuries are *not* random and unpredictable events. Exposure to the risk of injury varies tremendously with demographic and other human factors, as well as economic, temporal, environmental, and geographic factors.

There is now a growing body of scientific literature on injury control that recognizes the importance of these and other differences affecting risk and that treats the subject of injury control with the same scientific rigor as disease control.[2-5] This, in turn, is leading to the development of effective

---

*In much scientific work these days, the descriptor, accident, is being replaced by more useful terms and concepts, such as "unintentional injury," descriptions of the injuries, and specifications of the event itself.

countermeasures for many injury problems, although the application of such measures often lags years, sometimes decades, behind their development.

Injuries result from acute exposure to agents, such as mechanical energy, heat, electricity, chemicals, and radiation. When these agents interact with the body in amounts or at rates exceeding the thresholds of human tolerance, the result is an injury. Most injuries, including almost all from motor vehicle crashes, falls, sports, shooting, and so on, are caused by mechanical energy. Other types of injury, for example, drownings and frostbite, result from the absence of essential agents, such as oxygen or heat (these have been referred to by Haddon[2] as "negative agents").

It is still not widely recognized that there are no basic scientific distinctions between injury and disease. In some cases the causal agents are identical. For example, mechanical forces cause *injury* to the spine when applied in large amounts over a short period of time, whereas smaller amounts over longer periods of time can produce lumbar disc *disease*. Primarily because the events leading to injuries are close together in time and the role of human behavior in those events usually is more obvious than in diseases, human behavior has been assumed to be more important in injury causation and prevention than in diseases. In fact, human behavior can be important to both.

Diseases are often defined in terms that describe their manifestations— e.g., cancer, hypertension, or gastroenteritis. Injuries, on the other hand, usually are classified on the basis of the events and behaviors preceding them. Fatal injuries, for example, are commonly divided into three major subdivisions—unintentional, homicide, and suicide. Although such distinctions may be useful in many instances, even when the injury-producing events are substantially different the causal agents and outcomes are the same. In illustration, the same injuries can result from both unintentional and intentional falls. Moreover, many of the basic preventive strategies are the same regardless of intent: adequate fences on high bridges can reduce unintentional and suicidal falls; improved fuel systems in U.S Army helicopters have virtually eliminated postcrash fire deaths, regardless of whether the crashes occurred during training or as a result of combat.[6] These are examples of injury control based on epidemiological approaches that are basically the same as those that long have been used successfully in controlling infectious and other diseases. Only relatively recently has it been recognized that such approaches can be applied to the control of injuries.[2,3]

INJURIES IN COMPARISON TO OTHER LEADING HEALTH IMPAIRMENTS

The number of deaths or death rates (per 100,000 population from specific causes) each year are often used to compare various health problems. In 1980, there were 160,551 deaths from injuries in the United States. Of these, 105,718 were classified as unintentional, including 53,172 deaths from motor vehicle crashes and 52,546 deaths from other unintentional injuries. The 51,147 intentional deaths consisted of 26,869 suicides and 24,278 homicides. An additional 3,686 deaths were unclassified.[1]

Injuries are the fourth leading cause of death in the United States, with 71 deaths per 100,000 population in 1980, compared to 336 for diseases of the heart, 184 for cancer, and 75 for stroke. These simple mortality comparisons—useful as they are—tend to understate the injury problem. Unlike the other leading causes of death, injuries are much more likely to kill young people. For ages 1 through 44, injuries are the *leading* cause of death, and for ages 5 through 44, they kill more people than *all* other causes combined.[1]

Measuring and comparing the impact of nonfatal health conditions is more complicated because of the substantially different outcomes that can result. For this reason, many different methods of comparison have been used, including measuring the burden that nonfatal conditions place on health care systems by the number of visits to physicians and other contacts for treatment. In 1980, injuries were the leading cause of physician contacts—99 million compared to 72 million for diseases of the heart, the second leading cause of such visits, and 64 million for respiratory diseases, the third leading cause.[7] Another measure indicating some of the load on hospital facilities is the utilization of emergency rooms or hospital clinics. More than 25 percent of all such visits in 1980 were for the treatment of injuries.[7]

Another more comprehensive, but sometimes controversial, way of comparing health problems uses the dollar cost to society. Comparing health problems in economic terms must be treated with caution because effects such as pain, grief, and family and social disruption cannot be measured in dollars. However, economic comparisons do allow such important effects as lost productivity (indirect costs to society) and the use of medical and other resources (direct costs to society) to be compared. The societal costs of *all* injuries occurring in a given year have not been computed, but estimates have been made of the costs of the major subset of injuries, those resulting from motor vehicle crashes.

Injuries in motor vehicle crashes in 1975 were estimated to have cost society almost $15 billion (more than $20 billion in 1980 dollars).[8] These costs were compared to those for cancer, coronary heart disease, and stroke and were found to be second only to the costs of cancer. The direct costs resulting from motor vehicle crash injuries were approximately twice those for coronary heart disease. Even though fewer deaths were attributed to motor vehicle crash injuries than to the other three health conditions studied, their indirect costs ranked second, principally because the average age at which injuries occur is much lower than the corresponding ages for initial onset of the three diseases.

In a more recent, noncomparative study, the National Highway Traffic Safety Administration estimated that motor vehicle crash injuries in 1980 cost society over $25 billion.[9]

DATA SOURCES

Despite the obvious importance of injuries as a public health problem, there are serious inadequacies in most of the data sources currently available to study the epidemiology of the problem.

National mortality statistics on injuries are collected by the National Center for Health Statistics (NCHS) and are based on the International Classification of Diseases (ICD) codes. ICD codes for fatal injuries are subdivided, according to the apparent "intent" of the persons involved, into three basic categories—unintentional, homicide, and suicide. The codes specify various injury types, and so-called E-codes classify the events, circumstances, and conditions related to the cause of the injury.[10] These E-codes are seriously limited, for example, they do not specifically identify work- or recreation-related deaths. In addition, it is not possible to determine the location or time of injury; only the residence of the deceased and the time of death are specified. An additional problem with the NCHS mortality data is the delay (typically 2 to 3 years) in their availability.

More detailed and consequently more useful mortality data are collected on all motor vehicle crash deaths by the National Highway Traffic Safety Administration (NHTSA). These data, which summarize fatal crashes since 1975, are maintained in a computerized file referred to as the Fatal Accident Reporting System (FARS).[11] FARS data are collected by agencies in each state government under contract to NHTSA. The sources include police accident reports, state vehicle registration files, state driver licensing files, state highway department files, vital statistics, death certificates, coroner or medical examiner reports, hospital medical reports, and emergency medical services reports. The FARS file, which contains 90 different data elements for each fatal crash, has proved invaluable to researchers concerned with reducing motor vehicle crash deaths and injuries.

Routinely collected data on nonfatal injuries are very limited. As part of the ongoing National Health Interview Survey, the National Center for Health Statistics collects some information on injuries, but the sample sizes are relatively small and the data have limited utility for epidemiological research.

The Consumer Product Safety Commission collects data on injuries and fatalities associated with consumer products (excluding automobiles). This source, called the National Electronic Injury Surveillance System (NEISS), collects data from a sample of 73 hospitals. As with FARS, considerable information is collected on the circumstance and products involved and other important facts.[12]

NHTSA recently has begun another major data collection program, the National Accident Sampling System (NASS), which is designed to produce a nationally representative sample of all motor vehicle crashes. NASS uses specially trained traffic accident investigation teams located at 50 selected sites across the country to collect its data.[13] Each team is responsible for collecting detailed information about the people, vehicles, and environment involved in a sample of motor vehicle crashes. The resulting data are much more detailed than those available from police accident reports. They include specific information on the injuries involved as well as the deformation of vehicles—information that should be invaluable in relating vehicle crash forces to the injuries sustained. NASS teams are currently investigating about 10,000 crashes annually.

A number of states have computerized files of police accident reports that are available for research. The major shortcoming of these data is the limited information available about the injuries. Only four codes (A, B, C, and K) typically are used, with K as the code for death, and A, B, C the only codes covering the entire spectrum of nonfatal injuries. Moreover, police accident data typically underestimate, sometimes by substantial amounts, the numbers of injuries that occur.[14] Despite these limitations, for some purposes police accident data are valuable, especially when relatively large sample sizes are more important than detailed injury data.

For much injury research, especially involving nonfatal injuries, there is simply no alternative to special and often expensive data collection. It is hoped that as the importance of injuries as a public health problem is more widely recognized, there will be improvements in many of the present, somewhat inadequate data sources, as well as the development of new ones.

INJURY MORTALITY AND MORBIDITY

Injuries do not result from random and unpredictable events. The populations at risk of injury from different causes vary tremendously by age, sex, income levels, and various other human characteristics, by type of environment, and by geography, to mention just a few of the major factors. Identifying these variations among population subgroups enables appropriate countermeasures to be focused on those most at risk. It is only possible in this chapter to scratch the surface of these differences. A comprehensive description of injuries by cause, demographics, economics, geography, and time is given in a recent book by Baker et al.[15]

In 1980 there were 105,718 unintentional injury deaths, and 68 million persons were injured sufficiently to require either medical attention or one or more days of restricted activity.[16] Table 1 shows the 10 leading causes of unintentional injury deaths in 1980. These causes together accounted for about 85 percent of all such deaths. More than half of these deaths resulted from motor vehicle crashes on public roads and private property.* The second leading cause of unintentional injury death was falls, which accounted for almost 13 percent of all such deaths.

The best data on the incidence of nonfatal injuries come from a 1977 study of emergency department cases in 41 of the 42 acute care hospitals in a

---

*"Nontraffic" motor vehicle crashes not shown in the table were the eleventh leading cause of unintentional injury death, accounting for 1,242 deaths in 1980.

TABLE 1. THE 10 LEADING CAUSES OF UNINTENTIONAL INJURY DEATH, 1980

Rank	Cause	Number of Deaths	Percent of Total Sample
1	Motor vehicle crashes (traffic)	51,930	49.1
2	Falls	13,294	12.6
3	Drowning	7,257	6.9
4	Fires and burns	6,016	5.7
5	Poisoning by solids and liquids	3,089	2.9
6	Firearms	1,955	1.8
7	Aspiration of food	1,943	1.8
8	Airplane crashes	1,494	1.4
9	Machinery	1,471	1.4
10	Aspiration of nonfood material	1,306	1.2
	TOTALS	105,718	84.8

five-county region of northeastern Ohio.[14] The 10 leading causes of injury treated in these hospitals are shown in Table 2.

A comparison of Tables 1 and 2 indicates major differences in the incidence of fatal and nonfatal injuries by cause. Motor vehicle crashes are by far the leading cause of unintentional injury death, but in the case of nonfatal injuries, no single cause dominates to the same extent. Falls, which are the second leading cause of unintentional injury death, are the leading cause of injuries reported to hospital emergency departments. Motor vehicle crashes, which are the leading cause of death, are the fourth leading cause of nonfatal injuries. These differences reflect the greater average severity of injuries sustained in crashes.

TABLE 2. THE 10 LEADING CAUSES OF UNINTENTIONAL INJURY REPORTED TO HOSPITAL EMERGENCY DEPARTMENTS IN FIVE NORTHEASTERN OHIO COUNTIES, 1977

Rank	Cause	Percent of Total Sample
1	Falls	25.2
2	Cutting/piercing	14.9
3	Striking against/struck by objects or caught in between objects	14.5
4	Motor vehicle crashes	12.1
5	Overexertion/strenuous movements	8.5
6	Insect/animal bite/sting	4.4
7	Foreign body entering eye or other orifice	3.3
8	Other road vehicle	2.0
9	Contact with hot or corrosive substance	1.6
10	Poisoning	1.1
	TOTAL	87.6

TABLE 3. DEATH RATES PER 100,000 POPULATION FROM MOTOR VEHICLE CRASHES AND FALLS BY AGE AND SEX, UNITED STATES 1977–1979

	Motor Vehicle Crashes		Falls	
Age	*Males*	*Females*	*Males*	*Females*
Under 1 year	6.7	6.7	2.0	1.4
1–4 years	8.8	7.2	1.1	0.7
5–9 years	10.3	6.5	0.4	0.2
10–14 years	11.1	5.9	0.6	0.1
15–19 years	63.2	23.3	1.8	0.3
20–24 years	69.8	18.4	2.5	0.3
25–29 years	46.4	11.9	2.3	0.3
30–34 years	33.5	9.3	2.3	0.3
35–44 years	28.4	9.4	3.4	0.7
45–54 years	26.8	9.7	5.8	1.8
55–64 years	25.6	10.3	8.8	3.3
65–74 years	28.4	13.1	16.8	9.1
75–84 years	43.9	17.5	55.1	44.0
85 +	47.2	12.3	199.2	167.0
ALL AGES	33.5	11.8	6.5	5.5

Rates are computed from National Center for Health Statistics mortality data for 1977–1979 and 1980 census data.

The risk of injury from particular causes varies tremendously by age and sex. Table 3 shows by age and sex the death rates per 100,000 population from motor vehicle crashes and falls, the two leading causes of unintentional injury death. For both causes, the rates are higher for males than for females in all age groups. The death rates from motor vehicle crashes peak for males in the 20 to 24 age group, and for females in the 15 to 19 age group. The rates increase again for the elderly, but they do not reach the levels of the peaks for the teenage females or young adult males. The death rates for falls, on the other hand, are highest for the elderly, with only a small peak for males aged 20 to 24 years. All death rates begin increasing after age 35 for both sexes, and they continue increasing as people get older. Persons older than 85 years have especially high death rates from falls.

Within the group of injury deaths from motor vehicle crashes, there are also significant differences by age and sex depending on the type of crash. Table 4 shows by age and sex the death rates per 100,000 population for the three major classes of fatality from motor vehicle crashes—passenger vehicle occupants, pedestrians, and motorcyclists. As before, there are differences between the sexes in most age groups for each of the fatality categories, especially in the case of motorcyclists, where the rates for females relative to those for males are especially low. Death rates for passenger vehicle occupants and motorcyclists peak for males aged 20 to 24. Occupant death rates

TABLE 4. DEATH RATES PER 100,000 POPULATION FROM MOTOR VEHICLE CRASHES BY TYPE OF FATALITY AND BY AGE AND SEX, UNITED STATES 1979–1981

Age	Passenger Vehicle Occupants		Pedestrians		Motorcyclists	
	Males	*Females*	*Males*	*Females*	*Males*	*Females*
Under 1 year	5.2	4.8	0.3	0.1	0.0	0.0
1–4 years	3.6	3.8	3.5	2.3	0.0	0.0
5–9 years	2.4	2.3	4.5	2.6	0.2	0.0
10–14 years	3.4	2.8	2.3	1.5	1.2	0.2
15–19 years	42.3	18.0	4.9	1.9	8.4	1.2
20–24 years	48.5	14.7	6.0	1.6	13.3	1.2
25–29 years	33.2	10.5	5.2	1.3	9.3	0.7
30–34 years	25.3	8.3	4.2	1.3	6.0	0.4
35–44 years	20.2	7.9	4.1	1.4	3.2	0.2
45–54 years	17.3	7.5	4.8	1.6	1.3	0.1
55–64 years	16.4	7.6	5.2	1.9	0.7	0.1
65–74 years	17.7	8.7	7.1	3.3	0.5	0.0
75–84 years	24.8	10.2	14.2	5.8	0.3	0.0
85 +	25.1	7.2	20.4	5.0	0.1	0.0
ALL AGES	22.0	8.7	5.0	2.0	4.1	0.4

Rates are computed from 1979–1981 Fatal Accident Reporting System data, which permit specific identification of passenger vehicles. Census data for 1980 were used as denominators.

peak for females aged 15 to 19. The male pedestrian death rate exhibits two peaks, one for ages 5 to 9 years, and a second, much higher peak for males older than 85 years. The pedestrian death rates for females are lower than the male rates at all ages, especially among the elderly.

Intentional injury death rates also exhibit substantial differences by age and sex. The homicide and suicide rates for males are higher than the rates for females in all age groups (Table 5). For males the homicide and suicide rates peak for the 25 to 29-year-old age group. The suicide rate begins increasing again for males older than 55 years. The female homicide rate peaks for the 20 to 24-year-old age group. The female suicide rate peaks for the 45 to 54-year-old age group.

The substantial geographic variation in passenger vehicle occupant death rates is illustrated in Figure 1. For both sexes the highest rates are in the southwestern and northwestern states. The lowest death rates are in the Northeast. The states with the *highest male* death rates are Wyoming (61 per 100,000 males), Montana (46), New Mexico (45), and Nevada (39). The states with the *highest* passenger vehicle occupant death rates for *females* are Nevada (23), Wyoming (19), Montana (19), and New Mexico (18). For both *males* and *females,* the District of Columbia has the *lowest* passenger vehicle occupant death rates (5 deaths per 100,000 for males and 2 for females).

TABLE 5. DEATH RATES PER 100,000 POPULATION FROM HOMICIDE AND SUICIDE BY AGE AND SEX, UNITED STATES 1977–1979

	Homicide		Suicide	
Age	*Males*	*Females*	*Males*	*Females*
Under 1 year	5.2	4.4	0.0	0.0
1–4 years	2.6	2.4	0.0	0.0
5–9 years	1.1	1.0	0.0	0.0
10–14 years	1.6	1.2	1.4	0.4
15–19 years	13.6	4.7	13.5	3.2
20–24 years	26.1	7.2	26.7	6.4
25–29 years	27.5	6.5	25.9	7.3
30–34 years	23.3	5.2	21.4	7.8
35–44 years	21.8	5.2	21.2	9.4
45–54 years	17.0	3.9	24.4	11.6
55–64 years	11.2	2.8	26.3	9.4
65–74 years	7.8	2.8	31.8	7.6
75–84 years	6.3	3.2	42.7	6.4
85 +	6.9	3.5	46.7	4.5
ALL AGES	14.5	4.0	18.6	6.2

This is because the District of Columbia is an urban environment, where passenger vehicle occupant deaths are less frequent than they are in rural areas. Among the states, the *lowest* passenger vehicle occupant death rates for *males* are in Rhode Island (10), New York (13), and New Jersey (14). For *females,* the same three states, New York (5), Rhode Island (5), and New Jersey (5), had the *lowest* rates.

Figure 2, pedestrian death rates by state, shows different geographic patterns than Figure 1. For both males and females, the pedestrian death rates are highest in the South and Southwest and lowest in the northern states. The *male* pedestrian death rates are *highest* in New Mexico (13), Florida (10), and Arizona (9). The *highest* rates for *females* are in Florida (4), Arizona (4), and New Mexico (3). The *lowest* rates for *males* are in Iowa (2), Nebraska (2), and North Dakota (2). For *females* the *lowest* rates are in Alaska (1), Kansas (1), and Iowa (1).

Figure 3 shows death rates from falls by state and sex. This cause of death shows yet a different geographic pattern. The highest death rates from falls occur in the northern states, especially the northcentral states. The lowest rates are in the southwestern and western states. For *males,* the *highest* death rates from falls are in the District of Columbia (15 per 100,000 population), again reflecting the fact that this is an urban environment in which fall deaths are more common. Among the states, the *highest* death rates from falls for *males* are in Massachusetts (9), North Dakota (9), and Montana (9); for *females,* in Massachusetts (10), Montana (10), and Nebraska (10). For *males* the *lowest* rates are in Delaware (4), Hawaii (5), and South Carolina (5); for *females,* in Alaska (2), Hawaii (2), and Nevada (3).

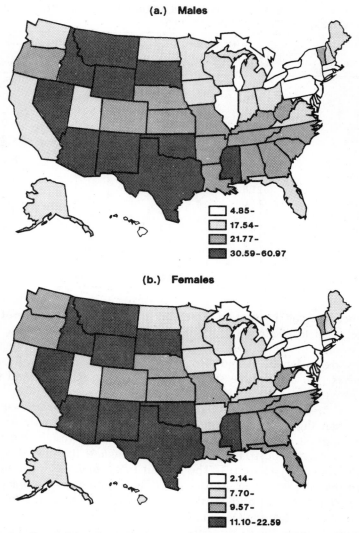

(a.) Males

☐ 4.85–
☐ 17.54–
▨ 21.77–
■ 30.59–60.97

(b.) Females

☐ 2.14–
☐ 7.70–
▨ 9.57–
■ 11.10–22.59

Figure 1. Passenger vehicle occupant death rates by state and sex, per 100,000 population, 1979–1981.

CONCLUSION

Injuries are a major burden to society, especially those involving young people. Contrary to popular belief, injuries do not result from random uncontrollable events. As this chapter illustrates, there are huge variations in injury death rates.

It is impossible within a single chapter to discuss the known variations

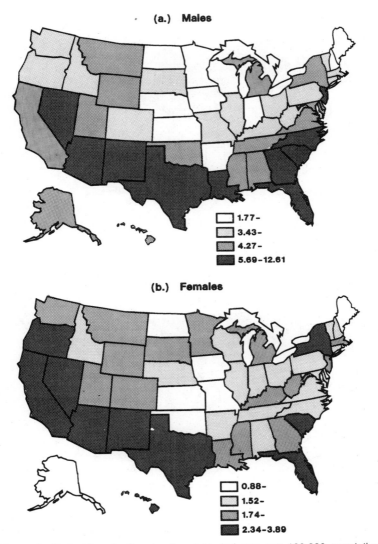

Figure 2. Pedestrian death rates by state and sex, per 100,000 population, 1979–1981.

in injury mortality and morbidity by cause, age, sex, and other relevant human factors (for example, osteoporosis), as well as the influence of economic, environmental, and geographic factors. Suffice it to say that by careful analysis of these and other factors, it is possible to identify groups that have especially high risks of sustaining particular types of injuries in order to target appropriate injury control measures.

For example, passenger vehicle occupant death rates are particularly

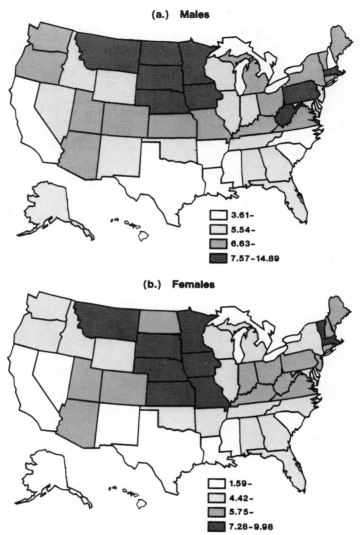

Figure 3. Death rates from falls by state and sex, per 100,000 population, 1977–1979.

high for teenagers and young adults, especially males. Yet younger people are among those least likely to wear automobile seat belts.[17] In fact, the groups of people most likely to be involved in automobile crashes, in general, also are those least likely to wear seat belts.[18] Thus, although seat belts are very effective when worn, their effectiveness as an injury control measure is limited because the high-risk groups are least likely to use them. (This finding, by the way, is true of belt use in countries where use is mandatory and 70 to 80 percent of occupants are belted. It is also true in the United States, where only about 14 percent of car occupants are belted. This illustrates the need for countermeasures that do not depend on behavior change by people in high-risk groups who have been shown to be resistant to such change.

With respect to the biomechanics of trauma, identification of high-risk population groups is a necessary prerequisite for much of the research on human tolerance to injury. DeHaven,[19] Stapp,[20] and other researchers have demonstrated that the human body can generally withstand substantial forces with little or no injury when properly packaged. There are, however, substantial biologic variations in such injury tolerance, and that is another reason why identification of high-risk groups becomes important.

Unlike the other leading causes of morbidity and mortality, there is a range of preventive measures for injury control that are effective, technologically feasible, and inexpensive in relation to potential societal benefits. Unfortunately, effective injury control measures often have not been used. Instead, ineffective efforts have been repeatedly tried—usually those aimed at changing behavior.

Injuries are increasingly being recognized as a major public health problem, one that is amenable to systematic scientific study to identify effective and ineffective countermeasures. Studying and understanding the biomechanics of trauma for the appropriate groups is an important component of this modern scientific approach. If this approach is followed, it eventually should lead to the same progress in controlling injuries that has been made this century in controlling infectious and other diseases, such as influenza/pneumonia and tuberculosis.

REFERENCES

1. National Center for Health Statistics: Advance report of final mortality statistics, 1980. *Monthly Vital Statistics Report.* August 1983, vol 32, no. 4 [Suppl].
2. Haddon W Jr: Advances in the epidemiology of injuries as a basis for public policy. *Public Health Rep* 1980; 95:411–421.
3. Haddon W Jr, Baker SP: Injury control, in Clark D, MacMahon B (eds): *Preventive and Community Medicine.* Boston, Little, Brown & Co, 1981.
4. Haddon W Jr, Suchman EA, Klein D: *Accident Research—Methods and Approaches.* New York, Harper & Row, 1964.
5. Robertson LS: *Injuries: Causes, Control Strategies, and Public Policy.* Lexington, Mass, Lexington Books, 1983.
6. Singley GT III: Army aircraft occupant crash-impact protection. *Army Research, Development and Acquisition Magazine* July–August 1981; 10–12.
7. National Center for Health Statistics: *Physician Visits: Volume and Interval Since Last Visit, United States, 1980.* Data from the National Health Survey. June 1983, Series 10, No 144.
8. Hartunian NS, Smart CN, Thompson MS: *The Incidence and Economic Costs of*

Major Health Impairments: A Comparative Analysis of Cancer, Motor Vehicle Injuries, Coronary Heart Disease and Stroke. Lexington, Mass, Lexington Books, 1981.

9. National Highway Traffic Safety Administration: *The Economic Cost to Society of Motor Vehicle Accidents.* January 1983, DOT HS 806 342.

10. Commission on Professional and Hospital Activities: *International Classification of Diseases, 9th Revision, Clinical Modification.* Ann Arbor, Mich, 1980, vol 1.

11. National Highway Traffic Safety Administration: *Fatal Accident Reporting System.* January 1982, DOT HS 806 0655.

12. Consumer Product Safety Commission: *NEISS Coding Manual.* Washington, DC, 1983.

13. National Highway Traffic Safety Administration: *The National Accident Sampling System.* September 1982, DOT HS 805 232.

14. Barancik JI, Chatterjee BF, Greene YC, et al.: Northeastern Ohio trauma study: I. Magnitude of the problem. *Am J Public Health* 1983; 73:746.

15. Baker SP, O'Neill B, Karpf R: *The Injury Fact Book.* Lexington, Mass, Lexington Books, 1984.

16. National Center for Health Statistics: *Current Estimates from the National Health Interview Survey: United States, 1980.* Data from the National Health Survey. December 1981, Series 10, No 139.

17. Williams AF, Wells JK, Lund AK: Voluntary seat belt use among high school students. *Accident Anal Prev* 1983; 15:161.

18. Williams AF, O'Neill B: *Seat Belt Laws: Implications for Occupant Protection.* No 790683. Dearborn, Mich, Society of Automotive Engineers, 1979.

19. DeHaven H: Mechanical analysis of survival in falls from heights of fifty to one hundred and fifty feet. *War Med,* 1942; and 2:539–546.

20. Stapp JP: Effects of mechanical force on living tissues: I. Abrupt deceleration and windblast. *J Aviation Med* 1955; 26:268.

CHAPTER 3

Anthropomorphic Models

Harold J. Mertz

INTRODUCTION

Anthropomorphic models are mechanical surrogates of the human body or body parts that are used to assess the potential for human injury of prescribed impact and/or acceleration environments. Such human surrogates are designed to mimic pertinent human physical characteristics (size, shape, mass, stiffness, articulation, energy dissipation) so that their mechanical responses (trajectory, velocity, acceleration, deformation) simulate human responses. Surrogate responses (head acceleration, neck load, chest acceleration and compression, leg load) are measured with transducers. Analyses of the time histories of these measurements are used to estimate the potential for various types and severities of injuries to humans, assuming that they were exposed to the same impact or acceleration environment.

There are many types and uses for anthropomorphic models. Whole-body models (commonly called "crash test dummies" or just "dummies") are used in the automotive industry to evaluate the effectiveness of occupant restraint systems of new car designs. Dummies also are used by the aircraft industry to evaluate ejection seat designs for high-speed aircraft. Anthropomorphic models of various body parts are used as subsystem test devices to evaluate the occupant protection potential of automotive steering assemblies and interior designs. Head/neck models are used by the helmet industry to assess the protective qualities of new crash helmet and sport helmet designs.

Anthropomorphic models are usually classified according to their physical size. For example, the height and weight of a 50th percentile adult male dummy (the most used dummy in automotive restraint system testing) approximates the median height and weight of the adult male population of the United States. Other adult dummy sizes used are the 5th percentile adult female dummy and the 95th percentile adult male dummy. Child dummies are classified according to the age of the child that they are to represent. Their height and weight approximate the median height and weight of children of the specified age. Three sizes of child dummies are currently used in the United States: infant, 3-year-old, and 6-year-old. A 10-year-old child dummy is available in Europe.

The efficacy of an anthropomorphic model for injury prediction is dependent on three factors: (1) the degree to which pertinent human physical characteristics are simulated (commonly referred to as biofidelity), (2) the measurement of appropriate mechanical responses, and (3) the ability to predict the likelihood of occurrence of injury types and severities based on analyses of the measured responses. A deficiency in any one of these factors will reduce the effectiveness of the anthropomorphic model as an injury-predictive surrogate. For example, if pertinent physical characteristics are not mimicked by the model, its responses to a prescribed acceleration or impact environment will not be representative of a human's response to the same environment. The credibility of any injury prediction based on the model's responses under this condition will be questionable. Obviously, if the model is not instrumented to make a measurement that can be related to a given injury or if the relationship between a measured response and associated human injury is not known, prediction of those injuries is not possible.

This chapter presents a summary of the biofidelity characteristics and injury-predictive measurement capabilities of some of the more advanced dummies and/or dummy parts used in automotive restraint system testing. Suggestions are given for improving dummy biofidelity and for additional response measurements. Correlations between injury-predictive measurements and types and severities of injuries will not be discussed. No discussion will be given to the construction of the transducers used to make the injury-predictive measurements. Both of these topics are subjects of other chapters.

WHOLE-BODY ANTHROPOMORPHIC MODELS

Most of the dummy development effort to date has been concentrated on developing 50th percentile adult male dummies for evaluating occupant protection countermeasures in frontal and side collisions. While some efforts have been made to develop an omnidirectional dummy, most of the effort has been directed toward developing separate dummies for frontal and side collision testing. These dummies are classified as "frontal impact dummies" and "side impact dummies," respectively.

Frontal Impact Dummies

Part 572 Dummy.[1-4] This is a 50th percentile adult male dummy specified by Part 572 of the Code of Federal Regulations[1] to be used for compliance testing of cars equipped with passive restraints. The Part 572 dummy is the Hybrid II dummy that was developed by General Motors in 1972 as a repeatable lap/shoulder harness test device. It was subsequently used for limited qualification testing of air cushion restraint systems.[4] The main features of the dummy are its good repeatability, durability, and serviceability. Its biofidelity is limited to its humanlike exterior shape, body weight, and range of motion of some of its articulated joints. Its response measurements are quite limited. Only orthogonal linear head and chest acceleration components and axial femoral shaft loads are measured. Because of its limited biofidelity and response measurement capabilities, the usefulness of this dummy as an injury-predicting surrogate is limited. However, the dummy does provide a basis for judging whether or not the repeatability and reproducibility of responses of other dummies are acceptable. The Part 572 dummy represents the state-of-the-art of dummy technology in the early 1970s.

Hybrid III.[4-6] The Hybrid III is a 50th percentile adult male dummy developed by General Motors in 1976.[4] The basis for the Hybrid III is the ATD 502, an advanced test dummy developed by General Motors in 1973 under a contract with the National Highway Traffic Safety Administration.[5] The ATD 502 featured a head with humanlike impact response characteristics for the hard surface forehead impacts.[7,8] A curved lumbar spine was used to achieve a more humanlike automotive seating posture. Constant-torque joints were incorporated in the knee and shoulder joints to improve repeatability and minimize the time required to set joint torques. The shoulder structure was designed to improve belt-to-shoulder interfacing, which was a problem with the Hybrid II (Part 572) shoulder design. The Hybrid III dummy retained these ATD 502 features, while design changes were made to improve the impact response biofidelity of its neck, chest, and knees. Transducers were incorporated into the Hybrid III design to measure the orthogonal linear acceleration components of the head and chest, the sagittal plane reactions (axial and shear forces and bending moment) between the head and the neck at the occipital condyles, the displacement of the sternum relative to the thoracic spine, the axial femoral shaft loads. Based on the testing of three prototype dummies, it was concluded that the Hybrid III repeatability was equivalent to that of the Part 572 dummy and that it appeared significantly more reproducible.[4]

Since the publication of the paper by Foster et al.[4] describing the Hybrid III, its response measurement capacity has been significantly increased.[6] Table 1 lists the measurement capacity of the fully instrumented Hybrid III. Note that the fully instrumented Hybrid III provides 44 response measurements for assessing occupant protection potential. The

**TABLE 1. MEASUREMENT CAPACITY OF FULLY
INSTRUMENTED HYBRID III DUMMY**

Measurement	Data Channels
Head	
Triaxial acceleration	3
Angular acceleration	1
Facial laceration	(Chamois)
Neck	
Axial load	1
Shear load	1
Bending moment	1
Chest	
Triaxial acceleration	3
Sternum acceleration	2
Deflection	1
Pelvis	
Triaxial acceleration	3
Anterior/superior iliac spine load	6
Upper Extremities	
Lower arm bending moment	4
Lower extremities	
Femur load	2
Femur/tibia translation	2
Tibia bending moments	4
Tibia axial load	2
Medial/laterial tibia plateau load	4
Lateral or fore/aft ankle bending moment and shear load	4
Knee laceration	(Chamois)
TOTAL DATA CHANNELS	44

From Mertz.[6]

only biofidelity improvements made to the Hybrid III since the Foster et al. paper have been to the knee and ankle joints. The current knee joint design allows the leg to translate relative to the thigh in a humanlike fashion.[9] The ankle joint allows lateral flexion.

The total body weight of the fully instrumented Hybrid III dummy exceeds the weight of the 50th percentile adult male by 4.5 kg. The benefits to be derived from the greatly expanded measurement capabilities of the Hybrid III dummy in terms of assessing the occupant protection potential of new car designs far exceed any shortcoming that might be associated with the increase in body mass. The Hybrid III dummy can be purchased commercially, except for the arm transducers. The dummy is used extensively by General Motors to assess the occupant protection potential of its new car designs.

Repeatable Pete.[10] This 50th percentile adult male dummy was developed by the Highway Safety Research Institute (now called University of Michigan Transportation Research Institute) in 1973 under a contract with the Motor Vehicle Manufacturers Association. Repeatable Pete features head, neck, and chest structures with humanlike impact response characteristics for a prescribed set of frontal impact conditions.[10,11] The head has been designed to have humanlike impact response for lateral impacts as well.[12] The dummy has flexible thoracic and lumbar spines that allow it to be placed in a humanlike, automotive seating posture. Constant-torque joints are used for major limb joints. The dummy is instrumented to measure orthogonal linear acceleration components of the head and chest, fore/aft chest compression, and axial femoral shaft loads. Based on the results of tests conducted with two prototype dummies, the repeatability and reproducibility were comparable to or better than the Part 572 dummy.[3-5] This is not a commercially available dummy.

OPAT Dummy.[13-15] The OPAT (Occupant Protection Assessment Test) dummy was developed by David Ogle Ltd. (a British dummy manufacturer) and MIRA (the Motor Industry Research Association of Britain) under a contract with the British government's Department of the Environment, Transport and Road Research Laboratory (TRRL) in 1972. The dummy was to be representative of the 50th percentile adult male in size and weight and was to provide humanlike behavior when used to evaluate lap–shoulder belt systems. The dummy features a humanlike clavicle and floating scapula design, and its rib cage mimics the shape of the human. These features make the OPAT dummy particularly useful in evaluating lap–shoulder belt systems. The chest structure has humanlike impact response for blunt frontal impacts. The dummy is equipped to measure orthogonal linear acceleration components of its head and chest and the axial loads in its femurs. Its repeatability is comparable to that of the Part 572 dummy. The dummy is commercially available.

Sophisticated Sam.[16,17] Sophisticated Sam was an experimental, frangible dummy that was developed by General Motors and Sierra Engineering (a dummy manufacturer) in the late 1960s. This 50th percentile adult male dummy featured frangible clavicles, humeri, radii, ulnas, femurs, tibiae, fibulae, and patellae. These structures were designed to fracture at the same static breaking load as their respective human counterpart. Dynamic testing of the dummy indicated that these design requirements were not sufficient to assure humanlike fracture response for more complex loading environments. Further development of a frangible dummy has not been pursued by these developers.

Side Impact Dummies

APROD Dummy.[18-21] APROD stands for Association Peugeot-Renault Omnidirectional Dummy. This dummy is a Part 572 dummy that was modified by Peugeot-Renault for lateral impact testing. Extensive modification was done to the chest and shoulder structure to provide humanlike response characteristics for lateral impacts.[18] The 1982 version of this dummy features head, neck, shoulder, chest, and abdominal structures with humanlike response characteristics. This version of the dummy incorporates transducers to measure the triaxial acceleration of the head, chest, and pelvis, the lateral displacements of the upper and lower rib sections relative to the thoracic spine, and go/no-go switches in the abdomen to indicate when injurious abdominal penetration has occurred. Impact performance of the dummy is discussed in papers by Maltha and Janssen[20] and Morgan et al.[21] Peugeot-Renault has an active side impact dummy development project, and the APROD dummy is continually updated with more desirable features.

DOT/SID Dummy.[20-24] DOT/SID stands for Department of Transportation/Side Impact Dummy. This is also a Part 572 dummy that has been modified for lateral impact testing by University of Michigan Transportation Research Institute under a contract with the National Highway Traffic Safety Administration of the Department of Transportation.[22,23] Subsequent modifications have been made by CALSPAN.[24] This dummy features a unique chest structure design that uses a hydraulic shock absorber to resist lateral loads and provides humanlike impact response. The shoulder and arm structures were eliminated, since the investigators felt that they interfered with obtaining repeatable responses and because cadaver test results indicated that these structures were not substantially involved in transmitting lateral impact loads. The thorax incorporates an array of 12 accelerometers to monitor the kinematic behavior of the ribs and spine during impact.[22] Triaxial accelerometers are located in the head and pelvis as well. Impact performance of the dummy is discussed by Maltha and Janssen[20] and Morgan et al.[21] A final design specification is being prepared by the National Highway Traffic Safety Administration.

MIRA Dummy.[20,25] This dummy is also a Part 572 dummy that was modified by the Motor Industry Research Association of Britain for lateral impact testing. The main features of MIRA dummy are its human-shaped rib cage and articulated shoulder structure. Transducers are provided to measure the individual loading of each rib. The dummy has a unique pelvis structure that is instrumented to measure triaxial loads applied to the ilium and lateral loads applied to the acetabulum and symphysis pubis.

ONSER Dummy.[20,26] This is a side impact dummy developed by the Organisme National de Securite Routiere (ONSER) in France. The ONSER dummy features a human-shaped thorax and flexible shoulder design. A transducer is incorporated to measure the lateral displacement of the simulated rib cage surface relative to the thoracic spine. The lateral impact performance of this dummy is summarized by Maltha and Janssen.[20]

ANTHROPOMORPHIC MODELS OF BODY PARTS

There are numerous anthropomorphic models of various parts of the body.The more germane models, with respect to biofidelity of response and injury-predictive measurement capabilities, are described in Tables 2 through 11. Each table gives a brief description of the design of the body part, its biofidelity attributes and deficiencies, and its injury-predictive measurement capabilities. References are given for those interested in obtaining more detailed information. Body parts for many of the frontal and side impact dummies are described in these tables.

DISCUSSION

Biofidelity
Most anthropomorphic models mimic the total weight and size of their human counterparts quite well because these human characteristics are easy to determine and to incorporate into the model design. In contrast, mass distribution is seldom mimicked. This is because metals are usually used for structural aspects of the models in order to provide adequate durability for testing of severe impact environments. It may be possible to obtain models with humanlike mass distribution and good durability if carbon filament and/or glass fiber composites are used for major structural elements instead of metals.

There is a need to greatly expand the impact-response requirements that are used to judge model biofidelity. In most cases, current requirements are specified for limited impact conditions. Anthropomorphic models designed to meet such limited conditions may not mimic human response over the broad range of impact conditions that can occur in automobile collision environments.

A major obstacle in specifying additional impact-response requirements for model development is obtaining those characteristics for the human. Current practice is to specify impact-response requirements for injurious impact environments. Since human volunteers cannot be subjected to such environments, indirect methods must be used to develop impact-response requirements.

The most frequently used method is to use human cadaver test results. There are two concerns with using cadaver data. First, different methods are used to infer how a live human would respond under similar impact conditions. Some investigators[18,20] have assumed that the cadaver and living

(*text continues on page 57*)

TABLE 2. ANTHROPOMORPHIC HEAD MODELS

Part 572 Head[1-4]

Description
  Aluminum shell covered by vinyl skin with facial features
  Trixial accelerometer package located at center of gravity

Biofidelity attributes
  Exterior size and shape of 50th percentile adult male

Biofidelity deficiencies
  Head-to-neck attachment location not humanlike
  Mass distribution not humanlike
  Center of gravity location not humanlike
  Hard surface impact response not humanlike

Injury-predictive measurements
  Facial laceration prediction possible if chamois covering used
  Brain injury and/or skull fracture predictions based on linear acceleration measurements
    poor for hard surface impact because impact response is not humanlike

Hybrid III Head[4-8]

Description
  Aluminum shell covered by constant thickness vinyl skin over cranium
  Vinyl facial features
  Triaxial accelerometer package located at center of gravity
  Sagittal plane angular acceleration measured

Biofidelity attributes
  Exterior size and shape of 50th percentile adult male
  Humanlike head-to-neck attachment location
  Humanlike mass and sagittal plane mass moment of inertia
  Humanlike response for deforming surface impacts
  Humanlike response for hard surface forehead impacts

Biofidelity deficiencies
  Mass moment of inertia may not be humanlike for other than sagittal plane
  Hard surface impact response may not be humanlike for side, top, and rear of head. *Note:*
    constant skin construction should give humanlike response, but response has not been
    evaluated

Injury-predictive measurements
  Facial laceration prediction possible if chamois covering used
  Brain injury and/or skull fracture predictions based on linear acceleration measurements
    possible for deforming surface impacts to front, top, back and side of head and for hard
    surface impacts to forehead
  Brain injury prediction due to sagittal plane, rotational effects possible

Repeatable Pete Head[10,12]

Description
  Solid skull of Uralite 3121 covered by a softer layer of Uralite 3110
  Cut out through rear of head for mounting triaxial accelerometer package
  No facial features

Biofidelity attributes
  Exterior size and shape of caphalic aspect of head that of 50th percentile adult male
  Humanlike response for deforming surface impacts
  Humanlike response for hard surface forehead and side of head impacts

TABLE 2. Continued

Biofidelity deficiencies
  Head-to-neck attachment location not humanlike
  Head mass too great
  Center of gravity location too low relative to top of head

Injury-predictive measurements
  Brain injury and/or skull fracture predictions based on linear acceleration measurement
    possible for deforming surface impacts to front, top, back, and side of head and for hard
    surface impacts to forehead

OPAT Head[13,14]

Description
  Aluminium shell covered by plastic skin with facial features
  Triaxial accelerometer package located at center of gravity

Biofidelity attributes
  Exterior size and shape of 50th percentile adult male

Biofidelity deficiencies
  Hard surface impact response not humanlike

Injury-predictive measurements
  Facial laceration prediction possible if chamois covering used
  Brain injury and/or skull fracture predictions based on linear acceleration measurements
    poor for hard surface impact because impact response is not humanlike

ITOH Head—3DGM-AM50-73[27,28]

Description
  Aluminum shell covered by plastic skin with facial features
  Triaxial accelerometer package located at center of gravity

Biofidelity attributes
  Exterior size and shape of 50th percentile adult male

Biofidelity deficiencies
  Hard surface impact response not humanlike

Injury-predictive measurements
  Facial laceration prediction possible if chamois covering used
  Brain injury and/or skull fracture predictions based on linear acceleration measurements
    poor for hard surface impact because impact response is not humanlike

Hodgson Head[29]

Description
  Self-skinning urethane foam skull covered by a silicon rubber skin with facial features
  Cranial cavity filled with silicon rubber
  Triaxial accelerometer package located at center of gravity

Biofidelity attributes
  Exterior head size chosen to be in the 7⅛ to 7¼ has size range since head model was
    developed for helmet testing
  Humanlike impact response when impacted through helmet

Biofidelity deficiencies
  Impact response characteristics without helmet unknown

Injury-predictive measurements   (*continued*)

TABLE 2. Continued

Brain injury and/or skull fracture predictions based on linear acceleration measurements possible for helmeted impacts

GMR Frangible Head[30]
Description
Polyester, glass fiber reinforced skull covered by rubber soft tissue simulation and a nylon fiber reinforced rubber skin
Cranial cavity filled with silicon rubber

Biofidelity attributes
Humanlike fracture loads for zygoma and frontal bone

Biofidelity deficiencies
Acceleration impact response unknown
Inertial properties may not be humanlike

Injury-predictive measurements
Zygoma and frontal bone fractures

APR Face Model[31]
Description
Aluminum shell with crushable metal honeycomb facial area
Facial area covered with artificial skin consisting of plastic foam covered by silicon rubber
Triaxial accelerometer package located at center of gravity

Biofidelity attributes
Exterior size and shape of 50th percentile adult male
Humanlike acceleration response for facial impacts
Humanlike lacerative response for facial skin covering

Biofidelity deficiencies
Hard surface impact response for forehead impacts not humanlike

Injury-predictive measurements
Facial laceration prediction possible
Facial bone fracture prediction possible from measured head acceleration and deformed facial honeycomb
Brain injury and/or skull fracture predictions based on linear acceleration measurements poor for hard surface forehead impacts

Load-Sensing Faceform[32]
Description
Part 572 head modified with four load-sensing elements in facial region: nose, both zygoma−suborbital regions and maxilla−mandible area
Triaxial acceleration package located at center of gravity

Biofidelity attributes
Exterior size and shape of 50th percentile adult male

Biofidelity deficiencies
Same as Part 572 head
Acceleration response not humanlike for facial impacts

Injury-predictive measurements
Same as Part 572 head
Facial bone fracture prediction possible based on measured facial loads

TABLE 3. ANTHROPOMORPHIC NECK MODELS

Part 572 Neck[1,4,33]

Description
　Monolithic butyl rubber cylinder with braided wire cable through center and attached to endplates

Biofidelity attributes
　None

Biofidelity deficiencies
　Bending response not humanlike

Injury-predictive measurements
　None

Hybrid III Neck[4]

Description
　One-piece structure consisting of four asymmetric butyl rubber segments bonded to thin aluminum disks and two endplates. Braided wire cable passing through center of neck and attached to endplates. Top endplate is a single-pivot nodding joint
　Load cell at nodding joint measures sagittal plane shear and axial forces and bending moment

Biofidelity attributes
　Humanlike fore/aft bending response

Biofidelity deficiencies
　Lateral bending response may not be humanlike
　May be too stiff in axial compression

Injury-predictive measurements
　Measured neck loads provide basis for predicting neck injuries

GMR Neck[33,34]

Description
　Split ball-jointed structure with four asymmetric polymeric elements. Single-pivot nodding joint at top
　Load cell at nodding joint measures sagittal plane shear and axial forces and bending moment

Biofidelity attributes
　Humanlike fore/aft bending response

Biofidelity deficiencies
　Lateral bending response may not be humanlike
　May be too stiff in axial compression

Injury-predictive measurements
　Measured neck loads provide basis for predicting neck injuries

Repeatable Pete Neck[10,11]

Description
　Universal jointed structure with three asymmetric butyl rubber elements

Biofidelity attributes
　Humanlike fore/aft bending response

(*continued*)

TABLE 3. Continued

Biofidelity deficiencies
    Lateral bending response may not be humanlike
    May be too stiff in axial compression

Injury-predictive measurements
    None

OPAT Neck[13,14]

Description
    Monolithic rubber cylinder

Biofidelity attributes
    None

Biofidelity deficiences
    Bending response not humanlike

Injury-predictive measurements
    None

APROD 82 Neck[18,19]

Description
    Modified Hybrid III design using three asymmetric polymeric elements instead of four.
      Nodding joint replaced with spherical joint. Spherical joint added to base of neck
    No load transducer

Biofidelity attributes
    Humanlike fore/aft and lateral bending response

Biofidelity deficiencies
    May be too stiff in axial tension and compression

Injury-predictive measurements
    None

ITOH Neck[27,28]

Description
    Single rigid link attached to head with pin joint and to torso with rubber ball joint

Biofidelity attributes
    Humanlike head-to-torso trajectory

Biofidelity deficiencies
    Bending response not humanlike

Injury-predictive measurements
    None

TABLE 4. ANTHROPOMORPHIC THORACIC MODELS FOR FRONTAL LOADING

Part 572 Thorax[1,4,35]

Description
  Six steel ribs connected to rigid steel spine and leather sternum. Damping material attached
    to ribs. Rib cage covered by vinyl skin
  Triaxial accelerometer package located in thoracic spine

Biofidelity attributes
  Exterior size and shape of 50th percentile adult male
  Humanlike mass

Biofidelity deficiencies
  Rib cage not humanlike in shape
  Mass distribution not humanlike
  Force deflection response not humanlike

Injury-predictive measurements
  Thoracic injuries related to gross thoracic acceleration

Hybrid III Thorax[4,6]

Description
  Similar construction as Part 572, except rib size and dampening material selected to give
    humanlike force deflection response for distributed sternal loading
  Triaxial accelerometer package located in thoracic spine
  Transducer to measure sternal-to-spine motion
  Sternal accelerometer

Biofidelity attributes
  Exterior size and shape of 50th percentile adult male
  Humanlike mass
  Humanlike impact response for fore/aft compression due to distributed sternal impacts

Biofidelity deficiencies
  Rib cage geometry not humanlike
  Mass distribution not humanlike
  Lateral impact response not humanlike
  Concentrated load impact response not humanlike

Injury-predictive measurements
  Thoracic injuries related to gross thoracic acceleration, sternal acceleration, and gross
    fore/aft thoracic compression

Repeatable Pete Thorax[10,36]

Description
  Six steel ribs molded in urethane shell attached to three-segment steel spine. The steel
    spine segments are isolated from each other with butyl rubber pads. A steel cable holds
    spine together
  Triaxial accelerometer
  Displacement transducer to measure fore/aft chest compression

Biofidelity attributes
  Exterior size and shape of 50th percentile adult male
  Humanlike mass
  Humanlike impact response for fore/aft compression due to distributed sternal impacts

(continued)

TABLE 4. Continued

Biofidelity deficiencies
 Rib cage geometry not humanlike
 Mass distribution not humanlike
 Lateral impact response may not be humanlike
 Concentrated load impact response not humanlike

Injury-predictive measurements
 Thoracic injuries related to gross thoracic acceleration and fore/aft chest compression due to distributed sternal loads

OPAT Thorax[13-15]

Description
 Nine flexible steel ribs attached to steel spine and plastic sternum. Unsupported ribs simulated. Lead sheets interleaved with polyurethane foam used to simulate internal organs
 Triaxial accelerometer

Biofidelity attributes
 Exterior size and shape of 50th percentile adult male
 Humanlike rib cage geometry
 Humanlike mass
 Humanlike impact response for fore/aft compression due to distributed sternal impacts

Biofidelity deficiences
 Mass distribution not humanlike
 Lateral impact response may not be humanlike
 Concentrated load impact response not humanlike

Injury-predictive measurements
 Thoracic injuries related to gross thoracic acceleration

Ford Thorax[37]

Description
 A deformed rib cage simulation composed of three aluminum beams, each attached to two triaxial load cells and covered with foam and rubber jacket. Load cells mounted to steel spine

Biofidelity attributes
 Size and shape of rib cage simulates that of a deformed human chest due to belt loading
 Humanlike mass

Biofidelity deficiencies
 Energy dissipation due to shoulder belt loading not humanlike
 Mass distribution not humanlike

Injury-predictive measurements
 Thoracic and clavicle injuries due to shoulder belt loading related to load cell outputs.

APROD Thorax[18-20]

Description
 Six steel ribs, three each attached to laterally displacing piston rods
 Leather sternum
 Lateral displacement transducer
 Rigid steel spine
 Triaxial accelerometer package located in thoracic spine

TABLE 4. Continued

Biofidelity attributes
    Exterior size and shape of 50th percentile adult male
    Humanlike lateral force-displacement response
    Humanlike fore/aft force-displacement response only with special thin ribs. (Not demon-
        strated in references listed)

Biofidelity deficiencies
    Rib cage geometry not humanlike
    Mass distribution not humanlike
    Concentrated fore/aft load impact response not humanlike
    Humanlike lateral and fore/aft impact response not demonstrated in same design
    Oblique force-displacement response not humanlike

Injury-predictive measurements
    Thoracic injuries related to gross thoracic acceleration
    Thoracic injuries related to lateral chest compression

MIRA Thorax[25]

Description
    Six flexible steel ribs per side attached to rigid steel spine and flexible sternum
    Each rib instrumented to sense frontal and lateral loads
    Triaxial accelerometer located in spine

Biofidelity attributes
    Exterior size and shape of 50th percentile adult male
    Humanlike rib cage geometry
    Humanlike fore/aft and lateral force-deflection response

Biofidelity deficiencies
    Mass distribution not humanlike
    Oblique force-deflection response may not be humanlike

Injury-predictive measurements
    Thoracic injuries related to rib loads and/or deflections
    Thoracic injuries related to gross thoracic accelerations

Sierra Slanted Rib Thorax[38]

Description
    Five berillium copper ribs per side encased in a viscoelastic material
    Ribs free to rotate at attachment with rigid steel spine
    Leather sternum
    Transducer to measure sternal displacement relative to spine

Biofidelity attributes
    Humanlike fore/aft sternal motion

Biofidelity deficiencies
    Exterior size and shape not humanlike
    Oblique force-deflection response may not be humanlike
    Mass distribution not humanlike

Injury-predictive measurements
    Thoracic injuries related to sternal compression

TABLE 5. ANTHROPOMORPHIC THORACIC MODELS FOR LATERAL LOADING

APROD Thorax[18-21]

Description
  Six steel ribs, three each attached to laterally displacing piston rods that deform rubber
    elements contained in piston housings
  Leather sternum
  Rigid steel spine
  Lateral displacement transducer
  Triaxial accelerometer package located in thoracic spine

Biofidelity attributes
  Exterior size and shape of 50th percentile adult male
  Humanlike lateral force-displacement response
  Humanlike fore/aft force-displacement response only with special thin ribs. (Not demon-
    strated in references listed)

Biofidelity deficiencies
  Rib cage geometry not humanlike
  Mass distribution not humanlike
  Humanlike lateral and fore/aft impact response not demonstrated in same design
  Oblique force-displacement response not humanlike

Injury-predictive measurements
  Thoracic injuries related to lateral chest compression
  Thoracic injuries related to gross thoracic acceleration

DOT/SID Thorax[21-24]

Description
  Five interconnected steel ribs attached to steel spine through flexible coupling
  Shock absorber used to provide lateral displacement resistance
  Lead masses attached to ribs
  Accelerometers mounted to upper and lower ribs and upper and lower spine

Biofidelity attributes
  Exterior size and shape of 50th percentile adult male
  Humanlike lateral rib acceleration impact response

Biofidelity deficiencies
  Rib cage geometry not humanlike
  Static stiffness not humanlike
  Mass distribution not humanlike

Injury-predictive measurements
  Thoracic injuries related to upper and lower impacted rib accelerations
  Thoracic injuries related to upper and lower thoracic spine accelerations

TRRL Thorax[39,40]

Description
  Four steel ribs per side, each mounted to steel spine via load transducers
  Triaxial accelerometer located in spine

Biofidelity attributes
  Exterior size and shape of 50th percentile adult male

Biofidelity deficiencies
  Lateral load-deflection response not humanlike
  Mass distribution not humanlike

Injury-predictive measurements
  Thoracic injuries related to rib loads

TABLE 5. Continued

Thoracic injuries related to gross thoracic acceleration

MIRA Thorax[20,25]

Description
  Six flexible steel ribs per side attached to rigid steel spine and flexible sternum
  Each rib instrumented to sense lateral load applied to rib
  Triaxial accelerometer located in spine

Biofidelity attributes
  Exterior size and shape of 50th percentile adult male
  Humanlike rib cage geometry
  Humanlike lateral and fore/aft force-deflection response

Biofidelity deficiencies
  Mass distribution not humanlike
  Oblique force-deflection response may not be humanlike

Injury-predictive measurements
  Thoracic injuries related to rib loads and/or deflections
  Thoracic injuries related to gross thoracic acceleration

ONSER Thorax[20,26]

Description
  Foam used for rib cage with steel spine
  Transducer to measure lateral deflection
  Triaxial spine accelerometer

Biofidelity attributes
  Exterior size and shape of 50th percentile adult male
  Humanlike rib cage geometry
  Humanlike lateral force-deflection response

Biofidelity deficiencies
  Mass distribution not humanlike
  Oblique frontal force-deflection response may not be humanlike

Injury-predictive measurements
  Thoracic injuries related to rib cage deflections
  Thoracic injuries related to gross thoracic acceleration

Ford SIBB Thorax[41]

Description
  Laterally guided, padded rib cage mass that reacts with steel spine through a spring damper
  Load cell to measure force transmitted to spine
  Accelerometers on rib cage mass and spinal mass
  Displacement transducer to measure motion of rib mass relative to spine

Biofidelity attributes
  Humanlike lateral force-deflection responses

Biofidelity deficiencies
  Rib cage geometry not humanlike
  Mass distribution not humanlike

Injury-predictive measurements
  Injuries due to gross spine or rib accelerations
  Injuries due to gross lateral rib cage compression

TABLE 6. ANTHROPOMORPHIC LUMBAR SPINE MODELS

Part 572 Lumbar[1,4]

Description
 Circular cylinder made of rubber with braided steel cable passing through the axis and attached to circular aluminum endplates

Biofidelity attributes
 None

Biofidelity deficiencies
 Bending response not humanlike
 Does not provide humanlike sitting posture

Injury-predictive measurements
 None

Hybrid III Lumbar[4,5]

Description
 Circular curved rubber segment with two braided steel cables attached to endplates to provide lateral bending stability

Biofidelity attributes
 Provides humanlike sitting posture

Biofidelity deficiencies
 Bending response not humanlike

Injury-predictive measurements
 None

Repeatable Pete Lumbar[10]

Description
 Rubber bar with braided steel cable used to attach spine to thorax and pelvis

Biofidelity attributes
 Provides humanlike sitting posture

Biofidelity deficiencies
 Bending response not humanlike

Injury-predictive measurements
 None

TABLE 7. ANTHROPOMORPHIC ABDOMEN MODELS

Part 572 Abdomen[1]

Description
  Polyurethane foam-filled rubber bladder

Biofidelity attributes
  None

Biofidelity deficiencies
  Force-penetration response not humanlike

Injury-predictive measurements
  None

TNO Abdomen[42]

Description
  Rubber foam composite with embedded lead pellets
  Three go/no-go force-penetration switches

Biofidelity attributes
  Humanlike lateral force-penetration response

Biofidelity deficiencies
  Frontal force-penetration response may not be humanlike

Injury-predictive measurements
  Go/no-go force-penetration switches to indicate potential for abdominal injury for lateral
    armrest location

TABLE 8. ANTHROPOMORPHIC PELVIS MODELS

Part 572 Pelvis[1]

Description
  Aluminum casting of human pelvis shape covered with vinyl skin

Biofidelity attributes
  Humanlike pelvis shape

Biofidelity deficiencies
  Mass and mass distribution not humanlike

Injury-predictive measurements
  None

Hybrid III Pelvis[4-6,43]

Description
  Aluminum casting of human pelvis shape covered with vinyl skin (Part 572 pelvis shape)
  Three load transducers on each iliac sartorius[43]
  Triaxial accelerometer

(*continued*)

TABLE 8. Continued

Biofidelity attributes
  Humanlike pelvis shape

Biofidelity deficiencies
  Mass and mass distribution not humanlike

Injury-predictive measurements
  Load transducers to indicate lap belt loading of pelvis

APR Pelvis[44-46]

Description
  Modified Part 572 pelvis shape
  Load transducers attached to iliac crest to measure lap belt forces when submarining occurs

Biofidelity attributes
  Humanlike pelvis shape
  Humanlike lap belt/pelvis interaction

Biofidelity deficiencies
  Mass and mass distribution not humanlike

Injury-predictive measurements
  Load transducers to measure lap belt loading of abdomen when submarining occurs

Citroen Pelvis[47]

Description
  Aluminum casting of human pelvis covered with vinyl skin (Part 572 pelvis shape)
  Load transducers at anterior-superior iliac spines

Biofidelity attributes
  Humanlike pelvis shape

Biofidelity deficiencies
  Mass and mass distribution not humanlike

Injury-predictive measurements
  Load transducers to indicate lap belt loading of pelvis

MIRA Pelvis[25]

Description
  Four-piece structure with humanlike contours
  Triaxial load transducer to measure ilium loading
  Load transducer to measure acetabulum loads
  Load transducer to measure symphysis pubis loading

Biofidelity attributes
  Humanlike pelvis shape

Biofidelity deficiencies
  Mass and mass distribution not humanlike

Injury-predictive measurements
  Load transducers to measure ilium loading and lateral loading of the acetabulum and
    symphysis pubis

TABLE 8. Continued

SIBB Pelvis[41]

Description
  Metal simulation of iliac crest and greater trochanter covered with rubber
  Load transducer to measure lateral loading of greater trochanter
  Provision for load transducer to measure lateral loading of iliac crest

Biofidelity attributes
  Humanlike lateral projection of iliac crest and greater trochanter
  Humanlike lateral exterior shape of hip area

Biofidelity deficiencies
  Mass and mass distribution not humanlike

Injury-predictive measurements
  Load transducer to measure lateral loading of greater trochanter/acetabulum

TABLE 9. ANTHROPOMORPHIC LOWER EXTREMITY MODELS

Part 572[1]

Description
  Steel shafts used for femur and leg structures covered with vinyl skin
  Limited rotation ball joint at hip and pin joints at knee and ankle
  Twist joint in femur shaft
  Adjustable friction joints
  Axial sensitive load cell in femoral shaft

Biofidelity attributes
  Humanlike ranges of motions, except ankle joint

Biofidelity deficiencies
  Mass distribution not humanlike
  Ankle joint cannot bend sideways
  Joint resistances not humanlike
  Knee impact response characteristics not humanlike

Injury-predictive measurements
  Axial compressive femur load

Hybrid III[4-6]

Description
  Steel shafts used for femur and leg shafts covered with vinyl skin
  Knee area has butyl rubber pad inserted beneath vinyl skin
  Ball joints at hip and ankle
  Knee joint allows leg to rotate and translate relative to femur in sagittal plane
  Adjustable frictional rotational joints

(continued)

TABLE 9. Continued

    Load cells in femur and leg shafts, ankle and knee joints
    Displacement transducer at knee joints

Biofidelity attributes
    Humanlike ranges of motion
    Humanlike knee impact response

Biofidelity deficiencies
    Mass distribution not humanlike
    Joint resistances not humanlike

Injury-predictive measurements
    Axial compressive femur load
    Leg relative to femur translation
    Sagittal and lateral tibial bending moments
    Medial and lateral tibial plateau compressive loads
    Fore/aft ankle bending moment and shear load or lateral bending moment and shear load
        depending on transducer orientation
    Knee laceration potential using chamois technique

Repeatable Pete[10]

Description
    Metal shafts used for femur and leg structures covered with self-skinning urethane foam
    Adjustable frictional joints
    Spherical shape patella
    Femur load cell

Biofidelity attributes
    Humanlike ranges of motion

Biofidelity deficiencies
    Mass distribution not humanlike
    Joint resistances not humanlike
    Knee impact response may not be humanlike

Injury-predictive measurements
    Axial compressive femur load

Sophisticated Sam[16,17]

Description
    Polymeric femur and tibia fibula shafts
    Ball joints at hip and ankle
    Pin-jointed knee
    Covered with rubber flesh

Biofidelity attributes
    Humanlike breaking strength of femur and tibia/fibula shafts

Biofidelity deficiencies
    Joint resistances not humanlike
    Knee impact response not humanlike

Injury-predictive measurements
    Breaking of femur and tibia/fibula shafts

TABLE 9. Continued

OPAT[13–15]

Description
    Steel shafts used for femur and leg structures covered with vinyl skin
    Ball joint at hip
    Pin joints at knee and ankle
    Spherical patella attached to leg
    Femur load cell
    Friction joints

Biofidelity attributes
    Humanlike ranges of motion

Biofidelity deficiencies
    Mass distribution not humanlike
    Joint resistances not humanlike
    Knee impact response may not be humanlike

Injury-predictive measurements
    Axial compressive femur load

Daniel Legs[37]

Description
    Metal shafts used for femur and leg structures covered with vinyl/foam skin and flesh
    Ball joint at hip and pin joint at ankle
    Knee joint allows leg to rotate and translate relative to femur in sagittal plane
    Load cells in femur and leg shafts and knee and ankle joints

Biofidelity attributes
    Humanlike ranges of motions, except ankle joint
    Humanlike knee impact response

Biofidelity deficiencies
    Mass distribution not humanlike
    Ankle joint cannot bend sideways
    Joint resistances not humanlike
    Translation of leg relative to femur not humanlike when leg and femur are straight

Injury-predictive measurements
    Axial compressive femur load
    Lateral and fore/aft bending moments and torque of femoral shaft
    Knee shear load
    Axial compressive tibia load
    Lateral and fore/aft bending moments and torque of tibial shaft
    Ankle bending moment for hyperdorsiflexion of foot

TABLE 10. ANTHROPOMORPHIC SHOULDER MODELS

Part 572 Shoulder[1]

Description
Aluminum casting pin-jointed to spine
Fore/aft range of motion controlled by rubber resistor.

Biofidelity attributes
Geometry of clavicle simulated in area where the shoulder belt would load the clavicle

Biofidelity deficiencies
Mass distribution not humanlike
Load-deflection responses not humanlike

Injury-predictive measurements
None

Hybrid III Shoulder[4,5]

Description
Aluminum casting pin-jointed to spine
Board area provided for shoulder belt loading

Biofidelity attributes
None

Biofidelity deficiencies
Mass distribution not humanlike
Load-deflection responses not humanlike

Injury-predictive measurements
None

OPAT Shoulder[13,15]

Description
Steel clavicle and scapula simulations
Clavicle attached to sternum

Biofidelity attributes
Geometric simulation of clavicle
Clavicle loads transmitted to sternum
Humanlike range of motion
Humanlike mass

Biofidelity deficiencies
Load-deflection responses not humanlike

Injury-predictive measurements
None

Sophisticated Sam[16,17]

Description
Wooden shaft used for clavicle
Polymeric scapula

TABLE 10. Continued

Biofidelity attributes
    One end of clavicle attached to chest
    Geometric simulation of scapula
    Clavicle has humanlike bending strength

Biofidelity deficiencies
    Load-deflection responses not humanlike

Injury-predictive measurements
    Breaking of clavicle shaft

APROD Shoulder[18,19]

Description
    Modified Part 572 shoulder to allow lateral translation

Biofidelity attributes
    Humanlike lateral force-deflection response

Biofidelity deficiencies
    Mass distribution not humanlike
    Load-deflection responses not humanlike in other than the lateral direction

Injury-predictive measures
    None

TRRL Shoulder[39]

Description
    Spherical surface mounted rigidly to spine
    Load cell to measure lateral applied loads

Biofidelity attributes
    None

Biofidelity deficiencies
    Load-deflection responses not humanlike

Injury-predictive measurements
    Lateral shoulder loads

MIRA Shoulder[25]

Description
    Designed to allow lateral motion of shoulder
    Telescopic clavicle attached at one end to sternum

Biofidelity attributes
    Humanlike articulation of shoulder

Biofidelity deficiencies
    Load-deflection response may not be humanlike in all directions

Injury-predictive measurements
    None

TABLE 11. ANTHROPOMORPHIC ARM MODELS

Part 572 ARM[1]

Description
  Steel shafts used for upper and lower arms
  Frictional pin joints at shoulder, elbow, and wrist
  Shafts covered with vinyl skin

Biofidelity attributes
  Humanlike range of motion

Biofidelity deficiencies
  Mass distribution not humanlike
  Joint resistance not humanlike

Injury-predictive measurements
  None

Hybrid III[4-6]

Description
  Steel shafts used for upper and lower arms
  Frictional pin joints at shoulder, elbow, and wrist
  Shafts covered with vinyl skin
  Bending moment transducer located in lower arm

Biofidelity attributes
  Humanlike range of motion

Biofidelity deficiencies
  Mass distribution not humanlike
  Joint resistance not humanlike

Injury-predictive measurements
  Fore/aft and lateral bending moments

Sophisticated Sam Arm[17,18]

Description
  Polymeric shafts used for upper and lower arms
  Frictional pin joints at shoulder, elbow, and wrist
  Shafts covered with vinyl skin
  Twist joints at end of shafts

Biofidelity attributes
  Humanlike bending strength of shafts

Biofidelity deficiencies
  Joint resistance not humanlike

Injury-predictive measurements
  Breaking of upper and/or lower arm shafts

human respond in the same way and just specify the cadaver response data. Other investigators[4] have modified the cadaver data to include effects of muscle tone. A second concern is that different procedures are used to normalize the data in an effort to account for differences in response characteristics due to variations in sizes of the cadavers compared to the human size that the model is to represent. Much work needs to be done in determining how human cadaver data should be interpreted for specifying living human impact-response requirements for anthropomorphic models.

A second method for developing impact-response requirements for anthropomorphic models is to scale animal response data. This approach has severe limitations, since both the impact response of the animal and the impact conditions must be scaled for size and shape differences. This method will require extensive developmental efforts in order to be a viable approach.

A third method is to simply extrapolate human volunteer test results.[33] The need for the model to simulate human response in the injury domain is debatable. From an occupant-restraint development viewpoint, the model only needs to provide a known increasing response with increasing stimulus in the injury domain. Such a response will allow the designer to estimate the effects of proposed design changes required to reduce the model response to a noninjury level. Based on this rationale, it would be more important for the model to have impact-response biofidelity in the noninjury range than in the injury range, which is counter to current practice.

Injury-Predictive Measurements
The following are suggestions for injury-predictive response measurements that are not incorporated in any of the anthropomorphic models listed in Tables 2 through 11.

Head. To aid in assessing the potential for different types of brain injuries, head models should be instrumented with sufficient accelerometers to completely define their linear and rotational acceleration responses. Much work still needs to be done to develop transducers to measure the potential for facial bone fractures.

Neck. A six-element load transducer (three orthogonal forces and three orthogonal moments) needs to be incorporated between the head and the top of the neck and another one between the base of the neck and the top of the thoracic spine for assessing the potential for neck injuries. A transducer needs to be developed to indicate the potential for neck injury due to direct impacts to the neck area.

Clavicle. A transducer needs to be developed to indicate the potential for clavicle fracture due to shoulder belt loading. For side impact testing, a transducer needs to be developed for indicating the potential for clavicle fracture and/or dislocation.

Lumbar Spine. A six-element load transducer (three orthogonal forces and three orthogonal moments) needs to be incorporated between the lumbar spine and pelvis to indicate the potential for lumbar spine injuries.

Abdomen. A transducer needs to be developed to indicate abdominal load or penetration as a function of time. This type of data is needed to assess the injury potential of steering wheel rim, lap belt, and/or armrest interactions with the abdomen.

Femur. Transducers are required to indicate the potential for femoral neck fractures and hip dislocations.

REFERENCES

1. Code of Federal Regulations, Title 49, Chapter V, Part 572. Anthropomorphic Test Dummy. *Federal Register* 1973; 38 (August 1):147.
2. Hubbard RP, McLeod DG: Geometric, inertial and joint characteristics of two Part 572 dummies for occupant modeling. *Twenty-first Stapp Car Crash Conference,* SAE 770937, October 1977.
3. Daniel RP, Trosien KR, Young BO: The impact behavior of the Hybrid II dummy. *Nineteenth Stapp Car Crash Conference,* SAE 751145, November 1975.
4. Foster JK, Kortge JO, Wolanin MJ: Hybrid III—A biomechanically based crash test dummy. *Twenty-first Stapp Car Crash Conference,* SAE 770938, October 1977.
5. Tennant JA, Jensen RJ, Potter RA: GM-ATD 502 anthropomorphic test dummy—development and evaluation. *Proceedings of Fifth International Technical Conference on Experimental Safety Vehicles.* London, England, June 1974.
6. Mertz HJ: Measurement capability of modified Hybrid III dummy. ISO/TC22/SC12/WG5, Document No. 105, Society of Automotive Engineers, June 1983.
7. Hubbard, RP, McLeod DG: A basis for crash dummy skull and head geometry. *Human Impact Response—Measurement and Simulation.* New York, Plenum Press, 1973.
8. Hubbard RP, McLeod DG: Definition and development of a crash dummy head. *Eighteenth Stapp Car Crash Conference,* SAE 741193, December 1974.
9. Viano DC, Culver CC, Haut RC, et al.: Bolster impacts to the knee and tibia of human cadavers and an anthropomorphic dummy. *Twenty-second Stapp Car Crash Conference,* SAE 780896, October 1978.
10. McElhaney JH, Mate PI, Roberts VL: A new crash test device—Repeatable Pete. *Seventeenth Stapp Car Crash Conference,* SAE 730983, November 1973.
11. Melvin JW, McElhaney JH, Roberts VL: Improved neck simulation for anthropometric dummies. *Sixteenth Stapp Car Crash Conference,* SAE 720958, November 1972.
12. McElhaney JH, Stalnaker RL, Roberts VL: Biomechanical aspects of head injury. *Human Impact Response—Measurement and Simulation.* New York, Plenum Press, 1973.

13. Warner P: *The Development of U.K. Standard Occupant Protection Assessment Test Dummy.* SAE 740115, March 1974.
14. Haslegrave CM, Croke MD: *Performance Measurements on the OPAT Dummy.* Nuneaton, Warwickshire, England, MIRA Publication, January 1974.
15. Searle JA, Haslegrave CM: *Improvements in the Design of Anthropometric/Anthropomorphic Dummies.* Nuneaton, Warwickshire, England, MIRA Bulletin No. 5, 1970.
16. Bloom A, Cichowski WG, Roberts VL: Sophisticated Sam—A new concept in dummies. SAE 680031, 1968.
17. Cichowski WG: A third generation test dummy—Sophisticated Sam. *Proceedings of GM Automotive Safety Seminar,* Warren, Mich, July 11 and 12, 1968.
18. Stalnaker RL, Tarriere C, Fayon A, et al.: Modification of Part 572 dummy for lateral impact according to biomechanical data. *Twenty-third Stapp Car Crash Conference,* SAE 791031, October 1979.
19. Hue B: *Summary of the Technical Evolution of the APROD.* ISO/TC22/SC12/WG5, Document No. 77, Society of Automotive Engineers, May 1982.
20. Maltha J, Janssen EG: *EEC Comparison Testing of Four Side Impact Dummies.* Final Report. Delft, The Netherlands, Research Institute for Road Vehicles, TNO, March 1983.
21. Morgan RM, Marcus JH, Eppinger RH: Correlation of side impact dummy/cadaver tests. *Twenty-fifth Stapp Car Crash Conference,* SAE 811008, September 1981.
22. Melvin JW, Robbins DH, Benson JB: Experimental application of advanced thoracic instrumentation techniques to anthropomorphic test devices. *Seventh Experimental Safety Vehicle Conference,* Paris, 1979.
23. Robbins DH, Benson JB, Brindamour JS, et al.: *Assembly Procedures for HSRI Side Impact Dummy Thorax.* Final Report No. UMHSRI8023. Ann Arbor, Mich, University of Michigan Transportation Research Institute, 1980.
24. Donnelly BR: *Assembly Manual for the NHTSA Side Impact Dummy.* Final Report. Contract No. DTNH2282C07366, CALSPAN/NHTSA, March 1983.
25. Standinger R, Nada M, Haslegrave CM: *Development of Side Impact Dummy.* Final Report EEC Phase III, Project UK2. Nuneaton, Warwickshire, England, MIRA, October 1981.
26. Cotte JP: The ONSER 50 dummy—A research tool for safety. *Eighth Experimental Safety Vehicle Conference,* Wolfsburg, 1980.
27. Anthropomorphic test dummy—Model: 3DGMAM5073. Tokyo, Japan, Itoh Seiki Brochure, 1973.
28. Satoh S, Aibe T, Maeda T: Some dummy problems to evaluate vehicle safety. Tokyo, Japan, Nissan Motor Brochure, October 1972.
29. Hodgson VR, Mason WM, Thomas LH: Head model for impact. *Sixteenth Stapp Car Crash Conference,* SAE 720969, November 1972.
30. McLeod DG, Gadd CW: An anatomical skull for impact testing. *Human Impact Response—Measurement and Simulation.* New York, Plenum Press, 1973.
31. Tarriere C, Leung YC, Fayon A, et al.: Field facial injuries and study of their simulation with dummy. *Twenty-fifth Stapp Car Crash Conference,* SAE 81103, September 1981.
32. Warner CY, Niven J: A prototype load-sensing dummy faceform test device for facial injury hazard assessment. *Proceedings of the Twenty-third American Asso-*

ciation for Automotive Medicine Conference. Arlington Heights, Ill, AAAM, October 1979.

33. Mertz HJ, Neathery RF, Culver, CC: Performance requirements and characteristics of mechanical necks. *Human Impact Response—Measurement and Simulation.* New York, Plenum Press, 1973.

34. Culver CC, Neathery RF, Mertz HJ: Mechanical necks with humanlike responses. *Sixteenth Stapp Car Crash Conference,* SAE 720959, November 1972.

35. Lobdell TE, Kroell CK, Schneider DC, et al.: Impact response of the human thorax. *Human Impact Response—Measurement and Simulation.* New York, Plenum Press, 1973.

36. Stalnaker RL, McElhaney JH, Roberts VL, et al.: Human torso response to blunt trauma. *Human Impact Response—Measurement and Simulation.* New York, Plenum Press, 1973.

37. Daniel RP, Yost CD: The design and experimental use of a chest load-distribution transducer and force-indicating legs for the Part 572 dummy. *Twenty-fifth Stapp Car Crash Conference,* SAE 811012, September 1981.

38. Foster K: Analysis of a slanted-rib model of the human thorax. *Human Impact Response—Measurement and Simulation.* New York, Plenum Press, 1973.

39. Harris J: The design and use of the TRRL side impact dummy. *Twentieth Stapp Car Crash Conference,* SAE 760802, October 1976.

40. Lowne RW: *Modification to the TRRL Side Impact Dummy.* ISO/TC22/SC12/WG5, Document No. 21, Society of Automotive Engineers, March 1979.

41. Daniel RP, Koga MS, Prasad P, et al.: *A Force Measuring Mechanical Test Device for Estimating and Comparing the Energy Absorbing Characteristics of Vehicle Interior Side Panels.* ISO/TC22/SC12/WG5, Document No. 89, Society of Automotive Engineers, November 1982.

42. Maltha J, Stalnaker RL: Development of a dummy abdomen capable of injury detection in side impact. *Twenty-fifth Stapp Car Crash Conference,* SAE 811019, September 1981.

43. Daniel RP: Test Dummy Submarining Indicator System. United States Patent 3,841,163, October 1974.

44. Leung YC, Tarriere C, Fayon A, et al.: A comparison between Part 572 dummy and human subject in the problem of submarining. *Twenty-third Stapp Car Crash Conference,* SAE 791026, October 1979.

45. Leung YC, Tarriere C, Fayon A, et al.: An anti-submarining scale determined from theoretical and experimental studies using three-dimensional geometric definition of the lap belt. *Twenty-fifth Stapp Car Crash Conference,* SAE 811020, September 1981.

46. Leung YC, Tarriere C, Lestrelin D, et al.: Submarining injuries of three-point belted occupants in frontal collisions—Description, mechanisms and protection. *Twenty-sixth Stapp Car Crash Conference,* SAE 821158, October 1982.

47. DeJeammes M, Biard R, Derrien Y: Factors influencing the estimation of submarining on the dummy. *Twenty-fifth Stapp Car Crash Conference,* SAE 811021, September 1981.

CHAPTER 4

The Biomechanics of Restraint

Richard F. Chandler

INTRODUCTION

Injuries resulting from impacts associated with vehicles used for transportation have been recognized for more than 80 years. Snyder[1] has traced the development of restraint systems for automobiles to a 1903 French patent for a sophisticated system of seat belts and upper torso restraint integrated into a highback seat and mentions the modificiation of a luggage strap as a device to keep the pilot from falling out of US Army airplane No. 1 in 1910. Fryer[2] describes a 1912 advertisement by A.V. Roe for a safety belt apparently incorporating elastic cords, "As supplied to the Army Aircraft Factory," and mentions illustrations of Farman biplanes of that era that showed the provision of simple lap belts. By 1917 many fighter aircraft were equipped with cruciform type upper torso restraint systems, usually without seat belts or other lower torso restraint. However, most of these early devices were developed primarily to keep the occupant coupled to the vehicle during normal operations. Little concern was expressed regarding the preventable loss of life due to crash injury, and the deaths of pioneers in aviation were regretfully accepted as a natural consequence of operating the new vehicles.

It was not until World War I, when the belligerents recognized the immense possibilities of the airplane as a factor in battle, that the critical nature of replacing trained pilots forced a careful look at the reasons for these losses. One such study, reported in *Air Service Medical*,[3] concluded that only 2 percent of the flyers were lost as a result of action by the enemy, 8 percent were lost due to a fault of the engine or plane, and the remaining 90 percent were lost because of "failure of the flyer himself." These findings led to the emphasis on accident prevention efforts to reduce the 90 percent loss, with strong programs in airman selection, training, medical monitoring, and continual education.

Although primarily concerned with accident prevention, this experience also provided the first opportunities for a biomechanical study of crash injury. The results, while rudimentary by today's standards, were effective. For example, medical officers observed that, in one aircraft, more than half the injuries sustained in crashes were caused by the aviator striking his head against the cowl. After following a suggestion that the cowl be cut so as to allow 8 inches more clearance in front of the pilot, head injuries were practically eliminated. Likewise, a suggestion to use a simple shock absorber between the aircraft and the restraint system "decidedly reduced" the number and extent of injuries to the upper abdomen and ribs.

Unfortunately, the end of World War I also appeared to signal the end of concern over crash-related injuries. The following two decades showed little progress in reducing the causes of crash injury, although the importance of automotive advances, such as hydraulic brakes and laminated safety glass, should not be understated. In the field of aviation, suggestions that aircraft be designed "so that they crash well"[4] went largely unheeded, although early regulations specified requirements for seat belt installations.[5] Results of military aircraft accident investigations continued to show injuries to the face and head due to the inadequate upper torso restraint provided by the lap belt type safety restraint. In response to these findings, Lt. Colonel M.C. Grow of the US Army Medical Department initiated a movement to have a shoulder type of safety belt designed and adopted. Human volunteer tests of this restraint were accomplished by Armstrong[6,7] on a swing seat test device but were terminated at 15 g, the limit of the recording instruments available. Nevertheless, Armstrong estimated that one could live through a deceleration of somewhere between 30 and 50 g with the shoulder type (and lap) safety belt if the belt and seat did not fail. Armstrong also mentioned the possibility of inflated rubber seat backs (air cushions?) in transport aircraft, so that the back of the seat would act as an upper body support.

RESEARCH IN WORLD WAR II

The development of high-speed aircraft during World War II led to a new operational problem. The conventional method of leaving a disabled aircraft, climbing over the side, and bailing out by parachute became impractical. Not only did the increased air pressure make it difficult to climb over the side, but it became increasingly difficult to clear the empennage or the propeller (of a pusher type airplane). The concept of a catapult seat to throw the pilot clear of the aircraft was aggressively developed in Germany. Analytical studies and trial ejections had already been conducted at the Heinkel Aircraft Company in 1940,[8,9] at a time when most others were just considering the problem. Seventy-six tests using human subjects on an inclined track were accomplished without injury at accelerations up to 15 g. The success of the ejection seat concept led to a directive by the Air Ministry of Germany in the fall of 1944, instructing that all fighter aircraft, including prototypes, be provided with ejection seats.[10]

Questions raised regarding the tolerance of the pilot to the loads imposed by the ejection seat led to what was possibly the first research study in what could be termed "biomechanics of impact." Arno Geertz, of the Heinkel Aircraft Company, worked under the direction of Dr. G. Madelung and Dr. S. Ruff at the Technische Hochschule in Stuttgart. His doctoral thesis, submitted in November 1944, described his work investigating human tolerance to ejection stress.[11] After presenting an analytical discussion of the dynamics of ejection from aircraft and the development of methodology to measure acceleration in laboratory simulations of ejections, Geertz described his research on the biomechanics of the spinal column. Significant content included:

1. Recognition of differences in response between a living subject and a cadaver test specimen
2. The utility of cadaver specimens "immediately after death" in evaluating structural strength
3. Experimental data on fresh cadaver spinal column segments, including vertebral disks, describing the fracture energy level for the intact column and individual vetebral segments, with differentiation between initial injury and final destruction levels
4. Experimental data from one living subject describing spinal compression (deflection) under loads up to 60 kg
5. A tolerance curve with distinct limitations based on circulatory disturbance, static strength of bones, and dynamic strength of bones, for impact durations greater than 0.5 seconds, between 0.5 and 0.005 seconds, and less than 0.005 seconds, respectively
6. The use of mathematical analysis to compute compression of vertebral segments
7. Analysis of the adverse effect of high rates of onset for the impact pulse
8. Recognition of the possible effect of individual body structure on tolerance to acceleration
9. Indication of the influence of seat cushion construction on "reinforcement or damping" of acceleration, with some types of cushions showing reinforcement in excess of 20 percent

This work was later summarized by Ruff.[12] Ruff also recounted volunteer human tests on a swing seat, similar to that used by Armstrong, in which the primary purpose was to study the effect of seat belt anchorage spacing. Tests up to 18 g were tolerated without injury, at an impact duration of 0.1 second. Ruff concluded that crashes exceeding 20 g could be accommodated without injury with proper seat and restraint design. Among the design criteria recommended were a static load harness strength of at least 3,500 kg, with the load distributed over as large an area as possible (German belt plates had an area of about 550 sq cm), and mandatory use of a shoulder harness with fastening points at least 40 to 50 cm above the upper edge of the seat. Ruff also suggested the use of a shoulder harness locking reel with three latch positions: a low tension position to allow comfort and movement during normal operation, a locked position for use in bumpy weather, and a high strap tension, locked position for preloading the harness straps in case of danger. For vertical (seat to head) acceleration he suggested that the forces should be limited to 1,500 kg if they act for more than 0.005 seconds. If no shoulder harness is used, as in passenger aircraft, he recommended that at least 80 to 90 cm of free space be provided ahead of the belt fastenings to reduce head impact.

Schneider, writing in the same publication, described protective measures for spinal injuries encountered with aircraft having skid type landing gear.[13] While the analytical position of this brief report is concerned mostly with energy management in crashing aircraft, Schneider also discussed the importance of seat pan angle in determining spinal column loads and the experience of thoracic vertebrae fractures complicated by the stiff support provided by a backpack parachute and a tight shoulder harness.

These reports represent only a small portion of the findings that resulted from over 200 volunteer ejection seat tests, 60 successful ejections in operational aircraft, and numerous investigations of injuries sustained in crashed aircraft. Nevertheless, they serve to illustrate the significant beginning of biomechanical studies of seat and restraint systems in Germany in the 1940s. While many of these findings have been reinvented since that time, it is interesting to note that disagreement, then as now, was quick to come. The Royal Swedish Air Force, for example, in developing an ejection seat for the J21 airplane, concluded that by limiting the acceleration to 10 g or less during the ejection, special shoulder harness headstraps "and the like" would not be necessary.[10]

Similar studies were also conducted at the Royal Air Force Institute of Aviation Medicine in Great Britain. Initial tests conducted on a rocket-propelled sled in 1944 produced mean accelerations of about 12 g for the first 0.1 second, with a total impact duration of 0.175 seconds. Because it was felt that the resulting 6 foot acceleration distance was not representative of the ejection seat applications, additional work was carried out on three ejection towers at the Martin-Baker Aircraft Company and a fourth at the Armament Research Department.[14] Reportedly, hundreds of live tests were conducted on these devices. As a result, it was concluded that it was probably unwise to subject a person to accelerations greater that 5 g in the first 0.01 second or to final values greater than 25 g and that an intimate relationship exists between accelerations in the seat and on the person. Unfortunately, no mathematical analysis of these findings was reported. Several significant recommendations relative to restraint systems were also made:

1. Use of combined parachute and safety harness with inertia lock on shoulder attachments and tensioning gear for the lower attachments, and automatic release
2. Automatic foot (lower limb) retraction
3. Design of the seat "as a splint" to restrain the body from flailing injury in a windstream, with a head blind (curtain) to protect and position the head
4. Use of the seat to adequately spread high shock loads
5. Dynamic tests, either on the pendulum or the high-speed track, to be carried out if any combined harness is designed
6. Possibility of the need for additional head restraint beyond that provided by the Martin-Baker type of head blind

Captured German ejection seats formed the basis for early experiments in the United States.[15] A 30-foot ejection tower constructed according to directions from Dr. E.J. Baldes of the Mayo Clinic served as the facility for early tests with ballast or human subjects in the seat. Most significant was the use of vacuum tube acceleration transducers to obtain accurate acceleration measurements of the body. These were placed on human subjects on the top of the head, the acromium, and the crest of the ilium, as well as on the ejection seat. Arm rests were also instrumented to measure load. Tests with human subjects were completed in December 1945. These measurements and observations of high-speed motion pictures of the tests led to the realization that the dynamic response of the occupant must be considered in optimizing the thrust-time characteristics of the catapult. This task was assigned to the Frankfort Arsenal. Kroeger[16] described the initial results of the work. His analysis modeled a system with two elastically coupled masses, one mass representing the body of the pilot and the other representing the seat, with the elastic link representing the parachute pack, life raft, hip muscles, and so on. An electrical analog of this system was constructed to allow convenient prediction of body acceleration as a function of catapult thrust-time characteristics, and a full-scale mechanical analog was constructed for testing of catapults. Although rudimentary, these analogs predicted the common use of computer analysis and the development of anthropomorphic dummies with realistic dynamic response. (Dummies composed of rigid links of appropriate mass for representing body segments had been used previously, but this was apparently the first attempt to provide dynamic fidelity.) As a result of his analysis, Kroeger concluded;

1. Attempts should be made to increase the damping of the elastic coupling cushion between the body and the seat to reduce the problem of overshoot.

2. An increase in spring constant (more rigid coupling of body and seat) would be limited by the elasticity of the body.
3. The force-time characteristics acting on the seat should not contain "abrupt discontinuities," particularly during its initial rise to a maximum.
4. The initial acceleration of the body and seat will be opposite in phase.
5. Increasing the mass of the seat will tend to decrease the excitation of "vibrations" acting on the body.

These reports present the state-of-the-art in the biomechanics of restraint as it existed at the close of World War II. The concepts that have been suggested in these reports are summarized in Table 1. The importance of these concepts is apparent when it is observed that many of these approaches are being pursued today with increased sophistication in methods of analysis, materials and mechanisms, but toward a goal stated almost 10 decades ago.

POSTWAR RESEARCH

After World War II, the technology relating to escape from high-speed aircraft continued to develop, although at a slower pace. Seating and restraint system progress evolved as a factor integrated with larger programs directed either toward development of human tolerance data or toward

TABLE 1. BIOMECHANICS/RESTRAINT CONCEPTS PROPOSED OR DEVELOPED PRIOR TO THE END OF WORLD WAR II

Use of upper torso restraint in addition to the seat belt to reduce injury due to secondary impact

Use of "inflatable rubber seat backs" to reduce injury when upper torso restraint not feasible

Dynamic testing to evaluate system performance

Use of unembalmed cadaver specimens to define tolerance limits

Consideration of energy required for fracture of spinal column segments

Tolerance limits as a function of the impact pulse temporal characteristics

Effectiveness of energy dissipation, or damping, to reduce injury, including the use of load-limiting devices in the restraint

Importance in load distribution over a maximum body area

Utility of self-locking reels on shoulder belts and preloading devices on seat belts

Potential problems of head-neck injury due to differential motion relative to the torso

Analytical models of the body to aid in understanding the response to impact

Electric analogs of the analytical models to relieve computational drudgery practical hardware. While much of this progress was guided by empirical judgment, the application of analytical methods of biomechanics to restraint system design was also progressing.

Perhaps the most extensive of the postwar projects was directed by J.P. Stapp and his associates. Although directed primarily to answering questions regarding human tolerance to impact and windblast, it was immediately apparent that restraint systems exercised a great deal of influence on tolerance. While a summary of the extensive results of this important research is far beyond the scope of this chapter, an excellent synopsis of the major findings was compiled by Stapp,[17] and some findings related to restraint biomechanics should be mentioned.

The inverted-V leg straps were developed early in the program and formed an important part of the restraint systems used in the remainder of the research. Stapp found that loads from the shoulder belts, which were then attached to the center buckle of the seat belt in standard military practice, would raise the seat belt so that it lodged against the lower ribs and abdomen.[18] An inverted-V strap, originating at the corners of the seat passing at a 45 degree angle under the thighs, and then carefully fitted around the thighs so that it would loop over the buckle, opposing the loading from the shoulder belts and kept the seat belt in place. Stapp found that the inverted-V strap also carried 25 percent of the total restraint load in a forward crash. Functional equivalents of the inverted-V strap, often called a "negative g strap," can be found on many modern military restraint systems.

In 1954, Nichols used data from Stapp's research to complete a comprehensive parametric study of the factors influencing restraint system performance.[19] This study, one of the first in the field of biomechanics of restraint to make extensive application of a digital computer, again used the familiar simple spring-mass model to represent the restraint-occupant system. This methodology allowed Nichols to demonstrate the effects on nonlinear system elements on overall restraint system performance. He was able to analytically judge the significance of slack in the restraint, the effect of altered webbing properties, and the advantages and disadvantages of energy-absorbing elements in series with the restraint straps. While a major significance of his work was the demonstration of computer analysis as a functional tool for the restraint system designer, Nichols also provided practical guidelines for restraint system development. For example, his suggestion that the occupant be tightly coupled to the seat with stiff restraint webbing and that energy-absorbing elements be placed between the seat and the aircraft forms the basic design concept used today in successful crashworthy seats for military helicopters.[20]

About this same time, a seemingly unrelated analytical study was completed. Kornhauser, working to develop a method of predicting switch closure in artillery fuses, used a spring-mass model to define a sensitivity curve in an acceleration-velocity domain.[21] The position of the sensitivity curve was dictated by limiting the stretch of the spring in the model. It was recognized that a similarity could exist between the stretch of a spring to close a switch and the strain of an element in the human body as a limit to impact tolerance. Perhaps the first wide application of this concept was the dynamic response index (DRI) developed by Stech and Payne for the US Air Force.[22] Using data on vertebral breaking strength presented by Ruff[12] and Perey,[23] they were able to devise a model to represent a seated pilot exposed to $+g_z$ ejection seat acceleration that would predict the probability of injury. This model was subsequently incorporated into military specifications for ejection seats and capsules.[24,25]

AUTOMOTIVE STUDIES

While pursuing his research on acceleration and windblast tolerance, Stapp realized that the Air Force lost nearly as many men in fatal auto accidents as in aircraft crashes and began a car crash study using salvaged automobiles. In May of 1955, after an expression of interest from the Society of Automotive Engineers, he invited representatives of the Armed Services, universities, automobile manufacturers, research laboratories, traffic and safety councils, and medical institutions to visit Holloman Air Force Base in order to witness sled tests and crash tests and to discuss automobile design and safety features.[26] These informal meetings were repeated in 1956, 1957, and 1959, the latter being designated the "Fourth Stapp Car Crash and Field Demonstration Conference." No conference was held in 1960, but they have been held annually since 1961. The Society of Automotive Engineers assumed administrative responsibility for conducting the conferences and publishing the proceedings for the Tenth Stapp Car Crash Conference in 1966 and for all subsequent annual conferences. These proceedings, and the five independently published proceedings of earlier conferences, provide a readily available, annually updated source of data pertinent to crash injury tolerance, restraint system design and test methodology, and field investigations.

Of course, others were also working to improve automobile safety. In 1957, Stapp representatives of Cornell University and of the University of California, representatives of major automobile companies, and others testified before the Eighty-fifth Congress of the United States regarding methods to reduce automobile crash injuries.[27] Ultimately, Congress enacted the National Traffic and Motor Vehicle Safety Act of 1966,[28] with a goal of reducing accidents and deaths and injuries to persons resulting from traffic accidents. The regulatory consequences of this act, supported by federal research and the initiative of industry, resulted in a massive escalation of efforts to develop improved crash injury protection systems. A detailed recitation of the many advances that have been made are far beyond the scope of this paper. Fortunately, they have been well documented in the literature. The proceedings of the Stapp Car Crash Conferences, The

Technical Paper Series and Special Publications of the Society of Automotive Engineers, proceedings of the annual conferences of the American Association for Automotive Medicine, proceedings of the International Research Committee on the Biokinetics of Impact, proceedings of special conferences dealing with experimental safety vehicles and passive restraint, and the Highway Safety (HS) series of documents selected by the National Highway Traffic Safety Administration are but a few of the readily available resources for restraint systems designers.

In reviewing the literature, definite stages of advancing technology seem apparent. In the Stapp conferences for example, the earlier reports pertaining to restraint systems generally presented the results of laboratory tests or simply described restraint systems being marketed at the time. Mathematical analysis of system performance was infrequent and rudimentary by today's standards. This began to change after the development by McHenry of a computerized seven degree-of-freedom, nonlinear mathematical model of a human body and restraint system on a test car.[29-31] This development provided a valuable new tool for the restraint system designer. At last it was possible to economically complete parametric analysis of the entire seat-restraint-occupant system, with an output visually and numerically comparable to what might be expected in a controlled laboratory test. The development of computer modeling became a specialized activity in itself, with sophisticated three-dimensional, lumped mass, or finite element models well within the state of the art, with symposia to discuss the considerable progress being made.[32,33] (However, it must be recognized that the progress in developing mathematical models to represent the body has far outstripped our knowledge of human body passive and active dynamic properties that are necessary to validate the performance of the models relative to the human occupant. Without such validation, the modeling techniques must be regarded only as sophisticated scientific displays of the empirical judgment of their creator. They are extremely useful tools but are not yet final answers.)

These techinques were quickly adapted for the design of vehicle interiors[34,35] and for restraint system design.[36-39] One of the major advances resulting from the application of these computer models was their ability to provide a system approach type of analysis. Beginning with a projected crash scenario, the designer is able to calculate the deceleration environment of the passenger compartment and then determine the optimum combination of seat and restraint characteristics to make the best use of the available space surrounding the occupant so as to reduce injury. Passive restraint components, such as energy-absorbing steering assemblies or seat cushions, padded instrument panels, or airbag restraint systems, can be evaluated. Finally, the more critical performance conditions can be selected for dynamic testing in the laboratory to better assure effectiveness and efficiency in a test program.

AUTOMOTIVE RESTRAINT

In the United States, a 1958 study indicated that the universal use of lap belts alone could save 5000 lives per year through control of occupant ejection in automobile crashes.[40] While this goal was sufficient to initiate the installation of seat belts in automobiles, it soon became apparent that additional provisions would be required if significant reductions in injuries were to take place.[41] In particular, the continuing occurrence of head, neck, and upper torso injuries led to consideration of upper body restraint. The three-point restraint (lap belt and diagonal torso belt) was the most widely acceptable design because the upper torso restraint anchor position could be readily installed in most automobiles. Extensive development of this restraint system has taken place, and several accident investigation and reconstruction programs have demonstrated the success of this approach.[42–54] Typically, these studies confirm the excellent performance of the system, reporting, for example, a 65 to 70 percent reduction in fatalities in over 99 percent of all frontal crashes[40] and a 10-fold reduction of serious injuries compared to those of unrestrained occupants.[51]

Even with such good performance, serious injuries continue to be reported but are usually encountered in the more severe crashes. Improvement in crash injury protection may still be possible for some systems if the accident data are carefully evaluated. For example, Ryan[47] associated lacerations of the jejunum, spleen, and ascending colon and fracture of the lumbar spine with incorrectly adjusted belts, with loose lap loops so that the buckle (and diagonal) set more toward the center of the abodmen. Nine years later Leung et al.[54] described the advantages of the adoption of a retractor belt system with improved geometrical layout that eliminated risks associated with contact of the adjusting buckle against the abdomen and had approximately halved the share of abdominal injuries. Recommendations were then made for improved test procedures and criteria to assess the potential for submarining injuries in tests of new systems. These reports illustrate a common problem with attempts to make statements about performance of crash protective systems in general. Each installation depends on details of design, installation, and operator use for its overall protective effect. Moreover, each accident tends to be unique to some degree relative to crash environment and occupant characteristics. While general statistical treatments of accident data are useful for demonstrating population trends, improvements in a specific restraint system must be predicated on the understanding of the interaction of those details.

With that precaution in mind, several recommendations for restraint improvement can be considered. The seat itself can influence the performance of the restraint system, particularly in regard to submarining or lap belt-related injuries.[55–58] The inclusion of an energy-absorbing wedge along the front of the seat pan would appear beneficial in reducing submarining and controlling forward and downward motion of the pelvis. Modifications to webbing and anchorages are often proposed. While conventional emergency locking retractors have done much to reduce slack in the restraint system, special devices to preload the restraint at the time of the crash[57,59] could improve coupling of the occupant to the vehicle to take advantage of the ridedown effect. Svensson[57] suggested a webbing lock device on the B-pillar loop to prevent slack caused by packing the webbing on the retractor during a crash. Such a device would also reduce the head strike envelope by reducing the effective length of webbing in the diagonal belt. Suggestions that energy-absorbing elements be incorporated are often made. These are intended to limit the maximum loads on the body from the restraint system and have been effectively used in some installations. However, the reduction in load is achieved at the expense of an increase in displacement, with the risk of increased secondary impact of the occupant with the interior of the vehicle. All of these factors must be considered together with occupant size and anchorage point location to achieve optimum protection with the three-point restraint.[59]

While improvements such as these would probably increase the crash injury protection offered by the three-point restraint, it must be acknowledged that the system is already highly perfected and the overall increase in protection would be small compared to the level of protection currently available to the occupant who chooses to use the system. Greater protection could be provided by advanced integrated seat and restraint systems, as suggested by McElhaney et al.[56] and others, but the acceptance and use of these advanced systems by the average automobile occupant is questionable. Indeed, the largest deficiency in the conventional automotive three-point restraint system is not design but the lack of seat-belt use by the occupants. In an attempt to correct this deficiency, several passive restraint concepts have been suggested.

Passive restraint concepts are inherent in modern crashworthy design of vehicles. Load-limiting steering columns and wheels, windshields, instrument panels, seatbacks, door structure, and so on are all forms of passive restraint. Passive three-point belt systems function in the same manner as conventional systems in a crash. The diagonal belt and knee bolster restraint[60,61] use femur contact with an energy-absorbing bolster across the instrument panel to provide lower torso restraint. Viano and Culver[61] showed that peak femur loads in excess of 6 kN could be generated without acute injury to lower extremity ligamentous or skeletal structure if the bolster was carefully designed. States et al.[60] found only 4 serious injuries (AIS 3 or greater) in a study involving 72 occupants, with the only injury above AIS 3 occurring to an occupant who had disconnected the torso belt. The main limitation of this concept appears to be that it is limited to outboard seating positions adjacent to doors.

Inflatable passive restraint systems (airbags) are the other major restraint concept considered for automobiles. These devices were first seriously proposed by Jordanoff as early as 1952, with Clark et al. accomplishing the first full-scale crash tests with human subjects in the early 1960s.[62] The issuance, in 1969, of a notice of proposed rulemaking by the Department of Transporation for "the prompt development and installation of passive restraint systems" began one of the longest and most extensively argued programs in the history of restraint system development. It is far beyond the scope of this paper to reiterate this convoluted history, but it should be recognized that the arguments center on socioeconomic factors rather than technology. The airbag simply fills the space between the occupant and the vehicle interior with a low-pressure air cushion, distributing the crash loads over the maximum body area. The literature describing this concept is extensive and well known.[63-65]

CONCLUSION

The study of biomechanics relative to restraint system performance had its beginning during World War I, when observations of injuries related to restraint system performance in aircraft were made and deficiencies corrected. This study became a science in Germany during World War II, when the development of aircraft ejection seats required data on the tolerance of the human body to impact injury. During the course of this work fundamental concepts of restraint system performance were recognized, and these concepts have served as goals for restraint system designers to this date.

Progress since that time has permitted more realistic modeling of the crash event, both in the test laboratory and in the analytical sense. The evolution of computer-based models that provide dynamic response with three-dimensional spatial representation of the seat, restraint, occupant, and vehicle interior, often with mathematical injury prediction, provides extremely useful tools for the restraint designer, allowing a true systems approach to be implemented. The development of adequate data for the full validation of these models represents a major field for further research.

The three-point automotive restraint system in its fully developed form provides effective crash injury protection limited primarily by the refusal of occupants to use the system. Passive restraint systems can be made available if socioeconomic considerations warrant their use.

REFERENCES

1. Snyder RG: *A Survey of Automotive Occupant Restraint Systems: Where We've Been, Where We Are and Our Current Problems.* SAE 690243. New York, Society of Automotive Engineers, 1969.
2. Fryer DI: RAF Experience with safety harnesses. *Ann Occup Hyg* 1962; 5:113–127.
3. *Air Service Medical: Air Service Division of Military Aeronautics.* Washington, DC, Government Printing Office, 1919.

4. Doolittle JH: Problems in flying. Presentation at the Third National Meeting of the ASME Aeronautic Division, St. Louis, Mo, 1929, AER-51-24, pp 147–150.
5. Snyder RG: *General Aviation Aircraft Crashworthiness. An Evaluation of FAA Safety Standards for Protection of Occupants in Crashes.* UM-HSRI-81-10. Washington, DC, Aircraft Owners and Pilots Association, 1981.
6. Armstrong HG, Heim JW: The effect of acceleration on the living organism. Air Corps Technical Report No. 4362. Dayton, War Department, Air Corps Material Division, 1937.
7. Armstrong HG: *Principles and Practice of Aviation Medicine.* Baltimore, Williams & Wilkins, 1939.
8. Richter H: *Schussversuche mit dem Katapultsitz.* Heinkel, Ernst, Flugzeugwerke, Rostock, 1940.
9. Richter H: *Physiological Analysis of the Effects of Catapulting by an Ejection Seat.* Heinkel, Ernst, Flugzeugwerke, Rostok, 1940.
10. Lovelace WR II, Baldes EJ, Wulff VJ: The ejection seat for emergency escape from high speed aircraft. Memorandum Report TSEAL-3-696-74C. Dayton, Air Technical Services Command, 1945.
11. Geertz A: *Limits and Special Problems in the Use of Seat Catapults,* thesis. Technische Hochschule, Stuttgart, 1944.
12. Ruff, S: Brief acceleration: Less than one second. *German Aviation Medicine, World War II.* Washington, DC, Government Printing Office, 1950, vol I, chap VI-C.
13. Schneider, J: Protective measures for the prevention of injuries—especially spinal fractures—in aircraft on skids. *German Aviation Medicine, World War II.* Washington, DC, Government Printing Office, 1950, vol I, chap VI-E.
14. Stewart WK: Ejection of pilots from aircraft, a review of the applied physiology. Flying Personnel Research Committee Report 671. RAF Institute of Aviation Medicine, 1946.
15. Close P, Lombard CF, Bronson SD, et al.: Pathophysiology of "protected" guinea pigs in high transverse impact accelerations. *Proceedings of the Seventh Stapp Car Crash Conference.* Paper No. 38. Springfield, Il, Thomas, 1965, pp 503–518.
16. Kroeger, EJ: Internal vibrations excited in the operation of personnel emergency escape catapults. Memorandum Report MR-340. Frankford Arsenal Laboratory Division, 1946.
17. Stapp JP: Biodynamics of deceleration, impact and blast, in *Aerospace Medicine.* Baltimore, Williams & Wilkins, 1971, chap 8.
18. Stapp JP: Human exposures to linear deceleration. Part 2. The forward facing position and the development of a crash harness—and Appendix. AF Technical Report No. 5915, Part 2. Dayton, United States Air Force, Wright Air Development Center, 1951.
19. Nichols G: Dynamic response of restrained subject during abrupt deceleration. Technical Report No. NAI-54-585. Hawthorne, Calif, Northrop Aircraft Company 1954.
20. Shulman M: The US Navy approach to crashworthy seating systems. Paper No. 21. *Impact Injury Caused by Linear Acceleration: Mechanisms, Prevention and Cost.* AGARD-CP-322. Neuilly Sur Seine, France, Advisory Group for Aerospace Research and Development, NATO, 1982.

21. Kornhauser M: Prediction and evaluation of sensitivity to transient acceleration. *J Appl Mechanics* 1954; 21:371.
22. Stech EL, Payne PR: Dynamic models of the human body. AMRL TR 66-157. Wright-Patterson AFB, Ohio, Aero Medical Research Laboratory, 1966.
23. Perey O: Fracture of the vertebral endplate in the lumbar spine. *Acta Orthoped Scand* 1957; 25 (suppl):1–101.
24. *Military Specification.* General specification for aircraft upward ejection seat system. MIL-S-9479A. United States Air Force, 1967.
25. *Military Specification.* General requirements for capsule emergency escape systems. MIL-C-25969B. United States Air Force, 1969.
26. Stapp JP: Twenty-five years of Stapp Car Crash Conferences. *Proceedings of the Twenty-fifth Stapp Car Crash Conference.* SAE 811011. Warrendale, Pa, Society of Automotive Engineers, 1981, p 97.
27. *Hearings.* First Session on Crashworthiness of Automobile Seat Belts. April 30, August 5–8, 1957. *Hearings before a Subcommittee of a Committee on Interstate and Foreign Commerce.* House of Representatives, Eighty-fifth Congress. Washington, DC, US Government Printing Office, 1957.
28. *Act.* National Traffic and Motor Vehicle Safety Act of 1966. Public Law 89-563. Eighty-ninth Congress, S 3005. September 9, 1966.
29. McHenry RR: *Analysis of the Dynamics of Automotive Passenger Restraint Systems.* CAL Report No. VF-1823-R1. Buffalo, NY, Cornell Aeronautical Laboratory, May 31, 1963.
30. McHenry RR: Analysis of the dynamics of automobile passenger-restraint systems. *Proceedings of the Seventh Stapp Car Crash Conference.* Springfield, Il, Thomas, 1965, pp 207–249.
31. McHenry RR, Naab K: Computer simulation of the crash victim—A validation study. *Proceedings of the Tenth Stapp Car Crash Conference.* SAE 660792. New York, Society of Automotive Engineers, 1967, pp 126–163.
32. VonGierke HE: Symposium on biodynamic models and their applications. *Aviation, Space Environ Med* 1978; 49:109–348.
33. Hirsch A (ed): *Crashworthiness Modeling Workshop: Computer Simulation Programs.* Washington, DC, National Highway Traffic Safety Admin, Dept of Transportation, July, 1983.
34. Miley RC: The synthesis of the optimized impact environment. *Proceedings of the Ninth Stapp Car Crash Conference.* Minneapolis, Minn, Nolte Center for Continuing Education, University of Minnesota, 1966, pp 403–411.
35. Austin CE, Brauburger RA, Kansal SC: Computer-implemented design of automobile interiors. *Proceedings of the Tenth Stapp Car Crash Conference.* SAE 660791. New York, Society of Automotive Engineers, 1967, pp 116–125.
36. McHenry RR, Segal DJ, Deleys NJ: Computer simulation of single vehicle accidents. *Proceedings of the Eleventh Stapp Car Crash Conference.* SAE 670904. New York, Society of Automotive Engineers, 1969, pp 8–56.
37. McElhaney JH, Roberts VL, Melvin JW, et al.: Biomechanics of seat belt design. *Proceedings of the Sixteenth Stapp Car Crash Conference.* SAE 720972. New York, Society of Automotive Engineers, 1972, pp 321–344.
38. Robbins DH, Bowman BM, Bennett RO: The MVMA two-dimensional crash victim simulation. *Proceedings of the Eighteenth Stapp Car Crash Conference.* SAE 741195. Warrendale, Pa, Society of Automotive Engineers, 1974, pp 657–678.

39. Viano DC, Culver CC, Prisk BC: Influence of initial length of lap-shoulder belt on occupant dynamics—A comparison of sled testing and MVMA-2D modeling. *Proceedings of the Twenty-fourth Stapp Car Crash Conference.* SAE 801309. Warrendale, Pa, Society of Automotive Engineers, 1980, pp 377–415.
40. Tourin B: Ejection and Automobile fatalities. *Public Health Rep* 1958; 73:381–391.
41. Campbell BJ, Kihlberg JK: Seat belt effectiveness in the non-ejection situation. *Proceedings of the Seventh Stapp Car Crash Conference.* Paper 15. Springfield, Ill, Thomas, 1965, pp 177–188.
42. Bohlin N: A statistical analysis of 28,000 accident cases with emphasis on occupant restraint value. *Proceedings of the Eleventh Stapp Car Crash Conference.* SAE 670925. New York, Society of Automotive Engineers, 1969, p 20.
43. Nelson WD: Restraint system effectiveneess. *Proceedings of the Fifteenth Annual Conference of the American Association of Automotive Medicine.* Colorado Springs, Colo, 1971.
44. Nelson WD: *Lap-Shoulder Restraint Effectiveness in the United States.* SAE 710071. New York, Society of Automotive Engineers, 1971.
45. Bohlin N, Nordin H, Andersson A: A statistical traffic accident analysis. *Fourth International Experimental Safety Vehicle Conference.* Kyoto, Japan, March 1973.
46. Henderson JM, Wyllie JM: Seat belts—Limits of protection: A study of fatal injuries among belt wearers. *Proceedings of the Seventeenth Stapp Car Crash Conference.* SAE 730964. New York, Society of Automotive Engineers, 1973, pp 35–66.
47. Ryan GA: A study of seat belts and injuries. *Proceedings of the Seventeenth Stapp Car Crash Conference.* SAE 730965. New York, Society of Automotive Engineers, 1973, pp 67–79.
48. Patrick LM, Andersson A: Three-point harness accident and laboratory data comparison. *Proceedings of the Eighteenth Stapp Car Crash Conference.* SAE 741181. Warrendale, Pa, Society of Automotive Engineers, 1974.
49. Hartemann F, Thomas C, Henry C, et al.: Belted or not belted: The only difference between two matched samples of 200 car occupants. *Proceedings of the Twenty-first Stapp Car Crash Conference.* SAE 770917. Warrendale, Pa, Society of Automotive Engineers, 1977, pp 97–150.
50. Niederer PF, Walz FH, Zollinger U: Adverse effects of seat belts and causes of belt failures in severe car accidents in Switzerland during 1976. *Proceedings of the Twenty-first Stapp Car Crash Conference.* SAE 770916. Warrendale, Pa, Society of Automotive Engineers, 1977, p 73.
51. Beier G, Schuller E, Spann W: Risk and effectiveness of seat belts in Munich area automobile accidents. *Proceedings of the Twenty-fifth Stapp Car Crash Conference.* SAE 811023. Warrendale, Pa, Society of Automotive Engineers, 1981, pp 765–788.
52. Dalmotas DJ: Mechanisms of injury to vehicle occupants restrained by three-point seat belts. *Proceedings of the Twenty-fourth Stapp Car Crash Conference.* SAE 801311. Warrendale, Pa, Society of Automotive Engineers, 1980, pp 439–476.
53. Walz FH, Niederer PF, Thomas C, et al.: Frequency and significance of seat belt induced neck injuries in lateral collisions. *Proceedings of the Twenty-fifth Stapp Car Crash Conference.* SAE 811031. Warrendale, Pa, Society of Automotive Engineers, 1981, pp 131–146.

54. Leung YC, Tarriere C, Lestrelin D, et al.: Submarining injuries of three-point belted occupants in frontal collisions—Description, mechanisms and protection. *Proceedings of the Twenty-sixth Stapp Car Crash Conference.* SAE 821158. Warrendale, Pa, Society of Automotive Engineers, 1982, pp 173–205.
55. Severy DM, Brink HM, Baird JD, et al.: Safer seat designs. *Proceedings of the Thirteenth Stapp Car Crash Conference.* SAE 690812. New York, Society of Automotive Engineers, 1969, pp 314–335.
56. McElhaney JH, Roberts VL, Melvin JW, et al.: Biomechanics of seat belt design. *Proceedings of the Sixteenth Stapp Car Crash Conference.* SAE 720972. New York, Society of Automotive Engineers, 1972, pp 321–344.
57. Svensson LG: Means for effective improvement of the three-point seat belt in frontal crashes. *Proceedings of the Twenty-second Stapp Car Crash Conference.* SAE 780898. Warrendale, Pa, Society of Automotive Engineers, 1978, pp 451–479.
58. Adomeit D: Seat design—A significant factor for safety belt effectiveness. *Proceedings of the Twenty-second Stapp Car Crash Conference.* SAE 791004. Warrendale, Pa, Society of Automotive Engineers, 1979, pp 41–68.
59. Hontschik H, Muller E, Ruter G: Necessities and possibilities of improving the protective effect of three-point seat belts. *Proceedings of the Twenty-first Stapp Car Crash Conference.* SAE 770933. Warrendale, Pa, Society of Automotive Engineers, 1977, pp 795–831.
60. States JD, Miller SR, Seiffert UW: Volkswagen's passive seat belt/knee bolster restraint, VWRA: A preliminary field performance evaluation. *Proceedings of the Twenty-first Stapp Car Crash Conference.* SAE 770935. Warrendale, Pa, Society of Automotive Engineers, 1977, pp 863–910.
61. Viano DC, Culver CC: Performance of a shoulder belt and knee restraint in barrier crash simulations. *Proceedings of the Twenty-third Stapp Car Crash Conference.* SAE 791006. Warrendale, Pa, Society of Automotive Engineers, 1979, pp 107–131.
62. Clark C, Blechschmidt C, Gordon F: Impact protection with the airstop restraint system. *Proceedings of the Eighth Stapp Car Crash and Field Demonstration Conference.* Detroit, Mich, Wayne State University Press, 1966.
63. *Air Bag Restraint Systems, Special Bibliography 1967–Oct 1973.* HS-801033. Washington, DC, US Department of Transportation, 1973.
64. *Air Bag Restraint Systems, Special Bibliography 1964–Feb 1981.* PB 81-804197. Washington, DC, US Department of Transportation, 1981.
65. *Air Bag Restraint Systems, Special Bibliography 1970–Feb 1981.* PB 81-804205. Washington, DC, US Department of Transportation, 1981.

CHAPTER 5

Mathematical Models:
Animal and Human Models

Carley C. Ward, Gopal K. Nagendra

INTRODUCTION

The object of this chapter is to review the current generation of biodynamic models. In recent years a revolution in modeling technology has taken place. New powerful solution methods, coupled with inexpensive digital computers, have greatly expanded the modeler's capability. The most useful of these new procedures are related to the finite element method, which first became popular in the early 1970s. In this method, the biologic structure is divided into small blocks or pieces and is mathematically reassembled in the computer program.

Prior to 1970 only a limited number of discrete parameters were used; that is, most modelers used systems of lumped masses and linear springs to represent a biologic structure. Models of this type continue and have been greatly improved and expanded. Another type of model was used for those regions that could not be idealized with springs and masses. These regions were modeled as continuums, using the three-dimensional equations of elasticity. However, solution of these equations required that the structure be simplified to a uniform shape, such as modeling the skull as a closed spherical shell. Many times these gross simplifications produced misleading results. Because biologic stuctures are never simple or uniform, continuum models had little to offer and are no longer used.

Today, modelers can access powerful general purpose programs. The mathematical description of the physical phenomena is a system of differential equations. These can be solved by a variety of applied mathematical solution methods, e.g., eigenfunction series expansions, Laplace transformation methods, or numerical analysis using either a finite difference or a finite element approximation. The simultaneous solution of hundreds of nonlinear equations is possible. New digital solution procedures for large displacements, rotations, and nonlinear materials have also been implemented. Because these new solution procedures are common to many technologies, improvements are occurring at an accelerated rate. They provide the tools for great modeling advances in the near future.

CATEGORIES OF MODELS

Mathmatical models of human and animal structures can be divided into the following three groups: (1) lumped parameter, (2) distributed parameter, and (3) combined lumped and distributed parameter models. These idealizations can be used for static analysis, where forces are applied but the body does not move, and for dynamic analysis, where the forces and motions are changing with time. Since trauma involves motion and impact, trauma models are included in the latter group and are commonly referred to as "biodynamic models." Some biodynamic models concentrate on kinematics in that they describe the relationships between displacements and motions of the different body parts, while others emphasize both kinematics and kinetics. These latter models treat the forces applied to the biologic structure and their effect on the moving body parts. In the more detailed dynamic models, the distribution of force inside the tissue and bone is studied and internal stresses and strains are calculated.

Lumped Parameter Models
These models use common engineering analogs, such as springs and dampers, to represent the tissue, fluid, and bones (Fig. 1). Equations for each of these analogs are written, and the appropriate constants are selected. Springs, concentrated masses, concentrated inertias, and dashpots or dampers represent the stiffness, mass, inertia, and damping or viscoelastic effects, respectively. Linear or nonlinear components can be used. The model is developed by combining the component equations to represent the biologic structure. In general, fewer equations and degrees of freedom are used than in the other models. In these idealizations, nonlinearities are more easily handled, and simpler solution procedures are employed. If the primary purpose is to represent the anatomic kinematics, lumped parameter idealizations are excellent. They are commonly used for whole body simulations and articulated regions, such as the neck. However, these models can not calculate stresses in the tissues or trace distribution of force in the various internal body structures. Thus, they do not predict tissue failure. Only with extensive injury test correlation can the kinematics be related to trauma.

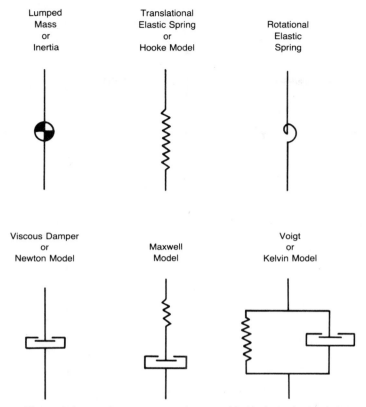

Figure 1. Lumped parameter analogs used in biodynamic simulation.

Distributed Parameter Models

Although some finite difference models have been developed, biodynamic models in this category are most often of the finite element type, since most of the advances in numerical methods are related to this method. Distribution of mass, material properties, and continuity of the structure are emphasized. Bones, soft tissue, fluids, and composite biologic tissue have all been modeled using this method. Structures to be analyzed are divided into small elements—blocks, tetrahedrons, plates, membranes—each having the continuum material properties of the host structure (Fig. 2). The mass is concentrated at points in the corners or along the sides of the elements. These points are referred to as "element nodes." Equations for nodes are developed and then combined to form a differential equation for the entire system. Using this technique unusual shapes and combined materials (bone, tissue, and fluid) can be analyzed.

This procedure requires many equations. Each node can have up to six degrees of freedom (six equations), and it is not uncommon to use more than 100 nodes. The volume of data and the number of calculations require

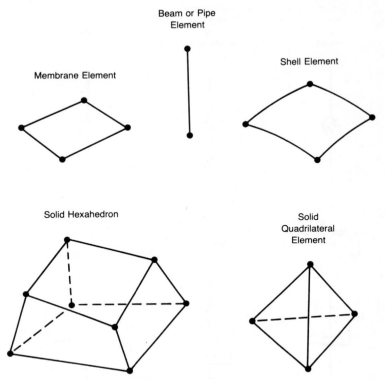

Figure 2. Finite elements used to represent biologic structures.

powerful data handling techniques and optimization procedures. Unfortunately, the cost of solution increases exponentially with the increase in the number of equations. Because of the size of the matrices, nonlinearities are not easy to handle; they increase cost and influence computational accuracy. Nonlinearities produced by large rotations and displacements are especially difficult to accommodate and require special numerical analysis techniques, which few finite element codes possess. For these reasons a finite element model of the entire body does not exist.

The value of finite element idealizations lies in their ability to calculate internal stresses and subtle internal motions and displacements. This information cannot be obtained by any other means. Using these idealizations, the researcher can obtain facts that could not have been anticipated. Since failures of tissue and bone can be predicted, these models are excellent for the study of trauma, provided the body part can be modeled.

Lumped and Distributed Parameter Models
These models are just beginning to make their appearance. To reduce the complexity of the finite element representation, regions of the body that are

remote from the area of interest are approximated with lumped parameter idealizations. For example, equivalent springs are used to represent the boundaries. In articulated structures the finite element calculations are done in local coordinates, while the lumped parameter equations are in the global system. The finite element solution procedure is modified to accept these approximations. Although they are more difficult to solve, these hybrid models avoid the disadvantages of lumped parameter idealization and make the finite element representation of a complex biologic structure possible.

ADVANTAGES

Mathematical models are valuable tools in the study of trauma. They can be used to predict body response to injury-producing conditions that cannot be simulated experimentally, and they can predict responses that cannot be measured in surrogate and animal experiments. It is unfortunate that in this field, where it is not possible to test directly on the subject, i.e., living humans, models have not been used more effectively. As stated by Liu,[1] when viewed as surrogate experiments, models offer an exceptional opportunity because they have absolute repeatability if done correctly. The investigator can vary any parameter in the smallest degree and measure the difference that particular change has on the final outcome.

Once developed, mathematical models are perhaps the most compact and economical of all research methodologies available. When combined with an experimental program, they provide a powerful means of extracting the major significance of experimental findings in animals and other surrogates. In addition, they provide a check on experimentally generated data records. Most important, mathematical modeling is the only means by which valid experimental animal and cadaveric data can be extrapolated to living man.

DIFFICULTIES

Some of the pitfalls of mathematical modeling are:

1. Oversophistication
2. Lack of validation
3. Lack of good physical properties data

It is often easier to make the model overly complex than to make limiting assumptions. This is especially true in finite elements, where it is easier to add elements than to make critical judgments. The result is a clumsy model that is expensive to use. The inclusion of unnecessary data can overshadow the valuable information, making the results difficult to interpret. In addition modelers tend to use their own programs, which are often undocumented codes, instead of the general purpose programs. When this is the case, the value of the model to the research community is limited.

The most important requirements for any biodynamic model is correlation with experimental test data. Modeling biologic systems requires many assumptions, and comparison with experiments is the only check on these assumptions. A model must be able to predict measured responses; otherwise it has no value and the solutions are mere mathematical exercises. In addition to being the most important requirement, correlation is the most difficult test for a model. Experimental tests are costly and data are difficult to obtain. Sometimes only a limited number of experiments are used for validation.[2] If the model's predicted response comes close to the measured results, the model is assumed validated, but this assumption is not necessarily correct, and the model's use in a different situation may be unsuitable. The mathematical idealization must be tested in many situations before it can be considered validated. Unfortunately, most mathematical analogs have limited use and survive only as long as their originator finds them interesting or the grant money lasts. As a result, few models have been so fully validated that they are considered reliable in all situations.

Properties of biologic tissue are difficult to define because they are nonlinear, nonelastic, and strain rate dependent. The body is composed of many different materials, each with its own mechanical properties. Physical properties data are usually screened from the literature, and that which is available is often meager, inaccurate, incomplete, or outdated.[2] In vivo soft tissue properties can be significantly different from those obtained from dissected specimens. The dynamic properties required for simulating impacts, events whose duration is measured in milliseconds, are especially difficult to obtain. It is sometimes necessary to assume material constants and verify them later through correlation with experimental data.

Other problems facing the modeler are the cost of the solution procedures and the computational accuracy of the results. Fortunately, these problems are being resolved by the rapid advancement in numerical methodology and computer technology. The advanced models of today will be commonplace in 5 years.

REVIEW OF BIODYNAMIC MODELS AND THEIR DEVELOPMENT

Many new mathematical idealizations are proposed every year. The object of this chapter is to review those models that can be applied to the study of trauma. Only those models that are in use or those developed since 1970 are included. For reviews of models developed prior to 1970 refer to King and Chou[3] and Liu.[1]

Grouping in this chapter is by body structure, whole body models being treated last. Because many models treat more than one body region, there is some overlap. Models that simulate the ribs are included in the torso section, although they may simulate the spine in detail. Spine models that treat only the cervical region are considered neck models. Femur models are included in the hip model section because the interaction between the femur and the hip joint is studied.

MODELS OF THE HEAD AND SKULL

Because of its shape, the head cannot be represented as a lumped parameter system. To study the motion of the head one might consider representing the skull with one mass and the brain with another, as did Roberts et al.[4] However, such a representation would be an oversimplification and would not provide information about head trauma. For this reason even the early head and skull models were of the continuum type, i.e., closed spheres and ellipsoids. Simple shapes had to be used in order to solve the equations. Because the skull has a unique shape and variable thickness, these early models had no value in the study of trauma.

Hardy and Marcal developed the first finite element model in 1971.[5] In this static model, actual skull geometry was approximated with triangular elements. Other finite element models of the skull followed. Some researchers continued to impose axial symmetry and to model the head as a sphere,[6,7] but others attempted to approximate the exact shape.[8] In the most complex model, Shugar[9] approximated the skull with three layers of eight-node brick elements, representing the inner and outer layers of dense bone and the center porous layer. This linear dynamic model was limited in application because rotations of the head were not permitted and translation of the head required a special version of the program. Chan and Ward[10] developed a simpler skull model using four-node shell elements. The thickness was varied to approximate the thickness distribution of an actual skull. The model was designed to study the effects of dynamic skull deformation on brain pressures and was restricted to head translations. Hosey and Liu[11] also developed a linear dynamic skull model composed of shell elements with uniform thickness. This model is also restricted to head translation.

Skull models have not been dynamically validated for the following reasons: (1) the neck, which is the boundary condition, is difficult to represent mathematically, (2) face and scalp effects are difficult to include, and (3) structural variations in the skull present difficult modeling problems. Skull stresses, strains, and displacements are exponential functions of skull thickness, and skull thickness varies from point to point and from skull to skull. Consequently, even small inaccuracies in skull thickness, due to the straight element edges, produce errors in the computed response.

Although skull models cannot be used to predict dynamic skull stresses and strains, they are used to study response trends. They are also useful in studying extreme conditions, such as estimating the maximum possible skull stress in a given event.

MODELS OF THE BRAIN

In most of the early models, the brain was treated as an incompressible fluid. It was completely enclosed inside either spherical or elliptical shells. These simplifications severely limited the usefulness and accuracy of these models. Because the human skull has openings and fluid flows into and out of the cranial cavity, modeling the brain as an incompressible material inside a closed shell is not an adequate representation. In these models, a small skull deformation, such as that which would result from a moderate impact, initiated high pressures in the simulated brain. Generally, pressure predictions were many times greater than the experimentally measured pressures.

The first finite element model approximating the exact geometry of the brain was developed by Ward in the early 1970s and was reported in 1974[12] (Fig. 3). The model was refined, and other models of test animal brains were developed from 1975 to 1978.[13-15] In these models, the skull is not simulated; only the internal shape of the skull is modeled, forming a container for the brain. The partitioning internal folds of dura, the foramen magnum, and opening for the cervical cord are represented. These are linear elastic models capable of handling large head translations and rotations. Special elements were developed to represent the nearly incompressible brain material, and a general purpose finite element computer program was modified to accommodate the large head rotations and translations. Later an attached skull and cervical cord were added.[16] Hosey and Liu, in 1979, added other head structures as well as a spine.[11]

The brain response in the Ward models is described as follows: The inertia of the brain produces stresses and strains in the tissue as the brain tends to lag the motion of the skull. Brain tissue compresses against the skull near the impact site and is in tension opposite the impact. The result is a pressure or stress gradient through the brain. When the head rotates, the brain again lags behind, producing stresses and strains on the brain–skull interface. The falx and tentorium help to keep the brain in position, and instead of a single rotational displacement inside the brain, separate rotational motions develop in each compartment. The rotations in the cerebrum and cerebellum produce a complex interaction with the brain stem and the interconnecting brain structure.

The Ward idealizations of the brain are perhaps the most extensively tested biodynamic finite element models.[16-19] In the simulation of brain pressure, experiments have accurately predicted brain pressures in more than 50 simulations correlating with hundreds of data traces. These models have also been useful in predicting brain injury in animal tests and other repressurized human cadaver subjects.

The results from these simulations show that the following design requirements must be met if the correct response is to be obtained:

1. The model must have approximately the same size and shape as the actual brain.

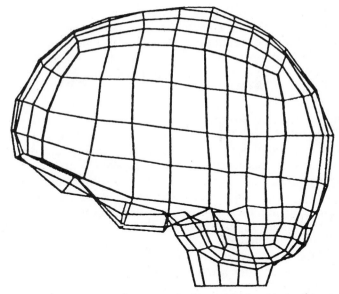

Figure 3. Finite element model of the human brain used for head impact simulations.

2. The opening in the base of the skull, the foramen magnum, acts as a pressure-release mechanism for the brain through which tissue and fluids move between the brain and spinal cord. This pressure-release mechanism must be simulated in the model.
3. The internal folds of dura, the falx, and the tentorium partition the cranial cavity and influence the brain response. These membranes must be represented.
4. The brain cannot be modeled as incompressible: the fluid flows into and out of the cranial cavity, and the movement of tissue through the foramen magnum provides some effective compressibility or volume elastance.

MODELS OF THE NECK

Models of the neck are primarily of the lumped parameter type and are essentially kinematic models. That is, they are designed to predict or approximate the motion of the neck. Springs and dampers are used to represent the effect of the disks and muscles and rigid masses represent the vertebrae. Large displacements and rotations are simulated.

Melvin et al.[20] in their cervical model used a three-parameter viscoelastic representation for each disk. Mertz and Patrick[21] developed equilibrium equations for the neck and correlated the results with human volunteer data and human cadaver injury experiments. In this study, tolerance levels for the neck are predicted. The Bowman model,[22] another simple idealization, has two ball-and-socket joints and an extensible neck. One joint is at the base of the cranium, and the other is at the seventh cervical vertebra. A Kelvin lumped parameter idealization is used for the neck. Joint stop ellipses are developed to limit motion at the joints simulating the anatomic limits. The effect of muscle contraction is also considered. Reber and Goldsmith[23] devised a somewhat more complex representation, using a total of 76 individual components and modeling the occiput to T3. Springs and dampers in parallel represent muscle, and axially and diagonally directed springs and dampers represent the disks. Suh,[24] using spring damper pairs and rigid bodies, developed a set of 96 first-order equilibrium equations to describe his neck idealization. Huston and Passarello[25] connected the rigid vertebrae with nonlinear springs and dampers to represent the disks, ligaments and muscles of the neck. Each rigid body has 6 degrees of freedom, the entire system having a total of 54 degrees of freedom. In the axial direction the disks are represented as two-parameter viscoelastic solids. In bending and shear, they are treated as linear elastic solids, approximating an elastic finite element. The ligaments are represented as nonlinear elastic bands capable of exerting force only in tension, and the muscles are represented as two-parameter, viscoelastic solids that only exert force when in tension.

Pontius and Liu[26] modeled the disk as a beam, elastic in shear and bending and viscoelastic in axial deformation. Neuromusculature is represented in detail with series and parallel springs. The contractile structure is represented using a dashpot in parallel with an isometric force generator. This model shows the effect of neuromusculature on the rotational stability at low levels of accelerations. Decreased bending is accompanied by increased compressive stress. Using experimental data directly to define a neck idealization for lateral impact, Wismans and Spenny[27] devised a simple linkage mechanism. The system has a lower pivot with one degree of freedom in the upper thorax and an upper pivot with two degrees of freedom at the occipital condyles. A simple four-mass model was developed by McElhaney et al.[28] to study vertex impacts. The neck is modeled with a single spring and damper, and the results have been correlated with neck injury data. The neck motions predicted by all of these models have been compared to experimental data either from human volunteer tests or cadaver head impacts. Good agreement between the calculated and predicted kinematics is claimed in every case.

Finite element modeling of the neck for arbitrary three-dimensional motion has not been done. The large rotations and displacements, and the many different materials included in the neck, make a finite element representation extremely difficult. A nonlinear solution procedure along with nonlinear elements would be required. Only Hosey and Liu[11] have used finite elements to represent the spine and cervical cord, but their model is restricted to small displacements and rotations, and the cervical portion of the model has not been experimentally verified. The only other finite element representation of a neck structure is a cervical cord model that was developed by Ward et al.[15] This model was designed to study the displacement interaction between the cord and brain stem in head impacts.

Because a complete, validated finite element model of the neck does not currently exist, stresses in the disks, ligaments, and bone cannot be correlated with injury. Neck tolerances can only be addressed indirectly using the lumped parameter kinematic models. A hybrid-type neck model combining the finite element representations with lumped parameters would be extremely useful.

MODELS OF THE SHOULDER

Modeling of the shoulder is difficult because of the lack of biomechanical data as well as the complexity of the region from an anatomic and kinesiologic point of view. Engin employed position vectors to describe the position of the arm relative to the body.[29] Using nonlinear matrix analysis and coordinate transformation, the forces, moments, and torques in the joint are studied. Computations are correlated with measured shoulder forces, torques, and moments obtained in tests on human volunteers.

MODELS OF THE SPINE

Modelers of single disk and disk–vertebra spinal units have made extensive uses of finite elements. These models have been useful in determining the load distribution and deflections in the disk anulus and nucleus pulposus and the distribution of force on the vetebral endplate. When treating the entire spine, it is not possible to treat the disk in as much detail because of the degrees of freedom that would be required. Usually in these whole spine models, single-beam elements represent the disk, rigid bodies represent the vertebra, and lumped parameter springs and dampers represent the muscles and connective tissue.

Vertebra and Disk Models
Liu and Ray[30] developed an interesting system identification scheme for the generalized relaxation modulus of the intervertebral disk subject to combined axial and shear loading. This is obtained by minimizing the error between the experimentally obtained displacement vector and the model-predicted deformation for each incremental time step. The scheme consists of constructing a two-dimensional linear viscoelastic finite element model of the disk using constant strain triangular elements. Kelly et al.[31] also, in a combined experimental and analytic investigation, modeled the viscoelastic

shear response of the disk using lumped parameters. A series of Kelvin unit idealizations were employed.

A three-dimensional linear finite element model of the intervertebral disk has been used by Veno et al.[32] to study the effects of geometry and material properties on the disk behavior due to axial compression. The model is sagittally and horizontally symmetrical, including a cartilaginous endplate (homogeneous, isotropic), anulus fibrosus (inhomogeneous, aniso-tropic), and nucleus pulposus (incompressible). Uniform axial displacements were applied on the superior surface of the model.

Hakim and King[33] have formulated a finite element model of a vertebra using a mesh generation technique. Three-dimensional, eight-node, isoparametric hexahedral elements represent the spongy bone. Thin plates simulate the endplates, while thin shell elements represent the cortex of the vertebral body. The common nodes between the hexahedral elements and plate or shell elements are restricted to the transfer of interelement shear loads only. The inferior intervertebral disk is represented by boundary elements, each of which is a one-degree of freedom spring acting at a node to resist axial, shear, or torsional loads. These boundary elements simulate the shear and torsional resistance of the intervertebral disk.

Yang and King[34] developed a finite element model of a functional spinal unit based on the single vertebra model of Hakim and King.[33] Newly acquired data on the stiffness of facet joints was utilized. The results show that the response of the facets was highly nonlinear and completely different in tension and compression. The facet joint is weak in tension, and the spinal column relies on the soft tissues between the processes and the spinal extensions to provide tensile resistance in spinal flexion. The compressive response is controlled by the bony contact of the inferior tip of the interior facet with the pars of the vertebra below it.

Complete Spine Models

A mathematical model of the complete spine considering axial, shear, and bending deformations was developed by Orne and Liu.[35] Each disk is essentially a finite element beam that is elastic in shear and bending and viscoelastic in axial deformation. Prasad and King[36] devised a 78-degree of freedom model of the spine, head, and pelvis. Twenty-six rigid links are used to represent the head, twenty-four vertebrae, and the pelvis. Intervertebral disks are represented by deformable, massless links that resist axial, shear, and bending deformation. The laminae and articular facets are treated as load-bearing members. In this two-dimensional model, eccentric loading of the spine and spinal curvature is included. King et al.[37] in a conceptually similar model include the effect of spinal musculature.

Privitzer et al.[38] developed combined lumped and distributed parameter models of human and baboon head–spine structures. These idealizations consist of three-dimensional assemblages of rigid bodies and deformable elements, for which the equations of motion are solved using a large-displacement, small-strain structural analysis program. The rigid bodies represent the inertial characteristics of the head, pelvis, sections of the torso, and some skeletal components, such as the ribs. Each vertebra is defined by a set of 13 points and is contained in a rigid body, which includes the inertial properties of a section or a portion of a section of the torso. Torso sections corresponding to a specific vertebral level are defined as the material bounded by parallel planes perpendicular to the vertical axis passing through the center of the inferior and superior intervertebral disks and by the torso wall. The rigid bodies interact through deformable elements (springs and beams) that represent the various connective tissues: the intervertebral disks, the spinal ligaments, the articular facets, the viscoelastic properties of the viscera–abdominal wall system, the costovertebral and costotransverse joints, the interchondral cartilage and intercostal tissues of the rib cage.

Jayaraman,[39] in his detailed spine models, determines the compression force and bending moment acting on the vertebra–disk interface. The induced tensile stresses in the anulus fibrosus is computed. Two approaches are used. In the first, the spine with its ligaments and back muscles is simulated as an elastic beam of two materials, with the plane of loading coinciding with the plane of the curved beam. In the second approach, a finite element model of the whole spine is used. The discretized elements of the elastic spinal structure have three degrees of freedom, and the back muscles can only take tension. The nodes of the spinal structure correspond to the intervertebral joint locations. The spine and back muscle finite element structure is then analyzed by applying the loading incrementally. Force and moment output from this model is applied to the vertebra–disk interface as distributed axial and shear forces plus a central moment on a finite element model of the disk made of two components: the nucleus pulposus cavity containing fluid and the surrounding anulus fibrosus elastic medium. The stress analysis of the anulus fibrosus is carried out.

A mathematical model using a lumped parameter torso, head, and helmet, which is capable of simulating displacement and time-dependent applied loads, has been developed by Sances et al.[40] Both slowly and rapidly applied axial loads were studied. This model is correlated with tests on isolated fresh cadaveric monkey cervical columns and on intact living and dead monkeys. Similar investigations were conducted on fresh human spinal columns.

MODELS OF THE THORAX

Inferior–Superior Acceleration Models
One of the better known analytic models is the dynamic response index (DRI) model, which is essentially a one-degree of freedom lumped mass

spring model of the head-spine torso.[41] It has been correlated with injury data and provides a useful criterion for evaluating the safety of an axial inferior–superior acceleration environment when the nonaxial response characteristics are minor.

When other than the axial response must be considered, such simple models as the DRI become less attractive because parameters, such as curvature of the spine, forward tilt, or asymmetric head mass, are not included. For these purposes, detailed idealizations are required.

Belytschko et al.[42] developed a three-dimensional large displacement model for the head–spine torso system consisting of rigid bodies interconnected by deformable elements. This was one of the first models to combine finite element with a lumped parameter representation. The deformable spring and beam elements represent the intervertebral disks, ligaments, and connective tissues.

Each pair of vertebrae is connected by seven spring elements and one beam element. The beam element represents the intervertebral disk, joining the endplate centers of adjacent vertebrae. The spring elements represent the following ligaments and connective tissues: the two intertransverse ligaments, the intraspinous and supraspinous ligaments, the ligamenta flava, and the articular facet joints. The facet joints are short, stiff elements that are primarily intended to represent the kinematic constraints.

Since the ribs themselves are rigid bodies, deformation of the rib cage results from the rotation of the ribs and the deformation of the costosternal cartilage. Each rib is connected to two vertebrae by three deformable elements representing the costovertebral joint. These elements were designed so that the directional properties of the anatomic joint, which exhibits large rotational flexibility about one axis, are reproduced.

The viscera are modeled by a set of confined hydrodynamic elements, and the loads transmitted through the viscera are transferred to the ribs at the T10 level. This simulates the transfer of loads to the spine through the diaphragm.

In addition to predicting body kinematics, the model predicts axial forces and moments in the body segments as functions of time. These forces and moments in the vertebral levels are converted to stresses.

Anterior–Posterior Acceleration Models

Viano[43] developed a lumped parameter model for the anterior–posterior thoracic impact response of the human thorax. His objective was to establish a comparison with available cadaver impact data and to develop curves of constant biomechanical response and injury potential that would be applicable over a wide range of frontal impact exposures. Three interconnecting lumped masses are used, and four differential equations (i.e., eight parameters and four degrees of freedom) derived from a mechanical analog formed from springs, masses, and dashpots. Equation parameters are adjusted to closely correlate the model response with established force-deflection response corridors suggested to be representative of human response based on cadaver and human volunteer tests.

Chen[44] developed a linear, dynamic finite element model of the thorax based on a modal synthesis technique. Rib geometry is modeled in detail using elastic 3-D beam elements. Linear elastic properties of the bone and cartilage are assumed. Although small-deformation theory was used in the ribs themselves, correlation with experimental chest deflection data is good. This model was later extended by Roberts,[45] and a similar model of the child's thorax was developed. The later Roberts models were solved by direct integration of the finite element matrix equations. A large-deflection model of the thorax was attempted by Reddi et al.[46] in the late 1970s. He developed a large-deformation element for the rib, and a rib cage composed of 32 elements was proposed. Some calculations for a single fourth rib were shown; however, the model was never completed.

MODELS OF THE HIP

Many finite element models of the hip have been formulated for the study of artificial joint replacement. Using these models, a complete stress analysis of the total joint replacement is possible. This type of analysis is essential in determining the strength and deformation characteristics needed for implant materials. Such analysis is also helpful in the selection of optimum prosthesis designs to prevent loosening fracture and concentrated loading on the bone. The loading on these models is usually static or quasi-static, which is characteristic of normal walking. However, if the loading times are shortened and the dynamics of the bone are considered, these models could simulate a traumatic event.

The first femur models were developed by Rybicki et al. in 1972,[47] Brekelmans et al.,[48] and Wood et al. in 1973.[49] Rybicki used two-dimensional elements for the femur, while Brekelmans used constant bone thickness. Wood, in a more detailed model, used isoparametric elements, nonhomogeneous material properties, and nonuniform thickness of the bone. The later models showed that a beam theory representation, used earlier, was not adequate. Wood[50] and Olofsson[51] analyzed the femur without implant, using three-dimensional finite elements. Valliappan et al.,[52] using a three-dimensional analysis, compared their results with experimental values. Additional three-dimensional finite element investigations were reported by Vichnin and Batterman[53] and Harris et al.[54]

Analysis of the combined femur and prosthesis began in 1975, when Andriacchi et al.[55] performed a two-dimensional analysis of an implanted hip. Finite element analysis of hip prosthesis continued in 1975 and 1976 and includes work by McNeice and Amstutz,[56] Forte,[57] Bartel and Ulsoy,[58] Bartel and Desormeaux,[59] and Bartel and Samahyek.[60] In 1977 Svensson et al.[61] were the first to investigate the influence of interface conditions by incorporating a slip and no tension criteria in their finite element analysis. Scholten et al.,[62] in 1978, used a three-dimensional model of nonhomogeneous material properties and a slip criteria on the interface. Other three-dimensional analyses of the combined femur and implanted prosthesis were performed by Crowninshield et al.,[63] Tarr et al.,[64] and Huiskes and Slooff.[65]

In 1980 Valliappan et al.[66] reported results on the influence of anisotrophy and the placement of the prosthesis. They used a three-dimensional model with 360 nodes, which simulated the bone, cement, and prosthesis. Also in 1980, Rohlmann et al.[67] computed stresses and strains while including the influence of forces in the iliotibial tract. This work was correlated with an experimental program where an isolated femur was instrumented with three strain gauges. Rohlmann's is one of the most detailed finite element models in existence. He uses 1950 volume elements with 2532 nodal points, resulting in 7188 equations (degrees of freedom). Tarr et al.[68] evaluated prosthesis designs and the effect of their stiffness. They also use a detailed three-dimensional model with 896 elements and 1215 nodes that has been verified experimentally.

Hampton et al.[69] investigated the stress distributions in an implanted femoral stem. Three-dimensional hexahedron elements are used to analyze the structure. Intraelement material property variations are achieved by assigning specific material property characteristics at each integration point within the element during its stiffness formulation. The hexahedrons are assigned the material properties of bone to the integration points in the regions of the element representing bone, and the material properties of cement to the region of the element representing cement.

MODELS OF THE LEG

The transmission of impact forces in a straight leg, from the foot to the level of the greater trochanter, has been investigated by Mizrahi and Susak,[70] using a two-degree of freedom linear damped spring model. A linear mathematical model is employed to describe the mechanical behavior of both bone and soft tissue of the leg as well as the whole body.

Hight et al.[71] developed a dynamic, nonlinear finite element model of the leg. The model accommodates large, three-dimensional displacements and rotations to accurately reflect the nonlinear stiffness characteristics of the knee joint. The undamped, passive model considers two types of nonlinearities: the geometric nonlinearity to account for the large displacements that a human leg can undergo and the nonlinearity within the joints. A beam element with six degrees of freedom per grid is used, and an incremental solution procedure is employed with force equilibrium at each load step.

MODELS OF THE KNEE

Although the knee is frequently injured, there are only a few models of this joint, and most of these are for static or quasi-static analysis. Kinematic models of the knee were developed by Huson[72] and Menschik[73] using four bar mechanisms. Later Andriacchi et al.,[74] in a combined lumped parameter and finite element model, analyzed the distal portion of the femur and proximal portion of the tibia, the interconnecting ligaments, joint capsule, joint surface, and meniscus. Bony portions are considered rigid, while ligaments, soft tissue, and joint capsules are represented by 28 uniaxial springs. The menisci are represented with 2 disklike shear beams, and the joint surfaces are represented with 12 hydrostatic elements that permit rolling contact. Force distributions in the knee joint during the walking cycle are calculated.

Treating the same quasi-static interactions, Wismans et al.[75] in 1979 developed a three-dimensional model that took into account the geometry of the joint surfaces. These surfaces are approximated with polynomials, and the ligaments and capsules are represented with nonlinear springs. For applied forces and moments at various flexion, extension angles for the following are calculated: location of the contact points, the magnitude and direction of the contact forces, and ligament elongation and forces.

In 1980 Moeinzadeh et al.[76] developed a two-dimensional nonlinear dynamic model of the knee joint in which profiles of the joint surfaces are represented with polynomials. Joint ligaments are modeled as elastic springs. Dynamic loads are applied to the center of mass of the tibia, and the resulting motion is studied.

MODELS OF THE WHOLE BODY

Many lumped parameter models for the prediction of gross body motion are described in the literature, with a history of their developments given by King and Chou.[3] Although they are primarily for automobile crash victim simulations, they have been extended to other injury conditions. The models are essentially computer programs that have been developed to solve the large-displacement and rotation equations generated in such simulations. The dimensions and mass of the body points can usually be changed to represent an accident victim. Hinges or rotational springs are used to connect the body segments, as shown in Figure 4. Initial conditions are input, and forces and/or motion are mathematically imposed on the body. The earlier versions were two-dimensional models designed for front or rear impacts. Later versions have full three-dimensional capability. Impressive graphics programs have been interfaced with these programs so that body motion can be visualized throughout the event.

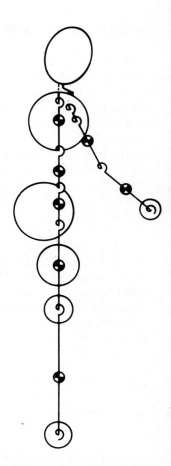

Figure 4. Typical articulated whole body model for two-dimensional motion.

Calspan Simulator,[77] a widely used whole body idealization, was developed for the National Highway Traffic Safety Administration. The model is flexible and modular in design so that the complete range of anthropometric variation, weight distribution, moment of inertia of segments, and joint limiting angles can be handled effectively. The occupant can be modeled by up to 20 segments and connected by 19 joints. Tension elements and spring dampers facilitate the representation of muscles and ligaments. Omnidirectional impact and dynamic initial conditions may be specified.

Validation of the model has been reported.[78,79] The model has also been modified for special applications, one of these being the simulation of aircraft pilots. Seat belt interaction and windblast forces have been added. The graphic representation has also been dramatically upgraded to generate computer plots of the body surface.

A model developed by Huston et al.[80] uses 12 rigid bodies to represent the human limbs, thorax, abdomen, and head. Body segments are represented by elliptical cylinders and ellipsoids. They are connected with simulated ball-and-socket joints having limit angle stops. The simulated body is positioned on a seat represented with springs and viscous damper stops. Neck stretch is modeled with a spring and damper at the attach point. Thirty-one degrees of freedom are used.

A 12-mass, 31-degree of freedom model was also developed by Young et al.[81] Body segments are represented by ellipsoids that are connected with ball-and-socket and hinged joints. This model was used to simulate belted and unbelted drivers in frontal impacts and to simulate a pedestrian in a vehicle–pedestrian impact. Good correlation with experimental data is reported.

The Highway Safety Research Institute (HSRI) at the University of Michigan began with a two-dimensional gross motion simulator. It was first reported in 1970[82] and consisted of eight segments. This model was similar to the two-dimensional Motor Vehicle Manufacturer's Association model. Validation was performed at HSRI using an anthropomorphic test dummy. The model was extended to three dimensions and at first consisted of three masses, representing the head, torso, and, legs. It was later extended to six masses and 14 degrees of freedom, with 20 ellipsoids representing the body segments.

MADYMO,[83] a crash victim simulator, has been used in many studies and experimentally validated. It is a computer program that can simulate crash victims in either two or three dimensions. Kinematic and dynamic behavior of the body is simulated. The number of linkages and number of elements in each linkage can be specified. Children as well as adults can be simulated. There is also flexibility in the treatment of force interactions between elements and the surrounding structure. MADYMO has been used for biomechanical crash research as well as for the development and optimization of crash safety devices, such as seat belts, child seats,[84] and vehicle padding.[85]

REFERENCES

1. Liu YK: *Biomechanics and Biophysics of CNS Trauma.* Central Nervous System Trauma Research Status Report, National Institute of Health, 1979, pp 36–52.
2. Panjabi M: Validation of mathematical models. *J Biomech* 1979; 12(3): 238.
3. King AI, Chou CC: Mathematical modeling, simulation and experimental testing of biomechanical system crash response. *J Biomech* 1976; 9(5): 301–317.
4. Roberts SB, Ward CC, Nahum AM: Head trauma—a parametric dynamic study. *J Biomech* 1969; 2:397–415.
5. Hardy CH, Marcal PV: *Elastic Analysis of a Skull.* Technical Report No. 8, Office of Naval Research, Contract No. N00014-67-A-0191-0007, Division of Engineering, Brown University, 1971.
6. Khalil TB, Goldsmith W, Sackman JL: Impact on a model head-helmet system. *Int J Mech Sci* 1974; 16:609–625.
7. Khalil TB, Hubbard RP: Parametric study of head response by finite element modeling. *J Biomech* 1977; 10:119–132.

8. Merchant HC, Crispino AJ: A dynamic analysis of an elastic model of the human head. *J Biomech* 1974; 7:219–228.
9. Shugar TA: Transient structural response of the linear skull-brain system. *Proceedings of the Nineteenth Stapp Car Crash Conference.* SAE 751161. Warrendale, Pa, Society of Automotive Engineers, 1975, pp 581–614.
10. Chan M, Ward C: Relative importance of skull deformation. *Proceedings of the ASME Biomechanics Symposium,* June 22–24, 1981.
11. Hosey R, Liu YK: A homeomorphic finite-element model of impact head and neck injury. *Proceedings of the International Conference on Finite Elements in Biomechanics,* Tucson, Ariz, 1980, pp 829–850.
12. Ward CC: *A Dynamic Finite Element Model of the Human Brain,* PhD dissertation, University of California, Los Angeles, 1974.
13. Ward CC, Thompson RB: The development of a detailed finite element brain model. *Proceedings of the Nineteenth Stapp Car Crash Conference.* SAE 751163. Warrendale, Pa, Society of Automotive Engineers, 1975, pp 641–674.
14. Ward CC: Analytical brain models for head impact. *Proceedings of the International Conference on Impact Trauma,* IRCOBI, Berlin, Germany, Sept 1977.
15. Ward CC: Finite element modeling of the head and neck. *Impact Injury of the Head and Spine.* Springfield, Ill, Thomas, 1983, p 421.
16. Nusholtz GS, Axelrod J, Melvin J, et al.: *Comparison of Epidural Pressure in Live Anesthetized and Post-Mortem Primates.* Highway Safety Research Institute Report No. UM-HSRI-79–90, 1979.
17. Nahum AM, Smith RM, Ward CC: Intracranial pressure dynamics during head impact. *Proceedings of the Twenty-first Stapp Car Crash Conference.* SAE 770922. Warrendale, Pa, Society of Automotive Engineers, 1977, pp 337–366.
18. Nahum AM, Ward CC, Smith R, et al.: Intracranial pressure relationships in the protected and unprotected head. *Proceedings of the Twenty-third Stapp Car Crash Conference.* SAE 791024. Warrendale, Pa, Society of Automotive Engineers, 1979, pp 613–636.
19. Nahum A, Ward C, Schneider D, et al.: A study of impacts to the lateral protected and unprotected head. *Proceedings of the Twenty-fifth Stapp Car Crash Conference.* SAE 811006. Warrendale, Pa, Society of Automotive Engineers, 1981, pp 241–270.
20. Melvin JW, McElhaney JH, Roberts VL: Improved neck simulation for anthropomorphic dummies. *Proceedings of the Seventh Annual Stapp Car Crash Conference.* SAE 720958. Warrendale, Pa, Society of Automotive Engineers, 1963, pp 45–60.
21. Mertz HJ, Patrick LM: Strength and response of the human neck. *Proceedings of the Fifteenth Stapp Car Crash Conference.* SAE 710856. Warrendale, Pa, Society of Automotive Engineers, 1971, pp 207–255.
22. Bowman BM, Robbins DH: Parametric study of biomechanical quantities in analytical neck models. *Proceedings of the Sixteenth Stapp Car Crash Conference.* SAE 720957. Warrendale, Pa, Society of Automotive Engineers, 1972, pp 14–144.
23. Reber JG, Goldsmith W: Analysis of large head-neck motions. *J Biomech* 1979; 12:211–222.
24. Suh CH: Dynamic simulation of the cervical spine with use of differential displacement matrix. *Proceedings of the Third International Conference on Impact Trauma,* IRCOBI, Berlin, Germany, Sept 1977, pp 355–365.

25. Huston RL, Passerello CE: *Multibody Dynamics Including Translation Between the Bodies—with Application to Head-Neck Systems.* Technical Report under Office of Naval Research Contract N00014-76-C-0139, 1978.
26. Pontius UR, Liu YK: Neuromuscular cervical spine model for whiplash. *Mathematical Modelling Biodynamic Response to Impact.* Michigan, Society of Automotive Engineers, Oct. 18–22, 1976, pp 21–30.
27. Wismans J, Spenny CH: Performance requirements for mechanical necks in lateral flexion. *Proceedings of the Twenty-seventh Stapp Car Crash Conference.* SAE 831613. Warrendale, Pa, Society of Automotive Engineers, 1983.
28. McElhaney J, Roberts V, Paver J et al.: *Etiology of Trauma to the Cervical Spine. Impact Injury of the Head and Spine.* Springfield, Ill, Thomas, 1983, pp 41–71.
29. Engin AE: On the biomechanics of the shoulder complex. *J Biomech* 1980; 13(7):575–590.
30. Liu YK, Ray G: A systems identification scheme for the estimation of the linear viscoelastic properties of the intervertebral disc. *Symposium on Biodynamic Models and their Applications, Wright-Patterson A.F.B., Ohio,* Feb 15–17, 1977, pp 53–57.
31. Kelly BS, Lafferty JF, Bowman DA, et al.: A viscoelastic model of the shear response of the intervertebral disc. *Biomechanics Symposium.* ASME, AMD, 1981, Vol 43, pp 145–147.
32. Veno K, Takahata T, Liu YK: Finite element analysis of an intervertebral disk. *Proceedings of the Eighteenth Midwestern Mechanics Conference,* University of Iowa, Iowa City, May 16–17, 1983, pp 481–484.
33. Hakim NS, King AI: A three-dimensional finite element dynamic response analysis of a vertebra with experimental verification. *J Biomech* 1979; 12:4.
34. Yang KH, King AI: Parametric study of a model of a functional spinal unit. *Proceedings of the Eighteenth Midwestern Mechanics Conference,* University of Iowa, Iowa City, May 16–18, 1983, pp 477–480.
35. Orne D, Liu YK: A mathematical model of spinal response to impact. *J Biomech* 1971; 4(1):49–71.
36. Prasad P, King AI: An experimentally validated dynamic model of the spine. *J Appl Mech* 1974; 41:546–550.
37. King AI, Nakhla SS, Mital NK: Simulation of head and neck response to $-Gx$ and $+Gx$ impact. *AGARD Conference Proceedings No. 253: Models and Analogues for the Evaluation of Human Biodynamic Responses, Performance and Protection.* Paris, France, Nov. 6–10, 1978, pp A7-1 to A7-13.
38. Privitzer E, Hosey RR, Ryerson JE: Validation of a biodynamic injury prediction model of the head–spine system. AGARD-CP-322. April 26–29, 1982, pp 30.1–30.10.
39. Jayaraman G: Biomechanical modelling of spine. *Third International Conference on Mathematical Modelling.* Los Angeles, Calif, July 29–31, 1981, p 55.
40. Sances A Jr, Myklebust J, Houterman C, et al.: Head and Spine Injuries. AGARD-CP-322, April 26–29, 1982, pp 13.1–13.34.
41. Payne PR: Some aspects of biodynamic modelling for aircraft escape systems. *Proceedings of a Symposium on Biodynamic Models and Applications.* Wright-Patterson AFB, Ohio, 1972.
42. Belytschko T, Schiver L, Privitzer E: Theory and application of a three dimensional model of the human spine. *Symposium on Biodynamic Models*

and their Applications, Bergamo Center, Dayton, Ohio, Feb 15–17, 1977, pp 38–42.

43. Viano DC: Evaluation of biomechanical response and potential injury from thoracic impact. *Symposium on Biodynamic Models and their Applications,* Bergamo Center, Dayton, Ohio, Feb 15–17, 1977, pp 15–19.
44. Chen PH: Finite element dynamic structural model of the human thorax for chest impact response and injury studies. *Symposium on Biodynamics Models and their Applications,* Bergamo Center, Dayton, Ohio, Feb 15–17, 1977, pp 27–31.
45. Roberts SB: The dynamic response of a child's thorax to blunt trauma. *Proceedings of an International Meeting on Biomechanics of Trauma in Children.* Lyon, International Research Committee on the Biokinetics of Impacts, 1974, pp 237–245.
46. Reddi MM, Tsai HC, Ovenshire L: Computer simulation of human thoracic skeletal response. *International Conference on Finite Elements in Biomechanics.* Tucson, Ariz, University of Arizona Press, 1980, pp 871–888.
47. Rybicki EF, Simonen FA, Weiss EB: On the mathematical analysis of stress in the human femur. *J Biomech* 1972; 5:203–215.
48. Brekelmans WAM, Poort HW, Slooff TJJH: A new method to analyze the mechanical behavior of skeletal parts. *Acta Orthop Scan* 1972; 43:301–317.
49. Wood R, Valliappan S, Svensson NL: Stress analysis of human femur, in Yamada Y, Gallagher RH (eds): *Theory and Practice in Finite Element Structural Analysis.* Tokyo, University of Tokyo Press, 1973, pp 461–478.
50. Wood RD: *Stress Analysis of the Femur,* PhD thesis. University of New South Wales, Australia, 1975.
51. Olofsson H: *Three-dimensional FEM Calculation of Elastic Stress Field in Human Femur,* thesis. Technikum, Inst of Technol, Uppsala Univ, Uppsala, Sweden.
52. Valliappan S, Svensson NL, Wood RD: Three-dimensional stress analysis of the human femur. *Comput Biol Med* 1977; 7:253–264.
53. Vichnin HH, Batterman SC: Three-dimensional anisotropic stress analysis and failure prediction in a femur with a proximal prosthesis. *Thirtieth Annual Conference on Engineering Medicine and Biology,* Los Angeles, 1977.
54. Harris LJ, Chao R, Bloch R, et al.: A three-dimensional finite element analysis of the proximal third of the femur. *Trans Orthop Res Soc* 1978; 3:16.
55. Andriacchi TP, Galante JO, Belytschko TB, et al.: A stress analysis of the femoral stem in total hip prosthesis. *J Bone Joint Surg* 1976; 58A:618–624.
56. McNeice GM, Amstutz HC: Stresses in prostheses stems and supporting acrylic—a finite element study of hip replacement. *Proceedings of the Twenty-second Annual Meeting of the Orthopaedic Research Society,* New Orleans, January 1976.
57. Forte MR: Structural analysis: Consideration in the design of the total hip prosthesis. *Proceedings of the Twenty-first Annual Meeting of the Orthopaedic Research Society,* San Francisco, 1976.
58. Bartel DL, Ulsoy GA: The effects of stem length and stem material on stresses in bone-prosthesis system. Paper presented at Orthopaedic Research Society, San Francisco, 1975.
59. Bartel DL, Desormeaux SG: On design objectives and testing techniques for femoral stems. *Twenty-third Annual Orthopaedic Research Society,* Las Vegas, 1977.

60. Bartel DL, Samahyek E: The effect of cement modulus and thickness on stresses in bone-prosthesis systems. *Trans Orthop Res Soc* 1976; 1:5.
61. Svensson NL, Valliappan S, Wood RD: Stress analysis of human femur with implanted Charnley prosthesis. *J Biomech* 1977; 10:581–588.
62. Scholten R, Rohrle H, Sollbach W: Analysis of stress distribution in natural and artificial hip joints using the finite-element method. *S Afr Mech Eng* June, 1978, Vol 28, p 220.
63. Crowninshield RD, Brand RA, Johnston RC: An analysis of femoral stem design in total hip arthroplasty. *Proceedings of the Twenty-fifth Annual Meeting of the Orthopaedic Research Society,* San Francisco, Calif, Feb. 1979, p 33.
64. Tarr RR, Lewis JL, Jaycox D, et al.: Effect of materials, stem geometry, and collar-calcar contact on stress distribution in the proximal femur with total hip. *Proceedings of the Twenty-fifth Annual Meeting of the Orthopaedic Research Society,* San Francisco, Calif, Feb. 1979, p 34.
65. Huiskes R, Slooff TJJH: Mechanical properties and stresses in intramedullary prostheses. *Proceedings of the Twenty-fourth Annual Meeting of the Orthopaedic Research Society,* Dallas, Tex, Feb. 1978.
66. Valliappan S, Kjellberg S, Svensson NL: Finite element analysis of total hip prosthesis, in Simon BR (ed): *Proceedings of an International Conference on Finite Elements in Biomechanics.* Tuscon, Ariz, University of Arizona Press, 1980, Vol 2, pp 527–548.
67. Rohlmann A, Bergmann G, Kolbel R: The relevance of stress computation in the femur with and without endoprostheses, in Simon BR (ed): *Proceedings of an International Conference on Finite Elements in Biomechanics.* Tuscon, Ariz, University of Arizona Press, 1980, Vol 2, pp 549–566.
68. Tarr R, Lewis J, Ghassemi F, et al.: Anatomic three-dimensional finite element model of the proximal femur with total hip prosthesis. *Proceedings of International Conference on Finite Elements in Biomechanics.* Tucson, Ariz, University of Arizona Press, 1980, pp 511–525.
69. Hampton SJ, Andriacchi TP, Galante JO: Three-dimensional stress analysis of the femoral stem of a total hip prosthesis. *J Biomech* 1980; 13(5):443–448.
70. Mizrahi J, Susak Z: In vivo elastic and damping response of the human leg to impact forces. *J Biomech Eng* 1982; 104:63–66.
71. Hight TK, Piziali RL, Nagel DA: A dynamic nonlinear finite element model of a human leg. *J Biomech Eng* 1979; 101:176–184.
72. Huson A: Biomechanische Probleme des Kniegelenks. *Orthopade* 1974; 3:119–126.
73. Menschik A: Mechanik des Kniegelenks. 1 Teil. *Z Orthop* 1974; 112:481–495.
74. Andriacchi TP, Mikosz RP, Hampton SJ, et al.: A statically indeterminate model of the human knee joint. *Biomech Symp AMD* 1977; 23:227–229.
75. Wismans J, Veldpaus R, Janssen J, et al.: A three-dimensional mathematical model of the knee joint. *J Biomech* 1980; 13: 677–686.
76. Moeinzadeh MH, Engin AE, Akkas N: Two-dimensional dynamic modelling of human knee joint. *J Biomech* 1983; 16(4):253–264.
77. Fleck JT, Butler FE, Volgel SL: An improved three-dimensional computer simulation of crash victims. Final Report for Contract. Vol 1-4 No. DOT HS-053-2-485, DOT Report No. DOT HS-801-507 through 510, NHTSA, 1975.
78. Frisch G, O'Rourke J, D'Auleris L: The effectiveness of mathematical models as

a human analog, in *Mathematical Modeling of Biodynamic Response to Impact.* SAE Paper No. 760774, 1976.

79. Frisch G, Cooper C: Mathematical modeling of the head and neck response to $-Gx$ impact acceleration—minimum articulation requirements. *Aviation Space Environ Med* 1978; 49(1):196–204.

80. Huston RL, Hessel R, Passerello C: A three-dimensional vehicle–man model of collision and high acceleration studies. Paper No. 740275. Society of Automotive Engineers, Warendale, Pa, 1974.

81. Young RD, Ross HE, Lammert WF: Simulation of the pedestrian during vehicle impact. *Proceedings of Third International Congress on Automotive Safety,* 1974, Vol II, Paper No. 27.

82. Robbins DH: Three-dimensional simulation of advanced automotive restraint systems. *International Automotive Safety Conference Compendium.* SAE 700421. P-30. Warrendale, Pa, Society of Automotive Engineers, 1970.

83. Wismans J, Maltha J, van Wijk JJ, et al.: MADYMO—A crash simulation computer program for biomechanical research and optimization of designs for impact injury prevention. *Impact Injury Caused by Linear Acceleration: Mechanisms, Prevention and Cost.* AGARD-CP-322, 1982, pp 24-1 to 24-11.

84. Wismans J, Maltha J, Melvin JW, et al.: Child restraint evaluation by experimental and mathematical simulation. *Proceedings of the Twenty-third Stapp Car Crash Conference.* SAE 791017. Warrendale, Pa, Society of Automotive Engineers, 1979.

85. Wismans J, Maltha J: Application of a three-dimensional mathematical occupant model for the evaluation of side impacts. *Proceedings of the Sixth IRCOBI Conference on the Biomechanics of Impacts.* Amsterdam, Vrije Universiteit, 1981, pp 331–341.

CHAPTER 6

Measurement Techniques and Experimental Methods

Rolf H. Eppinger

INTRODUCTION

The motivation of all science is successful prediction, and the common thread through all scientific development is the process of observation, postulation, and verification of relationships between observable phenomena. This process has been called the "scientific method." Measurement techniques and experimental methods are inextricable from this process and are, therefore, best discussed and developed by considering their relationship with the other aspect of the scientific method, the process of prediction.

THEORIES AND MODELS

The mechanism for accomplishing the process of prediction is commonly referred to as the "theory" or "model." The creation of a model or theory can be based on previous knowledge or observations, or it can be hypothesized without the consideration of any data. Theories or models may take on any form from physical models to abstract mathematical formulations, but they ultimately have the same purpose, namely, the prediction of a response (output) that occurs as a result of a set of imposed conditions (input).

Once a theory or model is postulated, the type, quality, and quantity of both the input and output variables and their expected relationships are invariant. It then becomes the task of measurement techniques and experimental methods to provide additional observations to determine the validity of the hypothesized theory or model. If the observations differ from predictions offered by the model, if confidence in the observations persists, and if a need to predict the particular output as a function of the input variables continues to exist, the model must be modified to accommodate this new information.

It can be seen that this iterative process suggests that models and theories are never absolute or permanent but are evolutionary and subject to revision. Thus scientific knowledge can be seen to be cumulative and built upon a foundation of successive models and observations.

Any model or theory only predicts or explains how certain things relate to one another but does not explain why they relate as they do in their described relationships. The generality of a model or theory is based upon how many specific kinds of observations can be predicted assuming the expounded theory to be correct. This concept of generality can be illustrated by the following example: Kepler, using more accurate observational data, extended the Copernicus circular orbit model of the solar system by postulating three laws of planetary motion: (1) each planet moves in an elliptical orbit around the sun and the sun is located at one focus of the ellipse, (2) a line from the planet to the sun sweeps out equal areas in equal times, and (3) the period of each planet around the sun is proportional to the cube of the planet's greatest distance from the sun. These three relationships provided accurate predictions of all planetary motion, both into the past and into the future, but they did not explain why the planets behaved in this manner nor were they able to predict other observed phenomena.

Later Newton, using other observational evidence and great insight, formulated certain relationships among forces, motion, and the mass of objects (his three laws of motion and the law of gravitation). Again, these models do not explain why these relationships behave as they do, only how they behave. However, Newton took his relationships and showed that Kepler's laws could be derived from his. That is, Kepler's laws could be understood and explained in terms of Newton's relationships. The fact that many other observations could also be explained using Newton's relationships gave them greater generality. In fact, Newtonian mechanics was able to provide answers to mechanical problems for the next 150 years. It was only when detailed examinations of atomic phenomena became available that a new theory (relativity) was necessary to explain them.

The continued desire and need to expand scientific knowledge requires that no theory will continue to exist without experiments, and, more importantly for this discussion, no experiment should exist without a theory. It is the theory being extrapolated to conditions where no observations exist that is being tested, and, therefore, only with a thorough knowledge of these conditions, their expected variations, and the quantity and quality of the variables that need to be observed can the best measurement technique and experimental methods be properly determined.

MEASUREMENT TECHNIQUES

It becomes the task of the measurement technique to obtain information about identified parameters in sufficient quantity and quality to address the hypothesis set forth by the theory or model under scrutiny.

The flow of that information from the thing being measured to the device recording the information requires a concomitant transmission of energy through each stage of the measurement chain. The device that is the basic component of all measurement systems is the *transducer,* and it is its observable responses to environmental stimuli that allow the information gathering process to begin.

The fact that a transmission of energy is required to initiate the flow of information causes the act of measuring a parameter to distort or modify both the system and the parameter being observed. It becomes the task of the measurement technique to minimize the distortion of the system and to document that what it presents as the value of the parameter being measured is close to the true value of the parameter were it not being measured at all. Several examples can illustrate this point clearly.

A strain gauge alters its electrical resistance as a function of how much it is strained from its nominal length. The gauge material has a certain nominal length and modulus of elasticity, and, therefore, a certain amount of force is required to strain it. The force times the strain determines the energy input required to alter the gauge from its nominal condition. If the gauge is now applied to a large steel beam that requires many orders of magnitude more energy to distort the local gauge area, the addition of the gauge will have little effect on the local strain. It will therefore provide a good estimate of the true strain. However, were the same gauge applied to a very thin foil, its presence would locally stiffen the foil, reduce strain in that area, and provide an extremely inaccurate estimation of the true strain.

Another example is the measurement of acceleration. This is usually accomplished by monitoring the effects an inertial mass has on either piezoelectric or resistive elements. The fact that this additional mass is attached to the structure, that it requires a force (an energy input provided by the structure) to accelerate it, causes the mass to apply a reaction force to the structure and distort the structure at the point of attachment. Depending on the nature of the structure at that point, this resulting distortion could significantly alter the true acceleration.

In both cases, the strain gauge and the accelerometer are indicating the correct values of the parameter being observed, but, in some cases, the presence of the transducer has distorted the observed value from the true value and thus made the measurement unacceptable.

Every transducer is constructed of materials and configured in such a manner that a particular environmental factor will alter its state, and it is this alteration that is monitored by some process to obtain the desired information. It is, however, an unfortunate fact that every transducer will respond in every way in which it can to every factor in the environment. It is the responsibility of the measurement technique to assure that only the desired environmental factor emerges at the measuring system output and that all other environment-response syndromes have been adequately suppressed.[1]

The strain gauge can again be used to illustrate this concept. As was stated previously, the environmental response being exploited in the strain gauge is characteristic of its changing resistance as a function of strain. It also changes its resistance as a function of temperature.

Therefore, if the temperature of the guage varies during the time of observation, part of the resistance change will be due to strain and part due to temperature change. A good strain measurement technique in a temperature-varying environment must recognize this and nullify the effect. Interestingly, if a temperature-sensitive resistive element were desired, a strain gauge could perform the task if it were located in a strain-free position within the temperature field.

Since many of the material characteristics used in the transducers are dependent upon the exposure history of the particular transducer, it is highly advisable to expose a transducer to the maximum environmental range it will experience in service before it is calibrated. This exposure should not be limited to only the parameter the transducer is designed to measure but should also be applied to all other expected environmental parameters. Using the strain gauge as an example again, the gauge factor, or more precisely, the change in resistance per unit strain, is dependent on the entire history of the gauge since it was manufactured. In fact, strain gauges go through a complicated heat-treat cycle to obtain their properties. Therefore, environmental conditions, such as additional temperature, electric field, magnetic field, and strain exposures since manufacture, could alter the gauge factor from its original value. If this shift occurs after calibration, all subsequent measurements could be faulty.

Calibration procedures should also be performed under conditions as close to the experimental environment as possible. This will insure that the calibration is correct for the nominal levels of the environmental factors.[2] Any effect a shift in these factors would have on the calibration is then proportional only to the deviation from the nominal level.

Another aspect of measurement technique should be addressed. It is the case when no transducer exists to directly measure the variable of interest. One such parameter is the rotational acceleration of a rigid body during impact-type collisions. This type of measurement has been sought by investigators of brain injury, since one of the injury mechanisms postulated deals with the effects that rotational motions have on the production of strains within the brain tissue.

Avoiding the details of rigid body mechanics, it is sufficient for this discussion to state that six accelerations, three orthogonal translational and three orthogonal rotational, completely define the motion of a rigid body in space. Examining the generalized equations of motion, it was determined that the output of an array of six translational accelerometers attached to the rigid body could be processed to determine the rotational accelerations. These calculations also required information on the three instantaneous rotational velocities of the rigid body. Since they were not directly available, they were determined from the integral of the individual rotational accelerations. As a result, if an error in the rotational accelerations calculation exists, it causes an error in the rotational velocity calculation, which in turn affects the next rotational acceleration calculation. Thus, a cumulative feedback error grows rapidly in the step-by-step solution technique and causes the solutions to become unstable.

Continued examination of the governing mechanical equations revealed that if three additional unique translational acceleration measurements were made, the calculation of the rotational acceleration levels no longer required information on the instantaneous rotational velocities, and it was stable over the entire time period of interest.

This example illustrates two necessary conditions for deducing variables from other measurements. The first is that the number of independent measurements made in an experiment must be equal to or greater than the number of unknowns to be determined. The six acceleration measurements provided sufficient data to determine the six rigid body accelerations when the rotational velocities were obtained by calculation. However, when it was desired to eliminate the calculation of the three rotational velocities, three additional measurements had to be made. Stated another way, information cannot be created, it can only be modified, and if one wishes to determine the actions of 10 individual independent variables, one has to provide at least 10 observers—otherwise, something is not being watched.

The second condition is obvious but should still be stated: a valid theory relating the experimental measurements to the desired variables must exist.

It is worthwhile at this point to reiterate the two most important axioms of measurement techniques. They are:

1. Introduction of a transducer and measurement process into a system will always modify the observed system and the quantity to be measured.
2. A transducer and its response are always modified by the total environment to which it is exposed.

As a result, the experimentalist must assure himself that his measurement technique always minimizes the alteration of the system being observed and that the effect of every environmental factor except the one of interest has been nullified.

EXPERIMENTAL METHODS

Experimental methods bring the individual measurement techniques into the context of the theory or model that is being verified or tested. It is with the experimental method that the proposed relationships of the output variables to the combination of input variables is explored, the objective being to obtain a series of observations to sufficiently satisfy oneself that they could have occurred governed by the hypothesized theory. This process is limited, since it can never prove the hypothesis true. Rather, it can only indicate the existence of a quantity of supportive data for the hypothesis or disprove it.

In the design of an experiment, careful consideration must be given to the observations being made in order to be assured that they are relevant and capable of providing any unambiguous answer. The strategy of how the input parameters are varied must be developed in concert with the analysis process to assure meaningful results.

Consideration must also be given to obtaining the results in the most efficient manner, since the economics of experimentation cannot be ignored. The time, money, and labor the experimentalist is allotted is often limited. Without prior and sufficient development of the experimental design *and* the analysis process, the available resources may not achieve the desired goal.

Rather than elaborate on various experimental designs, the reader is referred to some of the many good texts that deal with the subject.[3-5] These references, along with the references cited within them, are a rich resource and should provide either a technique that is directly applicable or suggest an amalgamation of methods to accomplish many experimental goals.

Sometimes the experimentalist is faced with the problem that, even with his best efforts toward the development of a transducer and measurement technique, the amount of desired information is only a very small proportion of the total signal being measured. Here, if the object is to get the data, certain circumstances will allow a variety of experimental techniques to extract the desired information, but at the expense of conducting additional measurements. These additional measurements are done by either measuring more things in each experiment or maintaining the same number of measurements and increasing the number of experiments.

An interesting example of how additional observations allow the extraction of a weak signal that is immersed within a strong signal is the process employed to obtain cerebral-evoked responses (either visual, auditory, or somatosensory).

What is desired in this case is the characteristic electrical response of an area of the brain resulting from a specific stimulus to another area of the body. This is either a short flash of light in the eye for the visual stimulus, a short tone in the ear for the auditory stimulus, or an electrical pulse to a nerve in the finger or toe for the somatosensory input.

The difficulty in capturing the desired response lies in the fact that it is only a very small signal superimposed upon the large and complex signal that constitutes normal brain activity. Since this normal activity appears to be the sum of a large number of unrelated stimuli, it has characteristics of a random signal. That is, the average of the signal voltage over long periods of time is zero or a constant, and the probability that the signal is either positive or negative at any instant in time is equal.

Therefore, if the continuous time signal is cut up into sequential, 1-second intervals, overlayed and divided into 100 equal time increments, the average of the signals at any signal time increment within the 1-second window will tend to zero. Theory indicates that scatter of the individual averages is proportional to 1 over the square root of the number of samples.

Now, if the segmentation of the continuous time signal is synchronous with the stimulus, the evoked response will present itself at the same points within the 1-second window. This means that the evoked response signal becomes additive at each time point and grows proportionally to the number of samples. Therefore, as more samples are taken, the amplitude of the desired signal grows, while the amplitude of the undesired, ansynchronous signal decreases.

The longer period of time required to obtain the necessary repetitive samples also requires that the evoked response not change during the total period of observation. If it does, two alternatives are available. Either maintain the current number of samples and realize that the resulting signal may be distorted or use fewer samples and have more undesired signal present.

This example illustrates that information is a commodity that is only obtained through an expenditure of time and effort, and, if the present quality is not sufficient, either time or effort must be increased. This increase, however, does not insure success because it imposes additional constraints upon the observation process. The expansion of the time necessary to observe a system limits the observations to systems that do things more slowly. The addition of effort, such as duplication of instrumentation, increases the probability of modifying the system being measured and distorting the information from its true form. A compromise must be struck, and ultimately, it is determined solely by the needs of the user.

SUMMARY

One could postulate that every living organism with a memory capacity proceeds through life performing the elements of the scientific method by making, iteratively, observations and associations. The benefit to those who make the correct associations is survival.

Man's use of this process, whether done intentionally or purely as an innate response, has without argument dramatically increased both his ability to survive and the quality of his survival.

Because of its success, the scientific method itself has become a subject of study. These studies have revealed a variety of maxims that, if followed, will guide the components of the scientific method to produce the most reliable observations in a most efficient manner.

The purpose of this discussion was to qualitatively characterize some of these maxims, illustrate them with simple examples, and hopefully, instill them into the consciousness of the reader as an aid in assisting him in the exciting quest for knowledge and survival. It is hoped that this purpose has been at least somewhat accomplished.

REFERENCES

1. Stein K: A new conceptual and mathematical transducer model application to impedance-based transducers such as strain gauges. Presented at International Measurement Conference—IMEKO, The Hague, Netherlands, September 1971. VDI Bericht Nr. 176, Dusseldorf, Germany.
2. Stein K: Traceability—the golden calf. *Measurements Data* 1968; 4 (July-August): 97–105.
3. Kempthorne O: *The Design and Analysis of Experiments.* New York, Wiley, 1952.
4. Schenck H: *Theories of Engineering Experimentation.* New York, McGraw-Hill, 1961.
5. Box GEP, Hunter WG, Hunter JS: *Statistics for Experiments: An Introduction to Design, Data Analysis, and Model Building.* New York, Wiley, 1978.

CHAPTER 7

Mechanical Behaviors of Soft Tissues: Measurements, Modifications, Injuries, and Treatment

Savio L-Y Woo, Mark A. Gomez, Wayne H. Akeson

INTRODUCTION

The knowledge of the mechanical and structural properties of soft tissue is an essential prerequisite for any theoretical, numerical, or experimental approach to analyzing its physiologic function in the body. There are several classic writings on the subject of biomechanics (or biorheology) of soft tissues in the literature. Fung,[1] who pioneered the application of classical mechanics to this field, has rigorously derived the biorheologic properties of such tissues as skin and mesentery, as well as muscles in their passive state. His work includes nonlinear constitutive equations for these tissues during loading and unloading at a constant rate.

Another pioneer in this field, Viidik,[2] has written a summary of his work emphasizing the relationship of morphologic structure to the biomechanical properties of parallel as well as randomly oriented fibrous connective tissues. The geometric configuration of the constituent collagen fibers and the interaction of collagen fibers with the noncollagenous tissue components are the basis of the mechanical behavior of soft tissues. The readers are encouraged to read these classic works in order to gain a better historical perspective on the subject matter.

These pioneers have stimulated numerous investigations on the measurements of the biomechanical properties of soft tissues. For example, follow-up studies on the biaxial properties of human skin have been published by Tong and Fung[3] and by Schneider et al.[4] Furthermore, all work presented at the Fourth International Congress of Biorheology in Tokyo (1982) was solely concerned with the mechanical properties of living tissues, such as the vascular wall, cardiovascular tissues, red blood cells, as well as the ligaments and tendons.[5] The approaches used in these studies to obtain the constitutive relationships for these soft tissues were (1) to directly measure the stresses (σ) and strains (ε), i.e., one-dimensional forces and deformations (tension, bending, or shear), to form simple stress–strain relationships and (2) to use a continuum approach in which a pseudostrain energy density function is utilized:

$$W = W(\varepsilon_{ij})$$

where i,j = 1,2,3. Thus:

$$\sigma_{ij} = \partial W/\partial \varepsilon_{ij}$$

is used to represent a general three-dimensional formulation of stresses and strains. Other soft tissues, such as articular cartilage, have also received considerable attention. Over the past 10 years Mow and his associates have shown that large amounts of free-flowing water can be moved in and out of the articular cartilage matrix and have developed a new biphasic theory to describe the mechanical behavior of the solid and fluid portions of the tissue.[6] More recently, these investigators have shown the interaction between collagen and proteoglycans and its effect on the tissue properties during normal cartilage function.[7]

In this chapter we discuss the current refinements and improvements of techniques for studying the biomechanical behavior of normal and healing soft tissues, ligaments and tendons in particular. It is convenient to directly measure the stress–strain relationships (constitutive equations) of the tissues in question, since the majority of their collagen fibers are oriented in a parallel fashion. The major function of these tissues is to connect muscle to bone, or bone to bone in order to permit joint motion. Ligaments and tendons are capable of supporting very large forces with minimal deformation. Thus, uniaxial tensile tests and small strain theory are suitable to accurately describe their mechanical and structural properties. We will also discuss the modification of the mechanical properties of these soft tissues (both in vitro and in vivo) as well as soft tissue injuries, particularly relating to the effects of stress and motion on their mechanical and structural properties during the healing process.

THE QUASI-STATIC STRESS–STRAIN RELATIONSHIPS

One of the most common methods of evaluating the biomechanical characteristics of ligaments and tendons is the uniaxial tensile test. The load-

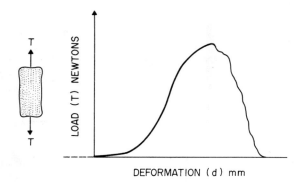

Figure 1. A typical load-deformation curve for soft tissues representing its structural properties.

deformation curves obtained from tensile testing are nonlinear and have concave upward shapes. Initially, the stiffness (or the slope of the curve) is very low. Gradually, the stiffness increases with the applied load until a linear region is reached. With further loading, the slope will begin to decline as a result of early failure of some fibers, and eventually, there will be complete failure. A typical load-deformation curve, as shown in Figure 1, represents the structural properties of these soft tissues.

The factors contributing to the structural properties of soft tissues are (1) mechanical properties, i.e., soft tissues being evaluated as a material, which are dependent upon the organization and/or orientation of their collagen fibers as well as the percentages of various constituent materials, and (2) the geometry of the soft tissues, i.e., cross-sectional area, length, and shape. On organization, it is well known that skin, for example, has a random and loose network of collagen fibers. Thus, during deformation, the largest percentage of strain, which could be as much as the first 70 percent, is to realign and to recruit the collagen fibers to the directions of the load. Tendons, on the other hand, have well-organized collagen fibers parallel along the length of the tissue. Therefore, there is little fiber realignment during stretch. This small amount of recruitment of collagen fibers in tendons results in a nonlinear stress–strain behavior. Viidik[8] has used a model of mechanical analogs to demonstrate the recruitment of collagen fibers or tendons by assuming the fibers are of different initial length (Fig. 2). Due to the large variation of collagen organization, the mechanical properties of various soft tissues cover a wide spectrum, with tendons on the one end and skin on the other (Fig. 3).

There exist large differences in the contents of constituent materials, i.e., the percentage of water, collagen, elastin, glycosaminoglycans (GAGs), and other noncollagenous proteins in the soft tissues. Examples of the constituent materials in several soft tissues are listed in Table 1. It can be seen that water occupies the largest percentage of weight in all soft tissues. In articular cartilage, the water content can be upward of 80 percent in weight. Collagen, a unique protein that has very high tensile strength and is mainly

112

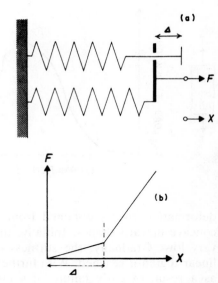

Figure 2. Mechanical analog model of collagen fibers and the effect recruitment of the fibers has on the nonlinearity of the load-deformation curve. (*From Viidik.*[8])

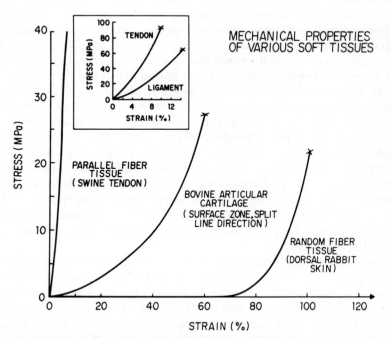

Figure 3. Typical stress–strain curves for various soft tissues, showing a wide spectrum of mechanical properties.

TABLE 1. APPROXIMATE PERCENTAGES OF BIOCHEMICAL CONSTITUENTS IN SEVERAL SOFT TISSUES

Tissue	Water (% of total wt.)	Collagen (% of fat free dry wt.)	Elastin	GAGs
Skin	60–65	65–70	5–10	1.5–2
Aorta	60–70	23–35	40–50	2–2.5
Tendon	65–70	75–80	< 3	1–1.5
Ligament				
Cruciate	65–70	75–80	< 5	2.5–3
Collateral	65–70	75–80	< 5	1–1.5
Articular				
cartilage	70–80	60–65	Trace	10–15

responsible for the mechanical strength of soft connective tissues, is the major solid component. On a fat-free, dry-weight basis, the percentage of collagen ranges from 60 to 65 percent in articular cartilage to 80 percent in ligaments and tendons. Elastin, another protein, is also an important constituent material in the solid matrix. Its stiffness is significantly lower than that of collagen, but it has a near linear elastic stress–strain behavior. There is practically no elastin in articular cartilage and only a small percentage in ligaments and tendons, 10 percent in skin, and 40 to 50 percent in aorta. The GAGs are noncollagenous macromolecules interwoven with collagen fibers and elastin. These macromolecules contribute only minimally to the stiffness properties but greatly affect the time- and history-dependent viscoelastic properties of soft tissues, particularly in articular cartilage. The percentage of GAGs is small in tendons and ligaments, yet it can be quite high (up to 15 percent in weight) in articular cartilage. GAGs, together with hyaluronic acid and a protein core, form a proteoglycan aggregate that has a centipedelike structure, and they occupy a large tissue space due to their water-binding capacity.[9] Such large variations in constituent materials in the soft tissues are the major contributors to the differences in the shapes of the middle and upper portions of the stress–strain curves (Fig. 3).

In the literature, the reported mechanical properties for each soft tissue can be widely divergent. For example, the ultimate strains for tendons can range from 9 percent to over 30 percent, and those for ligaments can range from 12 percent to over 50 percent.[10–16] Investigators who are involved in this area of research work are well aware of the factors, such as age, species, and type of specimens (obtained from different animals and from different anatomic locations), that will contribute to the variation in the experimental results. However, there appeared to be other and perhaps more important factors that contributed to this large variation of reported data, i.e., technical difficulties. These include (1) cross-sectional area measurements (for stress calculations), (2) clamping of the test specimens, (3) length–width (aspect) ratio for uniform stress, (4) deformation measurements (for strain determinations), (5) the test environment, and (6) other factors. To illustrate, soft connective tissues usually slip within their clamps during tensile stretch. Thus, the resulting strain measurements using the clamp-to-clamp deformation (a common practice in reporting strain data) are incorrect. Other soft tissues, such as ligaments, are usually too short (poor aspect ratio) to be tested in their isolated state. As a result, the bone–ligament–bone preparation is used as test specimen. In this case, the nonuniform properties between the ligament substance and its insertions to bone, as well as the complications in defining the original length of the ligament, are real technical difficulties.

In the past 10 years, our laboratory has been developing methodologies to minimize some of these technical problems. The readers are referred to Woo et al.[17-19] for some of the details. For example, to measure the thickness of the soft tissues, a micrometer instrument that applies negligible force on the tissue substance during its measurement should be used.[19] For more resilient soft tissue, such as the flexor digitorum profundus tendon, a device consisting of a slot of fixed width with semicircular ends to accommodate the tendon shape is needed. Thus, the cross-sectional area can be calculated by reading the width of the slot using a linear variable differential transducer (LVDT).[17]

Test specimen preparation and clamping must also be based on the geometric and physical dimensions of soft tissues. It is not always possible to fabricate standard dumbbell-shaped samples.[19] For tendons, it is possible to isolate long specimens from their muscle and bone attachments for tensile testing. For ligaments, it is necessary to prepare the test specimens with their bony attachments, i.e., a bone–ligament–bone composite. Therefore, clamping must be done in different ways for different tissues when performing tensile tests. For the free ends of the soft tissues it is possible to use clamps with pivoted joints or spring-loaded clamps so that there is a self-tightening effect on the soft tissue substance during the tensile pull to minimize slippage. For the high-strength digital flexor tendons or soft tissue specimens with a large cross-section, a coronal cut should be made, and a stainless wedge piece should be placed between the inner (cut) surfaces of the specimen and then inserted into the pivoted clamps. Thus, the specimen can be gripped at four surfaces rather than two (Fig. 4). For the bone–ligament–bone specimens, the bones should be fixed to clamps that permit minor adjustments. It is most important that the tensile load be applied directly along the longitudinal direction of the ligament with no bending (Fig. 5).

The experimental apparatus, such as shown in Figure 6, is suitable for measuring the mechanical properties of soft tissues. The tensile load during uniaxial stretch is measured with an electronic load cell. The tensile strain of the specimen is determined by using the video dimensional analyzer (VDA) system. This video system permits accurate measurement of the tensile strain of soft tissues within the gauge length marks.[17,19] The system has many

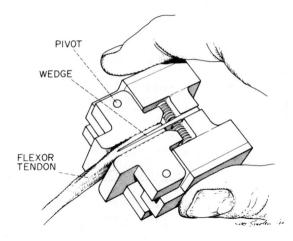

Figure 4. A self-tightening clamp with a wedge piece for gripping the free end of a tendon. After a 1 cm coronal cut is made in the tissue, it is inserted into the clamp. The design permits gripping of the tendon at four surfaces for tensile testing. (*From Woo.*[17])

advantages: (1) the optical/video system requires no external attachment (such as transducers or clip gauges) to the soft tissue specimen, as the strains are measured by the VDA automatically, (2) the strains obtained have no contribution from the specimen slippages at the clamps that normally occur and contribute to large experimental errors, (3) the video information is tape-recorded and can be used for repeated analyses or measurements of

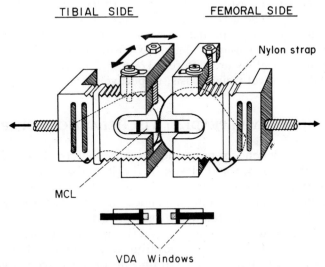

Figure 5. A diagram of the clamps designed for tensile testing of a bone–ligament complex. Note that the sample can be positioned so that the tensile load is applied along the longitudinal direction of the medial collateral ligament. Also depicted are the video dimensional analyzer windows that follow stain lines on the ligament for strain measurements. (*From Woo.*[17])

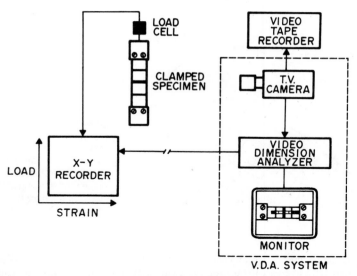

Figure 6. A flow chart detailing the apparatus used to acquire the load–strain properties of soft tissue. By utilizing the sample cross-sectional area, a stress–strain curve can be obtained.

strain variation along the soft tissues,[20] and (4) this system determines the strain with little or no human error contribution. This methodology has been successfully used to measure the mechanical properties of a wide range of soft tissues (Fig. 3). The accuracy of this system also permits the differentiation of the mechanical properties between similar soft tissues, e.g., digital flexors and their compensatory digital extensors (Fig. 7), as the collagen contents in these tendons are different (28.6 ± 2 percent vs 23.0 ± 4 percent, $P < 0.01$ on a wet weight basis, respectively.[21]

TIME- AND HISTORY-DEPENDENT VISCOELASTIC PROPERTIES

Time- and history-dependent properties are observed for all soft tissues. It is thought that the collagen and elastin in the soft tissue, together with the surrounding proteoglycans, contribute to this type of behavior. It is interesting to note that despite the time-dependent nature of these tissues, their stress–strain behaviors are relatively insensitive to the loading rate.[22] Recently, however, Haut[23] has demonstrated that at very high strain rates, i.e., 720 percent/second, the rat tail tendons from young animals show strain rate sensitivity to those tested at a low strain rate of 3.6 percent/second. This sensitivity decreases rapidly though during growth and sexual maturation of the experimental animals.

Several advanced theories of viscoelasticity are available to be used to describe time- and history-dependent properties of soft tissues. These the-

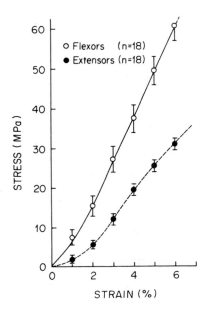

Figure 7. Stress–strain curves demonstrating the differences in mechanical properties between the extensor and flexor tendons of swine. (*From Woo.*[17])

ories include that of Lianis[24] and that of Bernstein, Kearsley, and Zapas (BKZ theory).[25] By far the most popular theory, however, is the quasi-linear viscoelastic theory of Fung.[22] This theory has been successfully applied by Chen on the rabbit mesentery,[26] Tanaka and Fung on aorta and the aortic tree,[27] Haut and Little[28] and Jenkins and Little[29] on parallel fibered tissues, and Woo et al. on articular cartilage and ligaments.[30,31]

The quasi-linear viscoelastic theory is relatively simple to use. The theory assumes that the stress relaxation function can be written as:

$$\text{(1)} \qquad \sigma[\varepsilon(t);t] = G(t)^*\sigma^e(\varepsilon)$$

where $\sigma^e(\varepsilon)$ is the nonlinear elastic response (function of ε only), and $G(t) = \sigma(t)/\sigma(o)$ is the reduced relaxation function (function of t only). The stress at time t, $\sigma(t)$, is thus the convolution integral of the reduced relaxation function and the rate of elastic stress, or:

$$\text{(2)} \qquad \sigma(t) = \int_0^t G(t-\tau) \, \frac{\partial \sigma^e(\varepsilon)}{\partial \varepsilon} \, \frac{\partial \varepsilon}{\partial \tau} \, d\tau$$

Therefore, when $G(t)$, $\sigma^e(\varepsilon)$, and the strain history, $\varepsilon(t)$, are known, the time- and history-dependent stress, $\sigma(t)$, is completely described by equation 2. For soft tissues that are insensitive to strain rate, Fung[22] further proposed a generalized reduced relaxation function as follows:

$$\text{(3)} \qquad G(t) = \frac{[1 + \int_0^\infty S(\tau)e^{-t/\tau}d\tau]}{[1 + \int_0^\infty S(\tau)d\tau]}$$

where $S(\tau)$ is a continuous spectrum and has the following special form:

(4)
$$S(\tau) = \frac{C}{\tau} \text{ for } \tau_1 < \tau < \tau_2$$
$$= 0 \text{ for } \tau < \tau_1, \tau > \tau_2$$

Equation 3 can then be rewritten as:

(5)
$$G(t) = \frac{[1 + C\{E_1(t/\tau_2) - E_2(t/\tau_1)\}]}{[1 + C\ln(\tau_2/\tau_1)]}$$

where $E_1(y)$ is the exponential integral, i.e.:

$$E_1(y) = \int_0^\infty \frac{e^{-t}}{t} dt$$

Constants C, τ_1 and τ_2 are parameters that can be determined by solving three simultaneous equations obtained from $dG/d(\ell nt)$, $G(\infty)$, and $G(\bar{t})$, where $\bar{t}$ is an arbitrary time such that $\bar{t} >> \tau$, and $\bar{t} << \tau_2$.

For the nonlinear elastic response, $\sigma^e(\varepsilon)$, a convenient exponential expression is chosen:

(6)
$$\sigma^e = A[\exp(B\varepsilon) - 1]$$

Constants A and B can be obtained by using a quasi-static tensile test where $\sigma(t)$ and the constant strain rate, α [where $\alpha = \Delta\varepsilon/\Delta t$], and $G(t)$ are known. Hence equation 2 can be written as

(7)
$$\sigma(t) = AB\alpha \int_0^t G(t-\tau)\, e^{\alpha B \tau} d\tau$$

The constants A and B from equation 7 can be determined by using a nonlinear least square curve-fitting procedure.

To illustrate, we have performed an experiment on the canine medial collateral ligament (MCL). The sample is stretched to 2.5 percent strain at a rapid rate of 10 percent/second and then allowed to stress relax up to 16 hours (Fig. 8). From this stress relaxation data, $dG/d(\ell nt)$, $G(t = 16 \text{ hours})$ representing $G(\infty)$, and $G(\bar{t})$, where $\bar{t} = 120$ seconds, values were used in equation 5. Thus, we have three equations for the three unknown constants C, τ_1, and τ_2. Solving the equations simultaneously, we obtain values for the constants to be 0.099 second, 0.29 second, and 1.99×10^5 seconds, respectively. Thus, the reduced relaxation function for MCL can be rewritten as

(8)
$$G(t) = 0.749 - 0.042\, \ell nt$$

In a separate experiment, eight canine medial collateral ligaments are tensile tested at a strain rate of 0.01 percent/second. The stress–strain relationships are shown in Figure 9. Using the experimental data together with a nonlinear curve-fitting procedure described in detail by Woo et al.,[31] constants A and B in equation 7 are 0.193 and 161, respectively. Thus, the elastic response of the MCL as written in equation 6 is deduced to be:

(9)
$$\sigma^e = 0.193\, [\exp(161\varepsilon) - 1]$$

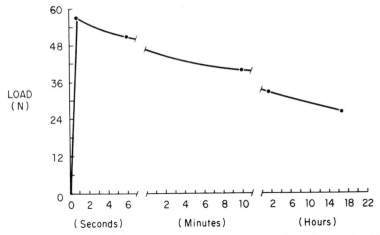

Figure 8. Typical long-term relaxation behavior of canine MCL. The experiments were carried out to 16 hours at 37°C.

It is necessary to further confirm the validity of the results shown in equations 8 and 9 by an independent set of experimental data, such as that from cyclic testing. The data of cyclic stretching of MCL between two strain levels, 1.6 and 2.4 percent, for 10 cycles at a strain rate of 0.1 percent/second are shown in Figure 10. Knowing the reduced relaxation function, $G(t)$ from equation 8, the elastic response $\sigma^e(\varepsilon)$ from equation 9, the strain rate $\partial\varepsilon/\partial t = 0.1$ percent/second, and the cyclic strain history, the peak and valley stresses, $\sigma(t)$, calculated from the loads shown in Figure 10 can be predicted by using equation 2. As can be seen from Figure 11, the calculated stresses using the quasi-linear theory match well with those obtained in the experiments. Thus, it is concluded that the $G(t)$ and the nonlinear $\sigma^e(\varepsilon)$ deter-

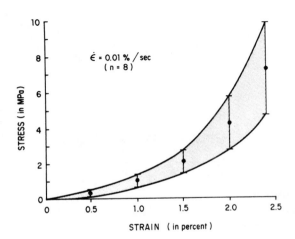

Figure 9. Stress–strain relationship of canine MCL.

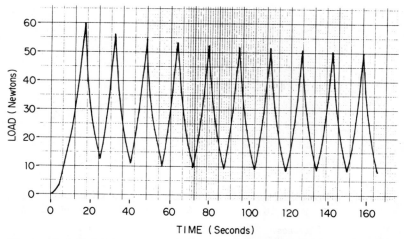

Figure 10. A load-time curve of canine MCL subjected to cyclic testing at a rate of 0.1 percent/second. Strain levels used were between 1.6 and 2.4 percent. (*From Woo et al.*[31])

mined for the canine MCL substance are reasonable and can be appropriately used to describe the time- and history-dependent viscoelastic properties of ligaments at the strain levels studied. Most recently, the sensitivity of the quasi-linear viscoelastic theory has been further evaluated by Sauren and Rousseau.[32] These authors find that C is the most important parameter to describe the viscous effects, while τ_1 and τ_2 govern the fast and slow viscous phenomena, respectively.

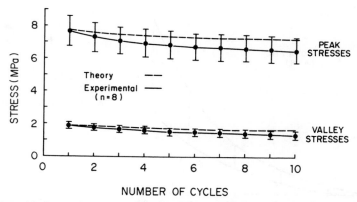

Figure 11. Curve showing confirmation of theoretically predicted peak and valley stresses by those obtained experimentally. (*From Woo et al.*[31])

MODIFICATION OF MECHANICAL PROPERTIES OF SOFT TISSUES

In Vitro and In Vivo Experiments

Since the biochemical compositions govern the mechanical properties of soft tissues, several investigators have studied the removal of matrix constituents in the soft tissue to examine the corresponding change in the mechanical properties. Minns and Soden[33] have used EDTA (a chelating agent) and α-amylase (an enzyme) to remove the ground substances and formic acid to remove both collagen and ground substances of human tendons and aorta and bovine ligamentus nuchae (contains a large portion of elastin). The use of EDTA and α-amylase induce a significant decrease in stiffness, stress relaxation, and hysteresis of all the tissues evaluated. The effective viscosity of interfibrillar matrix is reduced, and the interrelationship of collagen and GAGs is changed. The use of formic acid produces a marked change in the stiffness and time-dependent properties as well as a significant decrease in the stress and strain at failure due to the removal of collagen. Hoffman et al.[34] have studied enzymolysis effects on the canine aorta by using collagenase. The aortic tissue without a large portion of its collagen fibers has lost the nonlinear, concave, upward portion of the stress–strain curve and exhibits a more linear, elastic behavior, similar to that for elastin (Fig. 12).

On the other hand, Broom[35] has demonstrated the influence of glutaraldehyde fixation on the mechanical properties of the soft tissues. When the bovine mitral valve tissue is placed in 0.625 percent glutaraldehyde solution, collagen crosslinks are formed, and the stress–strain curve of the tissue becomes stiffened (Fig. 13). After 3½ to 4 hours of fixation, much of the initial elasticity, i.e., the lower portion of the curve called the toe region, has disappeared, presumably due to limited realignment of collagen fibers. Similar effects have also been reported for pericardium and the saphenous vein.[36]

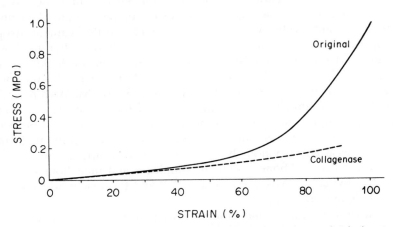

Figure 12. Curves showing the effect of collagenase on the mechanical properties of aortic tissue. (*From Hoffman et al.*[34])

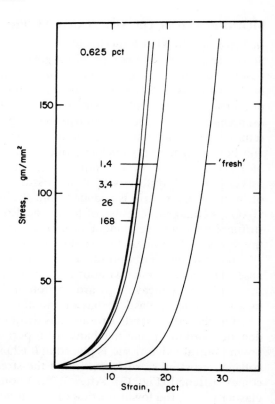

Figure 13. Curves demonstrating the effect of glutaraldehyde fixation on the stress–strain behavior of the bovine mitral valve. Exposure time in hours is given for each curve. (*From Broom.*[35])

Changes in mechanical properties of soft tissues also occur in vivo. There are significant homeostatic responses of ligaments and tendons in spite of their inert appearance. When a rabbit knee is immobilized for a period of 9 weeks, the mechanical properties of the collateral ligaments become significantly inferior, even though there is little ligament atrophy, i.e., reduction in cross-sectional area and weight.[18] On the other hand, when swine are subjected to exercise training at a schedule of 40 km/week of running at moderate speeds (6 to 8 km/hour), the mechanical properties of the digital extensor tendons do not change after 3 months. With periods of 12 months of training, however, the tendons from the exercised animals become hypertrophic and possess superior stiffness and strength properties in comparison with the sedentary controls. The collagen contents in these tendons increase correspondingly.[37] Excellent review articles on the structural characteristics of bone–ligament complexes and their homeostatic responses to physical activity are available,[38,39] and the reader will find these articles of significant interest.

Based on our studies and others in the literature, it is possible to hypothesize a highly nonlinear curve to represent the homeostasis of soft tissues (Fig. 14). With stress and motion deprivation, a rapid reduction of tissue properties and/or mass will occur, as the slope of the curve in this

HOMEOSTATIC RESPONSES OF BIOLOGICAL TISSUES

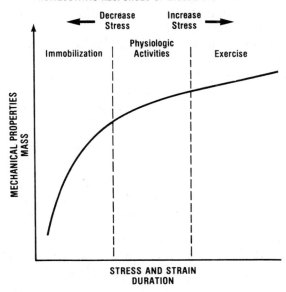

Figure 14. A hypothetical nonlinear curve representing the stress- and motion-dependent homeostatic responses of biologic soft tissues. (*From Woo.*[18])

region is the steepest. Hence, a short period of 9 weeks of immobilization of rabbit knees results in a profound reduction of ligament properties. On the contrary, with increased levels of stress and motion in tendon, the resulting changes will be small, as the slope of the curve in this region is relatively flat (Fig. 14). Hence, the changes of tissue properties and/or mass of tendons from the exercise-trained swine are negligible after 3 months and become significant only after 12 months of training.

TRAUMA OF SOFT TISSUE AND ITS TREATMENT

Soft tissues are frequently subjected to traumatic injuries. With currently used treatment procedures, the prognosis of many soft tissue injuries is less than promising. In the case of laceration of the flexor digitorum profundus tendon, healing is compromised by dense adhesions between synovial sheath and the repair site, leading to limited tendon excursion and poor digital function. Clinical and experimental efforts to improve the understanding of digital tendon healing mechanism are numerous. Early protective mobilization of the repaired flexor tendons have shown significantly improved digital function.[40-43] Our laboratory has demonstrated that immobilization is deleterious to the healing process following tendon repair.[44] The cellular activity at the repair site and within the sheath is promoted by early mobilization. Longitudinal vascular and cellular remodeling along the tendon is also improved by early motion. As a result, the angular rotation of the mobilized digits is increased during the excursion of the healed tendons.[44,45] Further, the strength and stiffness of the early mobilized tendons show significantly higher values than those that are immobilized. Delayed mobilization, beginning at 3 weeks yields only intermediate results.[44]

Much less is known about ligament healing following trauma. In spite of the very high frequency of ligamentous injury, only limited data on the structural properties of the bone–healing ligament complex are known to date. There is practically no knowledge of the mechanical properties of the ligament substance, as the precise area and length measurements of the ligaments are extremely difficult.

Loading Modes and Ligament Failure Mechanisms

In ligament trauma, it is recognized that the failure mechanisms are related to the loading modes, i.e., rate of loading. During slow rate of loading, avulsion failure at the ligament insertion to bone is common, whereas, during rapid rate of loading, midsubstance tear of the ligament is common.[46–48] The differences in failure modes can be explained based on the viscoelastic behavior of soft (ligament) and hard (bone) tissues. When examining the load-deformation curve for slow and rapid rates of loading of a bone–ligament complex, it can be seen that both curves are similar in the first half (up to the linear region) (Fig. 15). This portion of the load-deformation curves presumably represents the deformation of the ligament (soft tissue) by recruitment of its collagen fibers. With further loading, the second half of the curves and the deformations represent both the ligament and bone. Since the bone is more highly sensitive to the rate of loading than is the ligament,[22,49,50] avulsion failure and a lower failure load occur first at a slow rate of loading. With a faster rate of loading, the bone becomes stronger, and as a result, its failure load exceeds that for the ligament, resulting in midsubstance tearing and a higher failure load.

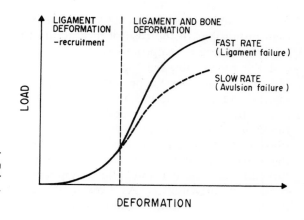

FAILURE MODES vs LOADING RATES

Figure 15. Curves showing the effect of different loading rates on the load-deformation behavior and the failure modes of a bone–ligament–bone complex.

Effect of Immobilization and Remobilization of Ligaments

The traditionally preferred clinical treatment of injured ligaments may be detrimental, since the procedure is conservative, i.e., immobilization by cast or splint. It has been repeatedly demonstrated by our laboratory,[18,51,52] as well as by Noyes et al.,[46,47] that immobilization significantly compromises the properties of the ligament and the bone–ligament–bone complex. Following 9 weeks of immobilization, the linear slope, ultimate load, and energy-absorbing capabilities of the rabbit MCL–bone complex during tension decrease to approximately one third of that of the contralateral nonimmobilized control.[53] The load-strain characteristics of the MCL substances become inferior. Further immobilization of up to 12 weeks causes additional degradation of the MCL substance (Fig. 16). These data are obtained with the aid of the video dimensional analyzer system, where the mechanical properties of ligament substance and structural properties of the bone–ligament complex can be simultaneously evaluated.[53]

This experimental method has also been used to study the effect of remobilization of the MCL substance and MCL–bone complex following a period of immobilization. Noyes et al.[46] have shown that the strength and failure mode of the anterior cruciate ligament (ACL)–bone complex following 8 weeks of immobilization will take as long as 1 year of daily conditioning to recover. Our work on the rabbit MCL demonstrated that after 9 weeks of remobilization following 9 weeks of immobilization the structural properties, i.e., the maximum load and energy-absorbing capability of the MCL–bone complexes, continue to inferior when compared to the con-

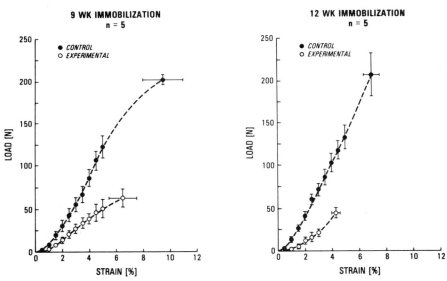

Figure 16. Comparison of load-strain curves between control and 9 and 12 week immobilized experimental rabbit MCLs. The highest point on each curve represents the mean values, maximum load, and maximum strain.

TABLE 2. COMPARISON OF THE MAXIMUM LOAD AND ENERGY ABSORBED AT FAILURE VALUES FOR RABBIT MCLs SUBJECTED TO 9 WEEKS OF REMOBILIZATION AFTER 9 OR 12 WEEKS IMMOBILIZATION

	Maximum Load (N)	Energy Absorbed at Failure (N-m)
9 week immobilization + 9 week remobilization		
Control	330 ± 20	1.00 ± 0.15
Experimental	250 ± 35	0.62 ± 0.20
12 week immobilization + 9 week remobilization		
Control	290 ± 20*	0.91 ± 0.11*
Experimental	200 ± 25*	0.49 ± 0.09*

*Significant difference between control and experimental $P < 0.02$.

trol knees (Table 2). These results are in agreement with Noyes et al.,[46] indicating that the strength at the bone–ligament junctions recovers slowly. However, the mechanical properties of the ligament substance, as represented by the load-strain curve from the experimental knee, are almost identical to those of a normal control (Fig. 17). The recovery of the MCL substance after 12 weeks of immobilization knee is not as complete. These findings suggest that it is necessary and important to differentiate the mechanical properties of the ligament substance from those of the structural characteristics of the bone–ligament functional unit, as their recovery rates are largely different. While the time course of recovery for bone–ligament junction is a slow one, the same may not be true for the ligament substance. Currently, collagen turnover studies of these ligaments are underway to gain further understanding of the ligament degradation and recovery processes secondary to immobilization and remobilization.

Healing and Repair of Ligaments: Immobilization vs Early Motion
In the case of severe trauma to the ligaments causing tearing of the midsubstance, there is no uniform clinical opinion with regard to the treatment

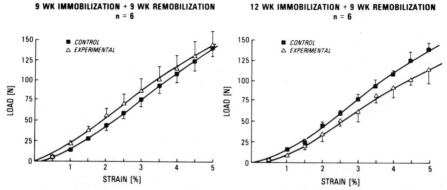

Figure 17. Average load–strain curves for the rabbit MCL subjected to 9 weeks of remobilization after 9 or 12 weeks of immobilization.

regimen. Clayton and Weir[54] found that suturing torn ligaments made them stronger than the comparable untreated tears for up to 9 weeks and reported that the ligaments healed by "union of ligament fibers" rather than by "a gap of fibrous tissues." O'Donoghue et al.[55] made similar observations, noting that suturing improved cellular organization and minimized scarring. He clearly distinguished between "scar formation" (without repair) and "dense collagenization" (with repair). However, at a recent consensus conference on knee ligament injuries (jointly sponsored by the National Institutes of Health and the American Academy of Orthopaedic Surgeons, April 1982, in Denver, Colorado),[56] most of the clinicians attending advocated discontinuing surgical repair for classes I and II MCL injuries and substituting in its place early protected motion and standard programs of rehabilitation.

Tipton et al.[38,57] are credited with the first systematic animal studies on the effects of physical activities on ligament healing following repair. Vailas et al.[58] recently published the maturational effects of exercise on ligament healing by biomechanical and biochemical methods. In our laboratory, we have been studying the management of medial–collateral ligament injuries by both unrepaired and repaired methods.

When the midsubstance defect of the medial–collateral ligament of the rabbit is created, the ligament will heal uneventfully via scar formation (unrepaired and untreated).[59] The time periods studied include 10 days and 3, 6, 14, and 40 weeks post injury. The structural properties of the healed bone–MCL–bone complex are persistently inferior to those of the contralateral controls despite the compensatory increases in scar mass and ligament size. The strength of the healing ligament appears to plateau at about 14 weeks postinjury, and there is no further improvement for up to 40 weeks (Fig. 18). The mechanical properties of the healing mid-ligament substance

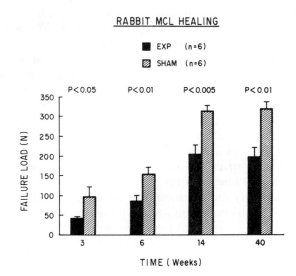

Figure 18. Histogram of strength properties of healing rabbit medialcollateral ligaments.

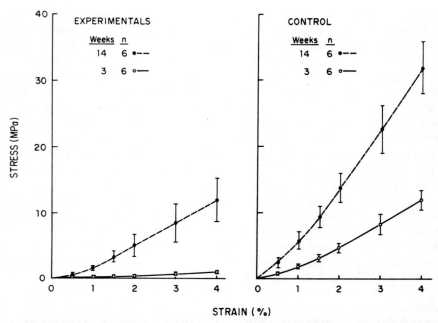

Figure 19. Curves showing the differences between the mechanical properties of healing and normal rabbit medialcollateral ligaments at 3 and 14 weeks post surgery.

also show a gradual recovery up to 14 weeks (Fig. 19). Furthermore, little change is seen in the stress–strain behavior from 14 to 40 weeks.

In a separate study, the MCL of the canine left knee is transected at the joint level, while the right knee is sham operated without disturbing its MCL. The experimental animals are divided into three groups following the methods of treatment. For group 1, the MCLs are repaired, and the knees are rigidly immobilized by internal fixation for 6 weeks. For group 2 the MCLs are repaired, immobilized for only 3 weeks, followed by removal of fixation, and allowed 3 additional weeks of cage activity. For group 3, the model is similar to that of the rabbit MCL study described above, and the animals are kept for 6 weeks with cage activity. At sacrifice, the varus–valgus knee laxities in terms of the change in angular deformation are measured at a torque of $\pm$ 0.2 N-m followed by tensile testing of the bone–MCL–bone unit to failure. Our findings are that all the experimental knees have greater laxity than have the controls (Fig. 20). Early motion will decrease the laxity of the experimental knee (group 2/group 1 = 61 percent; group 3/group 1 = 53 percent). Early mobilization also improved the mechanical properties of the healing ligament substance. The stress–strain curve for the group 2 ligaments was superior to that for group 1 (Fig. 21). The structural properties of the experimental bone–ligament–bone complexes from group 1 animals are similar to those published by Tipton et al.[57]

VARUS-VALGUS LAXITY

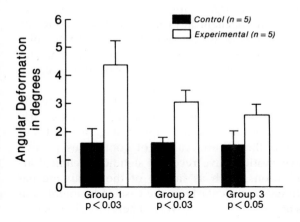

Figure 20. Varus–valgus laxity of dog knees. Laxity is defined by the angular deformation measured at an applied torque of ± 0.2 N-m.

These maximum load and energy failure values are approximately 40 to 50 percent of those of groups 2 and 3. In addition, there are no significant differences in strength characteristics between groups 2 and 3 (repaired with delayed motion vs nonrepaired with no immobilization). These preliminary results indicate that prolonged immobilization may be detrimental in the repaired MCL in terms both of the knee laxity and the mechanical properties of the ligament at the injury site. The ideal periods of immobilization and the question of repair vs nonrepair treatment of MCL injuries are subjects of continuing research.

SUMMARY

In this chapter, we have demonstrated the use of improved technology and experimental methodology to gain more accurate data on the mechanical

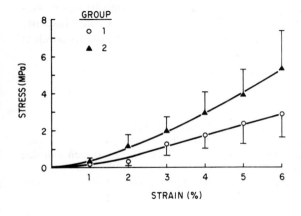

Figure 21. Comparison of the mechanical properties of surgically repaired ligaments. Group 1 ligaments are immobilized for 6 weeks, and group 2 ligaments are immobilized for 3 weeks, followed by 3 weeks of cage activity.

and structural properties of soft tissues. We have discussed the important contribution of tissue constituents and organization on their mechanical behaviors. It is possible to modify the mechanical properties of soft tissues by altering the contents and interactions of its constituents using in vitro biochemical techniques. It is also possible to measure the homeostatic changes in soft tissue properties in vivo by increasing or decreasing the levels of stresses and motion.

The data and techniques available today can be used to evaluate the success or failure in the treatment of soft tissue injuries. It has been clearly demonstrated that early mobilization is desirable for management of soft tissue trauma. As a result, the therapeutic values of continuous passive motion or intermittent passive motion have received significant clinical attention.[60,61] However, complications, such as failure of the repair mechanism and stretching of soft tissues, particularly in the case of ligament, can cause joint laxity if motion is applied too aggressively. Therefore, it is necessary to narrow the range of the spectrum of motion vs immobilization in the management of soft tissue trauma.

Obviously, more information on the functional forces and deformation of soft tissues in vivo is necessary. We believe, with this information together with the fundamental tissue properties, it will be possible to gain further understanding of the mechanisms of injury of soft tissues, as well as evaluation of the success of the clinical management. The technical information will be of significant importance to the prosthetic replacement of soft tissues as well. Thus, the goals of studying the biomechanics of soft tissues should be to provide the necessary information so that the decision of treatment of injuries can be made based on sound fundamentals and scientific information.

ACKNOWLEDGMENTS

The authors gratefully acknowledge their colleagues, Mr. D. Amiel and Dr. C. Frank, for their helpful discussions. We thank the Rehabilitation, Research and Development Department of the Veterans Administration, National Institutes of Health Grant AM 14918, and M. and D. Coutts Institute for Joint Reconstruction and Research for the financial support to conduct this research.

REFERENCES

1. Fung Y-C: Biorheology of soft tissues. *Biorheology* 1973; 10:139.
2. Viidik A: Biomechanical behavior of soft connective tissues. *Progress in Biomechanics*. NATO Advanced Study Inst, Ankara, July 10–21, Alphen aan den Rijn, Sijthoff and Noordhoff, 1978.
3. Tong P, Fung Y-C: The stress–strain relationship for skin. *J Biomech* 1976; 9:649.

4. Schneider DC, Davidson TM, Nahum AM: In vitro b sponse of human skin. *Arch Otolaryngol* 1984; 110:329–'
5. Hayashi K, Woo SL-Y (eds): Symposium on mechar tissues. *Biorheology* 1982; 19:397–408.
6. Mow VC, Kuei SC, Lai WM, et al.: Biphasic cree articular cartilage in compression: Theory and exp 1980; 102:73.
7. Mow VC, Myers ER, Rother V, et al.: Implications for collag interactions from cartilage stress relaxation behavior in isometric tension. *arthritis Symp* 1981; 22:41.
8. Viidik A: A rheological model for uncalcified parallel-fibered collagenous tissue. *J Biomech* 1968; 1:3.
9. Rosenberg LC: Structure of cartilage proteoglycans, in Burleigh PMC, Poole AR (eds): *Dynamics of Connective Tissue Macromolecules*. Amsterdam, North Holland, 1975, pp 105–128.
10. Abrahams M: Mechanical behavior of tendon in vitro. *Med Biol Eng* 1967; 5:433–443.
11. Yamada H: In Evans FG (ed): *Strength of Biological Materials*. Baltimore, Williams & Wilkins, 1970.
12. Viidik A: Biomechanical behavior of soft connective tissues, in Akkas N (ed): *Progress in Biomechanics*. Alphen aan den Rijn, Sijthoff and Noordhoff, 1979, pp 75–113.
13. Noyes FR: Functional properties of knee ligaments and alterations induced by immobilization: A correlative biomechanical and histological study in primates. *Clin Orthop* 1977; 123:210.
14. Kennedy JC, Hawkins RJ, Willis R, et al.: Tension studies of human knee ligament. *J Bone Joint Surg* 1976; 58:350.
15. Dorlot J-M, Ait ba sidi M, Tremblay GM, et al.: Load-elongation behavior of the canine anterior cruciate ligament. *J Biomech Eng* 1980; 102(3):190.
16. Woo SL-Y, Kuei SC, Gomez MA, et al.: The effects of immobilization and exercise on the strength characteristics of bone–medial collateral ligament–bone complex. *1979 Biomechanics Symposium*. ASME/AMD vol 32, pp 67–71.
17. Woo SL-Y: Mechanical properties of tendons and ligaments. I. Quasi-static and nonlinear viscoelastic properties. *Biorheology* 1982; 19:385.
18. Woo SL-Y: Mechanical properties of tendons and ligaments. II. The relationship of immobilization and exercise on tissue remodeling. *Biorheology* 1982; 19:397.
19. Woo SL-Y, Akeson WH, Jemmott GF: Measurement of nonhomogeneous directional mechanical properties of articular cartilage in tension using a video dimensional analyzer. *J Biomech* 1976; 9(12):785.
20. Woo SL-Y, Gomez MA, Seguchi Y, et al.: On the measurement of mechanical properties of ligament substance from a bone–ligament–bone preparation. *J Orthop Res* 1983; 1(1):22.
21. Woo SL-Y, Gomez MA, Amiel D, et al.: The effects of exercise on the biomechanical and biochemical properties of swine digital flexor tendons. *J Biomech Eng* 1981; 103(1):51.
22. Fung Y-C: Stress–strain history relationship of soft tissues in simple elongation, in Perrone N, Anliker V (eds): *Biomechanics, Its Foundations and Objectives*. 1982, pp 181–208.
23. Haut RC: Age-dependent influence on strain rate of a tensile failure of rat tail tendon. *J Biomech Eng* 1983; 105:296.

is G: Application of thermodynamics of viscoelastic materials with a fading
emory: Integral constitutive equations. *Int J Nonlinear Mechanics* 1970; 5:23.
Bernstein B, Kearsley EA, Zapas LS: A study of stress relaxation with finite
strain. *Trans Soc Rheology* 1963; 7:391–410.

26. Chen HY-L: *Rabbit Mesentery as Viscoelastic Material—An Approach to the Mechanical Properties of Soft Tissues,* PhD thesis. University of California, San Diego, Calif, 1973.

27. Tanaka TT, Fung Y-C: Elastic and inelastic properties of the canine aorta and their variation along the aortic tree. *J Biomech* 1974; 7:357.

28. Haut RC, Little RW: A constitutive equation for collagen fibers. *J Biomech* 1972; 5:423.

29. Jenkins RB, Little RW: A constitutive equation for parallel-fibered elastic tissue. *J Biomech* 1974; 7:397.

30. Woo SL-Y, Simon BR, Kuei SC, et al.: Quasi-linear viscoelastic properties of articular cartilage during stress relaxation. *J Biomech Eng* 1980; 102(2):85.

31. Woo SL-Y, Gomez MA, Akeson WH: The time and history-dependent viscoelastic properties of the canine medial collateral ligament. *J Biomech Eng* 1981; 103:293.

32. Sauren HJ, Rousseau EPM: A concise sensitivity analysis of the quasi-linear viscoelastic model proposed by Fung. A technical brief. *J Biomech Eng* 1983; 105:92.

33. Minns RJ, Soden PD: The role of the fibrous components and ground substance and the mechanical properties of biological tissues: A preliminary investigation. *J Biomech* 1973; 6:153

34. Hoffman AS, Grande LA, Park JB: Sequential enzymolysis of human aorta and resultant stress–strain behavior. *Biomater Med Devices Artif Organs* 1977; 5(2):121.

35. Broom ND: The stress/strain and fatigue behavior of glutaraldehyde-preserved heart-valve tissue. *J Biomech* 1977; 10:707.

36. Harjula A, Mattila S: The effect of glutaraldehyde tanning on elasticity of the human saphenous vein. *Ann Chir Gynaecol* 1980; 69:60.

37. Woo SL-Y, Ritter MA, Amiel D, et al.: The biomechanical and biochemical properties of tendons. Long term effects of exercise on the digital extensors. *Connect Tissue Res* 1980; 7:177.

38. Tipton CM, Matthes D, Maynard JA, et al.: The influence of physical activity on ligaments and tendons. *Med Sci Sports* 1975; 7(3):165.

39. Butler DL, Noyes FR, Grood ES: Measurement of the mechanical properties of ligaments, in Feinberg BN, et al. (eds): *CRC Handbook of Engineering and Medical Biology. Section B, Instruments and Measurements.* Palm Beach, Fla, CRC Press, 1978, Vol 1, pp 279–314.

40. Duran RJ, Hauser RG: *American Academy of Orthopaedic Surgeons Symposium on Tendon Surgery in the Hand.* St. Louis, Mosby, 1975, pp 105–114.

41. Kleinert HE, Kutz JE, Atasoy E, et al.: Primary repair of flexor tendons. *Orthop Clin North Am* 1973; 4:865.

42. Strickland JW, Glogovac SV: Digital function following flexor tendon repair in zone 2: A comparison of immobilization and control passive motion techniques. *J Hand Surg* 1980; 5:537.

43. Lister GD, Kleinert HE, Kutz JE, et al.: Primary tendon repair followed by immediate controlled mobilization. *J Hand Surg* 1977; 2:441.

44. Gelberman RH, Woo SL-Y, Lothringer KS, et al.: Effects of early intermittent passive mobilization on healing canine flexor tendons. *J Hand Surg* 1982; 7(2):170.
45. Woo SL-Y, Gelberman RH, Cobb NG, et al.: The importance of controlled passive mobilization on flexor tendon healing: A biomechanical study. *Acta Orthop Scand* 1981; 52(6):615.
46. Noyes FR, DeLucas JL, Torvik PJ: Biomechanics of anterior cruciate ligament failure: An analysis of strain-rate sensitivity and mechanisms of failure in primates. *J Bone Joint Surg* 1974; 56A:236.
47. Noyes FR, Torvik PJ, Hyde WB, et al.: Biomechanics of ligament failure. II. An analysis of immobilization, exercise and reconditioning effects in primates. *J Bone Joint Surg* 1974; 56A:1406.
48. Crowninshield RD, Pope MH: The strength and failure characteristics of rat medial collateral ligaments. *J Trauma* 1976; 16:99.
49. McElhaney J: Dynamic response of bone and muscle tissue. *J Appl Physiol* 1966; 21:1231.
50. Crowninshield RD, Pope MH: The response of compact bone in tension at various strain rates. *Ann Biomed Eng* 1974; 2:217.
51. Amiel D, Woo SL-Y, Harwood FL, et al.: The effect of immobilization on collagen turnover in connective tissue: A biochemical–biomechanical correlation. *Acta Orthop Scand* 1982; 53:325.
52. Akeson WH, Amiel D, La Violette D: The connective tissue response to immobility: A study of chondroitin 4 and 6 sulfate and dermatan sulfate changes in periarticular connective tissue in the control and immobilized knees of dogs. *Clin Orthop* 1967; 51:183.
53. Woo SL-Y, Gomez MA, Lothringer KS, et al.: The effect of changes in stress levels on the homeostasis of cortical bone, tendons and ligaments, in Wang JJ, Fung YC (eds): *The First China-Japan-US Conference on Biomechanics*. Wuhan, China, Huachung University of Science and Technology, May 1983, in press.
54. Clayton ML, Weir GJ: Experimental investigations of ligamentous healing. *Am J Surg* 1959; 98:373.
55. O'Donoghue DH, Rockwood CA, Zarnicznyj B, et al.: Repair of knee ligaments in dogs. I. The lateral collateral ligament. *J Bone Joint Surg* 1961; 43A:1167.
56. Finerman G (ed): *Proceedings of the Conference on Knee Ligament Injuries*. Sponsored jointly by the NIH and AAOS, April, 1982, Denver, Colo. St. Louis, CV Mosby, in press.
57. Tipton CM, James SL, Mergner W, et al.: Influence of exercise on strength of medial collateral knee ligaments of dogs. *Am J Physiol* 1970; 218:894.
58. Vailas AC, Tipton CM, Matthes RD, et al.: Physical activity and its influence on repair process of medial collateral ligaments. *Connect Tissue Res* 1981; 9:25.
59. Frank CB, Woo SL-Y, Amiel D, et al.: Medial collateral ligament healing. A multidisciplinary assessment in rabbits. American Orthopaedic Society for Sports Medicine Basic Science Award paper. *Am J Sports Med* 1983; 11(6):379.
60. Frank CB, Akeson WH, Woo SL-Y, et al.: Physiology and therapeutic values of passive joint motion. *Clin Orthop* 1984; 185: 113.
61. Salter RB, Simmonds DF, Malcolm BW, et al.: The biological effects of continuous passive motion in the healing of full thickness defects in articular cartilage. *J Bone Joint Surg* 1980; 62A:1232.

CHAPTER 8

Biomechanics of Bone

Dennis R. Carter

INTRODUCTION

In any physical activity, a complex pattern of forces is imposed on the bones of the skeletal system. These forces are of three types: (1) external forces acting on the body, (2) internal forces caused by muscle contraction or ligament tension, and (3) internal reaction forces between bones. The forces (also referred to as loads) cause small deformations of the bones on which they act. The mechanical response of a particular bone can be described by quantitatively assessing the relationships between the applied loads and the resulting deformations. The relationships between forces and deformations reflect the structural behavior of the whole bone. In moderate loading situations, bone deformations are only present while the loads are applied. When the loads are removed, the bone reassumes its original position and geometry. If the skeletal system is exposed to severe trauma, the loads imposed on a bone become extremely high, resulting in large deformations and possibly in bone fracture. The major factors that determine the deformation characteristics and fracture resistance of a bone are (1) the direction, magnitude, and rate of force application, (2) the size and geometry of the bone, and (3) the material properties of the tissue that comprises the bone.

Large forces obviously cause greater deformation and tendency toward fracture than smaller forces. Less well appreciated is the fact that the fracture behavior of a long bone when subjected to axial forces is much different from when it is subjected to transverse or torsional forces and that the bone can resist rapidly applied forces much better than slowly applied forces.

The anatomic differences among the bones of the skeleton tend to reflect the types of forces that act on the individual bones. Large bones are much better suited to resist forces than are small bones. Bones are specialized in their geometric configuration to resist forces in a particular direction.

The material properties of the tissue that comprises the bone are very important in determining the deformation and fracture properties of the bone. For example, an osteoporotic bone with the same geometry as a normal bone will experience greater deformation under loading and will fracture at lower force magnitudes.

The objective of this chapter is to present some of the basic mechanical concepts related to bone fracture. Important aspects of whole bone fracture will be provided. An attempt will then be made to relate these phenomena to the material properties and fracture behavior of bone tissue. This presentation is based to a large extent on that of Carter and Spengler.[1] Related topics are provided by Carter and Spengler[2] and by Hayes and Carter.[3]

WHOLE BONE FRACTURE

Influence of Bone Size and Shape

When the skeletal system is exposed to severe trauma, the bones are subjected to very high forces. Fracture occurs when the stresses (internal force intensities) in one region of the bone exceed the ultimate strength of the bone material. Bone fracture can therefore be thought of as an event that is initiated at the level of the material. The size and shape of the bone under loading determine the distribution of stresses throughout the bone. A large bone is more resistant to fracture simply because it distributes the internal forces over a larger volume of bone material. The stresses at any one location are therefore less than those in a smaller bone loaded under similar conditions.

A notable physical characteristic of long bones is that their diaphyses are tubular in shape. A tubular structure can more evenly distribute the stresses imposed by bending and torsional loading than can a solid cylindrical structure of equal mass.

A given structural member (bar) can better resist torsional and bending loads if the material that comprises that member is distributed at a distance from the central axis. The distribution of mass about the center of a structural member can be quantitatively described by the *second moment of inertia* of the cross-section. Tubular structures have larger second moments of inertia both in bending and torsion than cylindrical structures with the same amount of mass. Figure 1 illustrates the cross-sections of three bars composed of the same material. Bar A is cylindrical and has a cross-sectional area of one square unit. Bar B has the same mass and cross-sectional area as bar A. However, bar B has a hollow interior and, therefore, a greater moment of inertia. Bar C has twice the mass and, therefore, twice the cross-sectional area of bars A and B. Since the additional mass of bar C

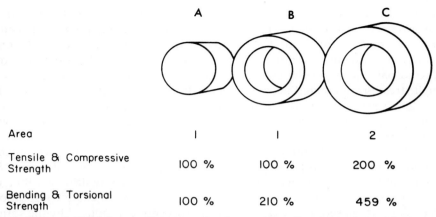

Figure 1. The influence of cross-sectional geometry on the structural strength of circular bars. (*From Carter and Spengler.*[1])

is at the perimeter of the bar (far from the center of the bar), this mass will result in a much greater second moment of inertia. The tensile and compressive strength of the three bars illustrated in Figure 1 is directly proportional to the cross-sectional area. Therefore, when exposed to tensile or compressive loading, bars A and B will have the same strength, and bar C will be twice as strong as A and B. In torsional or bending loading, however, the favorable distribution of mass in B and C makes these bars much stronger than bar A. In these loading modes, bars B and C will be 210 percent and 459 percent stronger than bar A, respectively.

The influence of the second moment of inertia on the structural strength of long bones is particularly relevant when the bone shape during fracture healing is considered. During the healing process, a fracture callus consisting of newly mineralizing bone tissue is formed at the fracture site. This callus bridges the gap between fractured bone ends to produce a cuff of bone with a significantly greater diameter than that of the normal bone. During healing the fracture callus is not as well mineralized as the normal bone and, therefore, has inferior material properties. The high moment of inertia offered by the fracture callus, however, tends to compensate for this lack of material integrity and is an important factor in establishing stability at the fracture site. As the fracture heals, the fracture callus becomes progressively more mineralized and gradually increases in material strength. The excessive bone material around the perimeter of the callus is then progressively resorbed. With time, the healing bone reassumes its normal material properties and geometry.

Bone Defects
Certain surgical procedures, as well as naturally occurring pathologic bone lesions, may create defects in the normal bone geometry that significantly affect the fracture resistance of whole bones. Specific examples of these

defects include screw holes, surgically excised bone slots, and bone cysts. Such defects reduce the strength of the bone by removing bone mass, causing the internal forces to be distributed over a smaller volume of bone tissue. Additionally, and often more importantly, bone defects tend to produce a pattern of poorly distributed stresses within the bone such that very high stresses are created near the defect. In such cases, the defect is said to produce *stress concentrations* in the bone when forces are applied, and the bone tissue material strength near the defect is exceeded under relatively low forces.

Burstein et al.[4] studied the effects of screws and screw holes on the torsional material properties of rabbit bones. A 70 percent decrease in energy-absorption capacity during fracture was created by drilling a hole in the bone diaphysis and inserting a screw. This decrease in energy-absorption capacity was caused by the stress concentration effect produced by the surgical procedure. Eight weeks after the insertion of the screw, the stress concentration caused by the hole and the screw had been alleviated by significant bone remodeling around the screw. When the screw was removed, however, the screw hole in the cortex again served as a stress concentrator, and the strength of the bone was again significantly reduced. With time, such screw holes are filled in with mineralizing tissue, and the normal bone material properties, geometry, and strength are restored.

Figure 2 illustrates a fractured femur that was treated by open reduction and internal fixation with a metal plate. After the initial fracture had healed,

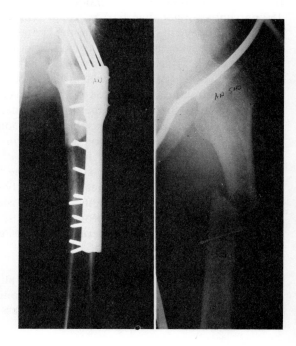

Figure 2. Refracture of the femur caused by the stress concentration due to the presence of a screw hole after the removal of the bone plate. (*From Carter and Spengler.*[1])

the metal plate and screws were removed, leaving the bone in a healed but significantly weakened condition. Refracture of the bone then occurred with relatively minor trauma. The fracture line can be seen to have begun at one of the screw holes, which served as a stress concentrator.

Fracture Patterns

In clinical practice it is common to evaluate patients who have sustained bone fractures but cannot recall the specific injury mechanism. Knowledge of the events that caused injury, however, is important in assessing the patient and suggesting appropriate fracture management. Some insight into the mechanism of injury can be gained from careful evaluation of the patient's roentgenograms.

Evaluation of roentgenograms can yield two important kinds of biomechanical information. First, the degree of fracture comminution tends to reflect the amount of energy imposed during fracture. For example, a highly comminuted fracture with multiple fragments indicates a large amount of energy dissipation and quite likely a very rapid loading rate, such as is seen in high-speed vehicular accidents (Fig. 3). In such fractures, a great deal of soft tissue injury generally accompanies the bone fracture. Second, the ori-

Figure 3. Severe fracture comminution indicating high-energy bone fracture. (*From Carter and Spengler.*[1])

entation of the fracture lines can provide information on the type of loading that caused the fracture (i.e., tension, compression, torsion, bending, or combined loading). The following section examines some common fracture patterns and relates these patterns to the mechanical events that caused the fracture.

Tension and Compression. Bone fracture caused by pure tensile loading is extremely uncommon in the long bone diaphyses. Directly applied tensile forces are, however, associated with avulsion fractures in cancellous bone. Avulsion fractures in metaphyseal areas are encountered at sites of major ligamentous and tendinous attachment (e.g., medial malleolus and the peroneal brevis insertion on the fifth metatarsal). An example of a predominantly tensile failure of a patella is demonstrated in Figure 4. This fracture, which was produced by forceful contraction of the quadriceps muscle rather than by externally imposed trauma, resulted in a transverse fracture on a plane of high tensile stresses.

A common site of compression fracture in cancellous bone is the vertebral body. Figure 5 illustrates a compression fracture resulting in significant shortening of a lumbar vertebra. This compression fracture of a vertebral

Figure 4. Tensile fracture through the patella. (*From Carter and Spengler.*[1])

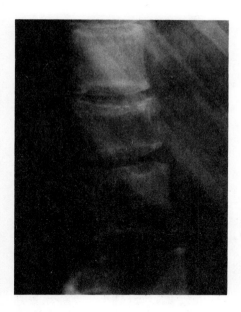

Figure 5. Compression fracture in a vertebral body. (*From Carter and Spengler.*[1])

body resulted in a much greater absorption of energy than the tensile fracture through the cancellous bone shown in Figure 4. The ability of cancellous bone to absorb significant energy during compressive fracture is one of the most important characteristics of the skeletal tissues.

Cortical bone fracture caused by compressive loading along the long bone axis is sometimes observed in the diaphyses of long bones. Axial compression results in high shear stresses on planes that are oblique to the long bone axis. Bone fracture due to axial compression occurs along these shear planes, producing an oblique fracture pattern in the diaphyses. Figure 6 illustrates a compression fracture in the tibia and fibula under axial loading conditions. Numerous oblique fracture lines are present.

Torsional, Bending, and Combined Loading. Torsion is a commonly encountered loading situation, which results in diaphyseal bone fracture with a spiral oblique appearance. Compare the roentgenograms of a human tibia (Fig. 7) and a human femur (Fig. 8). In both of these cases a spiral fracture pattern was created; however, the degree of comminution is different in each fracture. The degree of comminution reflects the relative degree of trauma and force imposed on each of the bones. The fracture illustrated in Figure 7 was produced with relatively minor trauma and was primarily a result of the weakening of the bone caused by the open section defect where a cortical graft was removed. Figure 8 shows slightly greater comminution, which suggests that the fracture was caused by significantly higher forces and greater energy.

Fracture in long bones is often caused by high bending forces that

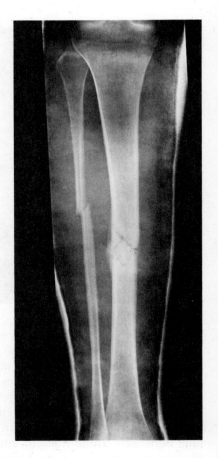

Figure 6. Axial compression fracture in the tibia and fibula. (*From Carter and Spengler.*[1])

create tensile stresses on one aspect of the cortex and compressive stresses on the opposite side. Figure 9 illustrates a fracture in the middiaphysis of the femur. A primarily transverse fracture is seen through one cortex, and a large butterfly fragment is seen on the opposite cortex. The butterfly fragment was formed by the propagation of two cracks on oblique fracture planes. Large butterfly fragments may be produced when an axial compressive load is superimposed on the bending load. If no compressive load is acting, the fracture may be essentially transverse through the entire diaphysis. The size of the butterfly fragment in Figure 9 suggests that, in this case, a significant axial compressive force may have been acting, as well as bending forces.

Most fractures seen clinically are not produced by simple loading mechanisms but are produced by a more complex loading situation. Figure 10 illustrates a fracture in the tibial diaphysis that consists of a spiral fracture surface through one cortex and a transverse fracture through the opposite cortex. Also present are several small oblique lesions in the proximal bone fragment. This fracture was probably caused by a combination of torsional,

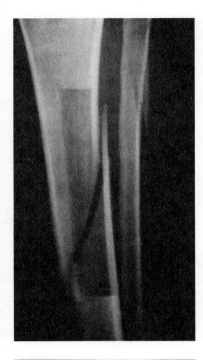

Figure 7. Torsional fracture through an open section defect that was surgically created in the tibia. (*From Carter and Spengler.*[1])

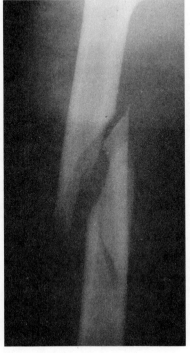

Figure 8. Torsional fracture in the diaphysis of the femur. (*From Carter and Spengler.*[1])

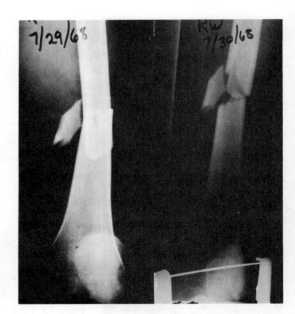

Figure 9. Disphyseal fracture caused by bending forces. A butterfly fragment was produced on the side of the bone subjected to compressive stresses. (*From Carter and Spengler.*[1])

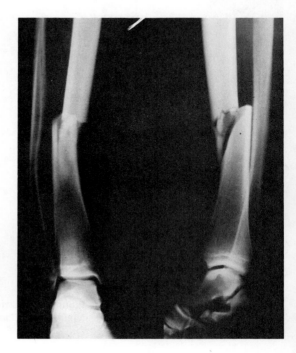

Figure 10. Tibial fracture produced by a combined loading situation, which probably included torsion, axial compression, and bending. Note the residual deformation of the fibula caused by yielding. (*From Carter and Spengler.*[1])

bending, and compressive forces. The transverse fracture through one cortex was probably caused by tensile stresses produced by bending forces. Torsional loading was probably responsible for the spiral oblique fracture surface. Fracture of the tibia and the resulting tibial shortening resulted in significant residual deformity of the fibula. This residual bending in the fibula suggests that the bone material of the fibula was loaded beyond its yield point but not sufficiently to produce complete fibular fracture.

Fatigue Fracture. Fatigue (or stress) fractures commonly occur but are often undocumented or simply not appreciated during clinical examination. Patients with these fractures generally present with complaints of pain and localized bony tenderness. Initial roentgenograms may be normal, although a bone scan often reveals locally increased metabolic activity that reflects a localized increase in bone turnover. Fatigue failure of bone in vivo occurs when the accumulation of bone microdamage caused by mechanical stress exceeds the bone's ability to repair that damage. Fatigue fractures have commonly been reported in military recruits, professional athletes, dedicated joggers, and "weekend" athletes. When a fatigue fracture occurs in the hip area, significant morbidity may result. Early recognition and treatment are essential to prevent complications. In most situations prompt treatment, consisting of elimination of weightbearing or immobilization, results in healing.

The fracture patterns created in fatigue fractures are similar to those previously described for the fracture of bone caused by a single traumatic episode. Figure 11 illustrates a fatigue fracture consisting of a transverse lesion in the midtibial diaphysis. Biomechanical studies have shown that during running, this area in the tibia is subjected primarily to tensile stresses along the bone axis. The production of a transverse lesion is consistent with the predicted fracture pattern caused by longitudinal tensile stresses. In other locations of the skeleton (e.g., the medial aspect of the femoral diaphyses) longitudinal compressive stresses are present during rigorous activity. In these locations, fatigue fractures generally present as oblique lesions that form on planes of high shear stresses.[5]

Pathologic Fractures. The presence of underlying tissue abnormalities can significantly alter the mechanical response of bones. Pathologic changes can affect the mechanical behavior by causing (1) a decrease in bone mass (e.g., osteoporosis, osteogenesis imperfecta), (2) an altered bone quality (e.g., osteomalacia, osteogenesis imperfecta), and/or (3) changes in the distribution of bone mass (e.g., acromegaly). Pathologic fractures generally have little comminution, since the preexisting abnormalities markedly impair the ability of the bone to absorb energy, and fracture occurs under fatigue or moderate loading conditions.

Figure 12 illustrates a bone fracture through an area of metastatic carcinoma. Note the extensive bony destruction and minimal comminution. The presence of distal cortical destruction involving the lateral cortical wall suggests the necessity of using an internal fixation device that will safely reinforce the entire area.

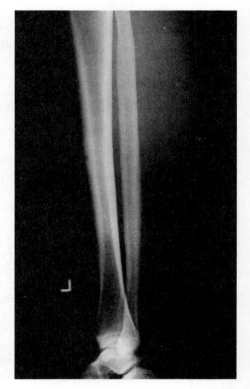

Figure 11. Fatigue fracture in the anterior tibial diaphysis. (*From Carter and Spengler.*[1])

Figure 13 illustrates multiple fractures in various stages of healing in a patient with osteogenesis imperfecta. Note the extreme loss of bone apparent density and the gross bone deformity. Osteogenesis imperfecta is a systemic disease that markedly impairs bone strength by altering bone composition and bone geometry.

Osetoporosis is a commonly encountered condition that impairs bone strength by increasing the bone porosity and decreasing the cortical thickness even though no alterations in bone mineralization occur. The loss of total bone mass is associated with an increase in the intramedullary area and periosteal diameter of long bone diaphyses. The redistribution of bone mass farther from the bone axis tends to maintain a high second moment of inertia, which partially offsets the decline in bone strength caused by the loss of bone mass.

BASIC MECHANICAL CONCEPTS

Stress and Strain

To understand bone fracture, one must have a basic understanding of the relationship between the forces applied to a bone and the stress and strain

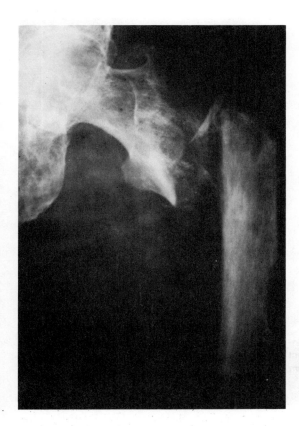

Figure 12. Pathologic fracture through a region of bone significantly weakened by metastatic carcinoma. (*From Carter and Spengler.*[1])

distributions that result. When forces are applied to any object, the object will be deformed from its original dimensions, and internal forces will be produced within the object. The deformations created at any point in the object are referred to as the *strains* at that point. The internal force intensities are referred to as the *stresses* at that point. These stresses and strains vary thoughout the bone in a highly complex manner.

Stresses = local force intensities (dimensions = force per unit area, MN/m^2 or MPa)
Strains = local deformations (dimensions = length per length)

The strains at any point in a bone subjected to forces are mathematically related to the stresses at that point. The quantitative relationships between stresses and strains are governed by the material properties of the bone tissue. If the whole bone is loaded with very high forces, the stresses and strains at one region may exceed the ultimate stresses or strains that the bone tissue can tolerate. Mechanical failure will then occur at that point, and bone fracture will result. If a bone is comprised of tissue with poor

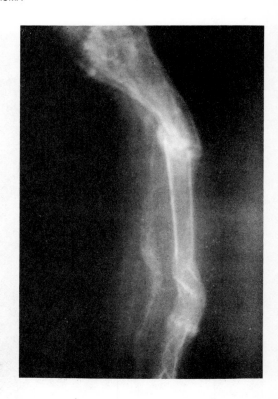

Figure 13. Multiple fractures in various stages of healing in a patient with osteogenesis imperfecta. (*From Carter and Spengler.*[1])

material properties (e.g., osteomalacic bone), the bone tissue at the region of high stresses and strains will fracture at lower force levels than will a bone consisting of normal tissue.

To fully characterize the stresses or strains at any one point, one must specify six stress values that correspond to normal and shear strains on each of three independent planes passing through that point. The concept of stresses and strains is considerably simplified if we restrict our consideration to stresses present on an imaginary plane through a point in the bone. Figure 14 demonstrates a human femur that is subjected to a pattern of applied forces. We wish to examine the stresses at point 0 in this bone that act on a plane passing through the diaphysis in a transverse direction. We can consider point 0 to consist of an infinitesimally small cube. The top face of this cube has an area that we will designate as A. Two types of internal forces may act on the top face of the cube. The first is a force that is perpendicular to the face (force F). This internal force results in a normal stress, σ, which is equal to F/A. The other type of force that may act on this surface is shear force, which we will designate as S. This shear force results in a shear stress, τ, which equals S/A. A normal stress can act either toward the face of the cube (compression) or away from the face of the cube (tension). The shear stress can likewise be oriented in any direction parallel

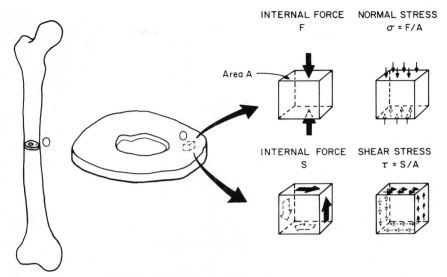

INTERNAL FORCE   NORMAL STRESS
F $\sigma = F/A$

Area A

INTERNAL FORCE   SHEAR STRESS
S $\tau = S/A$

Figure 14. Schematic representations of the stresses acting on a transverse plane through a point in the femoral diaphysis. (*From Carter and Spengler.*[1])

to the top face of the cube, depending on the loading condition imposed on the whole bone.

Normal stress = force per unit area acting perpendicular to a given plane (MPa)
Shear stress = force per unit area acting parallel to a given plane (MPa)

The two types of stresses illustrated in Figure 14 result in local deformations to the cube. A normal tensile stress will cause the front face of the cube to become longer and thinner. Normal compressive stresses cause the front face of the cube to become shorter and wider (Fig. 15). The normal strain in the cube can be defined as the ratio of the change in length (Δl) of the side of the cube to the original length of the side (l):

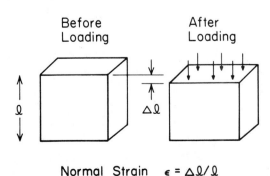

Before   After
Loading   Loading

l

Δl

Normal Strain $\epsilon = \Delta l / l$

Figure 15. Representation of the strain caused by compressive stresses. (*From Carter and Spengler.*[1])

Normal strain $\varepsilon = \Delta l/l$

The shear stresses imposed on the top face of the cube will cause the front face of the cube to be deformed from a square into a parallelepiped (Fig. 16). The shear strain is defined as the angular deviation of one side of the cube from its original right angle position. This angle (expressed in radians) is approximately equal to $\Delta l/l$. The amount of normal strain and shear strain that is experienced by the cube will be influenced by the magnitudes of the normal stress and shear stress on the top face of the cube as well as by the inherent material properties of the bone tissue. In general, if the bone tissue is very well mineralized, the bone tissue itself will be very stiff, and small strains will result from the application of stresses at that point. If the bone material is of poor quality, however, the same stresses will result in larger strains in the cube, since this bone material is softer and more compliant. It should also be noted that we have thus far considered stresses at point 0 only on a *transverse* plane through the whole bone (i.e., the top face of the cube). If we were to consider the stresses at point 0 acting on a plane that passes through the whole bone at an oblique angle, the stresses on that plane would differ from those on the transverse plane.

Tension and Compression

Bones of the skeletal system are geometrically complex and are exposed to complex force patterns. These factors lead to very complex patterns of stresses and strains throughout the bone tissue. Simplified structures that are loaded under well-defined conditions can be used to demonstrate some basic mechanical concepts. Figure 17 demonstrates a bar of length l and a constant cross-sectional area (A) that is subjected to pure tensile loading (F). This loading condition results in a uniform, homogeneous pattern of stresses and strains throughout the structure at any cross-section. As load is applied to the bar, the bar begins to stretch. The relationship between the applied force (F) and the increase in the length of the bar (Δl) can be demonstrated on a force-deformation curve (Fig. 18A). The initial portion of the force-deformation curve is essentially linear. This linear portion of the curve

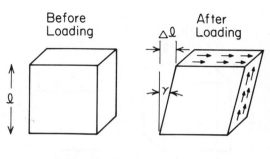

Figure 16. Representation of the strain caused by shear stresses. (*From Carter and Spengler.*[1])

Shear Strain $\gamma = \Delta l/l$

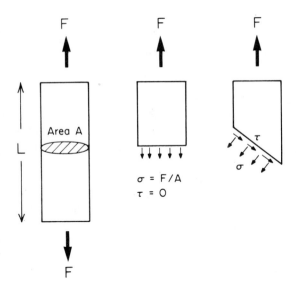

Figure 17. Schematic representation of the stresses acting on a
transverse and oblique plane
through a bar subjected to tensile
forces. (*From Carter and Spengler.*[1])

represents the *elastic* behavior of the structure. If load is applied in the
elastic region of the curve and is then released, the bar will return to its
original length. If sufficient force is applied, however, internal damage will
be created in the structure and the bar will begin to yield. Yield occurs at
point Y on the force-deformation curve of Figure 18A. Further loading
beyond the yield point will result in marked deformation until total fracture
occurs (point U in Figure 18A). The total energy absorbed by the bar during
the fracture process is represented by the area under the force-deformation
curve.

One can also consider the stresses and strains present on a transverse
plane through the bar at any point. Because of the simplicity of the loading

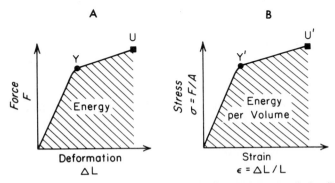

Figure 18. A. Force-deformation curve of a bar subjected to tensile loading. **B.**
The corresponding stress–strain curve of the material comprising the bar that is
subjected to tensile forces. (*From Carter and Spengler.*[1])

situation, the stresses on this plane are identical over the entire cross-section. A stress–strain curve can be constructed for any point on a transverse plane through the bar (Fig. 18B). In this loading configuration, the normal stress is equal to F/A and the normal strain is equal to Δl/l. No shear stresses or strains act on the transverse plane. The stress–strain curve is, in effect, a normalized curve that reflects the mechanical behavior of the material that comprises the bar. If one were to double the cross-sectional area and triple the length of the bar, the force-deformation curve would be altered significantly. Since the ultimate force (point U) is proportional to the cross-sectional area, the bar would be able to withstand twice as much force before it fractured. The bar would deform three times as much before it fractured, since ultimate deformation is proportional to initial length. The stress–strain curve of this new bar, however, would be indentical to that of the first bar. The yield strength and yield strain of the material (represented by point Y'), as well as the ultimate strength and ultimate strain of the material (represented by point U'), are independent of the size of the bar being loaded. The force-deformation curve is therefore a representation of the mechanical behavior of the *structure,* while the stress–strain curve is a mechanical representation of the behavior of the *material.* In the elastic region (from A' to Y') the stiffness of the material is measured by the slope of the stress–strain curve. This slope is called the "elastic modulus" or "Young's modulus" and has dimensions of force per unit area. The elastic modulus of steel is approximately 10 times greater than that of bone, and the ultimate tensile strength of steel is approximately 5 times greater than that of cortical bone.

It is important to realize that the stress in the bar as demonstrated in Figure 17 is dependent upon the plane under consideration. If one considers a plane oriented at an oblique angle to the applied load, shear stresses as well as normal stresses will be present on that plane. This fact becomes particularly important in considering the fracture behavior of cortical bone, since under tensile loading the bone tends to fracture on a transverse plane, where the normal tensile stresses are greatest. In compression, however, the bone tends to fail along oblique angles; this pattern corresponds to failure along planes of high shear stress.

Bending

A beam can be subjected to bending loads in two ways. These two types of bending are generally referred to as *pure bending* and *three-point bending.* Figure 19 demonstrates a simple beam that is subjected to pure bending loads. This loading situation can be thought of as if one were to grab each end of the beam and twist that beam in such a manner as to produce a convex surface on one side of the beam and concave surface on the other side. This action produces constant bending loading over the entire length of the beam. The material on the concave side of the beam will be subjected

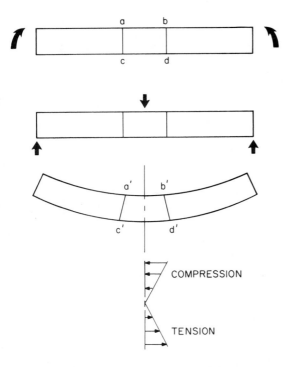

Figure 19. Representation of the deformation and stresses in a beam subjected to bending forces. (*From Carter and Spengler.*[1])

to compressive strains, while the material on the convex side will be subjected to tensile strains. The strains produced at any cross-section of the beam will result in stresses at that cross-section. The concave side of the beam will be exposed to high compressive stresses, while the convex side will experience high tensile stresses. If the bar is loaded so that all the material is stressed only in the elastic region of the stress–strain curve, the distribution of stresses at any cross-section through the beam will be proportional to the distance from the center of the beam. Notice that the highest stresses are experienced by the material on the surface of the beam. If the bending force is increased until the bar begins to fracture, fracture will be initiated at the surface of the beam, where the stresses are highest. Since pure bending is being applied, fracture may occur at any location along the length of the bar.

Pure bending rarely occurs in the skeletal system. Bending forces experienced by bones in vivo are better simulated by a single load applied to the beam, which is supported at the ends (three-point bending). In such a loading situation the bending moment at a section through the beam is greatest at the point where the load is applied, and failure of the beam in bending will occur at that point. Three-point bending also introduces, on transverse planes through the beam, shear stresses that would not be present in a pure bending situation. However, if the beam is much longer than it is thick, these shear forces will be negligible.

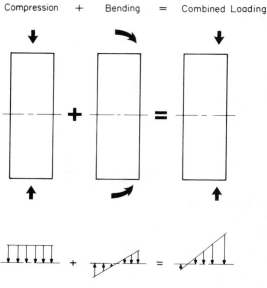

Figure 20. Stresses imposed on a transverse plane through a bar that is subjected to a combination of compressive and bending loading. (*From Carter and Spengler.*[1])

Transverse Stress Distributions

Combined Bending and Axial Loading

Long bones in vivo are often subjected to combined compressive and bending loads. This loading situation can be illustrated by a bar subjected to a compressive force that is not directed through the center of the bar (Fig. 20). The resulting stresses on a transverse section through the bar can be found by merely summing the stresses caused by the independent actions of axial force and bending moment. High compressive stresses will be created on one side of the bar, while the other side of the bar will experience either lower compressive stresses or tensile stresses, depending on the relative magnitudes of the axial force and the bending load.

Torsion

A circular bar subjected to torsional loading will tend to be twisted about its axis. This twisting is evidence of shear strain on any transverse section through the bar (Fig. 21). The shear strains are associated with shear stresses in a transverse and axial direction. The magnitude of the shear stresses and strains varies linearly with the distance from the central axis of the bar such that the greatest shear stresses are experienced by the material on the surface of the bar. In considering the stresses on the surface of a circular bar in torsion, one generally thinks of the shear stresses in the transverse and longitudinal planes. However, it can be demonstrated that significant tensile and compressive stresses are present on oblique planes through the bar. If a bar is subjected to torsional loading and fractures along

Torsional Loading

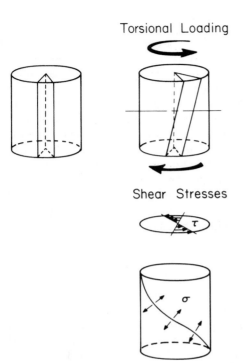

Shear Stresses

Tensile Stresses

Figure 21. Deformation and stress distributions in a cylindrical bar subjected to torsional loading. Shear stresses are imposed on the longitudinal and transverse planes, while tensile and compressive forces are imposed on planes that make an oblique spiral through the bar. (*From Carter and Spengler.*[1])

an oblique or spiral plane, it can be assumed that failure occurred primarily because of high tensile stresses on this plane.

MATERIAL PROPERTIES OF BONE TISSUE

The fracture behavior of whole bones is strongly dependent upon the material behavior of the tissue that comprises the bone. To determine the material properties of bone tissue, small specimens can be extracted from the whole bone. Specimens can be loaded under well-defined conditions in the laboratory in such a manner as to produce uniform, known stresses throughout the specimen. Specimen deformation can be measured and strains calculated. The material properties of the tissue that comprises the bone specimen can thus be determined for the specific loading conditions imposed. This general approach to materials testing has facilitated the documentation of bone material properties in tension, compression, bending, and torsion (shear).

Whole long bones are composed of bone tissue exhibiting two forms of structural organization. Cortical (or compact) bone tissue forms the diaphyses of long bones and the thin shell of the bone ends. Cancellous (or trabecular) bone in the metaphyses and epiphyses is continuous with the inner surface of the cortical shell and exists as a three-dimensional network of bony plates and columns. The trabeculae divide the interior volume of bone into intercommunicating pores of different dimensions, producing a structure of variable porosity and apparent density. The classification of bone tissue as cortical or cancellous is based on bone porosity, which is the proportion of the volume occupied by nonmineralized tissue. Cortical bone has a porosity of approximately 5 percent to 30 percent; cancellous bone porosity may range from approximately 30 percent to more than 90 percent. However, the distinction between very porous cortical bone and very dense cancellous bone is somewhat arbitrary. The chemical composition of cancellous bone is similar to that of cortical bone. The major difference in these two bone types is the high degree of porosity exhibited by cancellous bone. This porosity is reflected by measurements of bone tissue apparent density, which is the mass of bone tissue divided by the bulk volume of tissue (including nonmineralized tissue spaces).

Cortical Bone

Laboratory testing has shown that the material properties of cortical bone are dependent upon the rate at which the bone tissue is loaded or deformed. A specimen of bone tissue that is very rapidly subjected to forces will have a greater elastic modulus and ultimate strength than bone tissue that is loaded more slowly (Fig. 22). In addition, bone tissue that is exposed to very rapid loading will absorb considerably more energy than bone tissue that is loaded more slowly. To quantify the rapidity of deformation, one can refer to the *strain rate* (dimensions = strain per unit time) to which the tissue is exposed during the loading process. In normal activities bone is subjected to strain rates that are generally below 0.01 per second. In traumatic bone fracture, however, strain rates may exceed 10.0 per second. Materials, such as bone, whose stress–strain characteristics are dependent upon the applied strain rate are said to be viscoelastic materials. The elastic modulus and

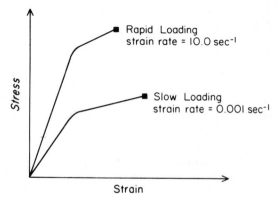

Figure 22. The influence of strain rate on the stress–strain characteristics of bone tissue. (*From Carter and Spengler.*[1])

ultimate strength of bone are approximately proportional to the strain rate raised to the 0.06 power.[6]

The stress–strain behavior of bone tissue is also strongly dependent upon the orientation of bone microstructure with respect to the direction of loading. Several investigators have demonstrated that cortical bone is stronger and stiffer in the longitudinal direction (direction of osteon orientation) than in the transverse direction. In addition, bone specimens loaded in a direction perpendicular to the osteons tend to fail in a more brittle manner, with little nonelastic deformation subsequent to yielding (Fig. 23). Long bones are, therefore, better able to resist stresses along the axis of the bone than across the bone axis. Materials, such as bone, whose elastic and strength properties are dependent upon the direction of applied loading are said to be anisotropic materials.

The viscoelastic, anisotropic nature of cortical bone distinguishes it as a very complex material. Because of these characteristics, one must specify the strain rate and the direction of applied loading when discussing bone material behavior.

Tension, Compression, and Shear Fracture. The most complete investigations of cortical bone elastic and strength characteristics were conducted by Reilly and co-workers.[7,8] These studies also provided detailed information on the preparation and testing of standardized bone specimens machined from whole long bones. Mechanical testing was conducted at strain rates between 0.02 and 0.05 per second. The resulting stress–strain curve showed that cortical bone behaves in a manner similar to other engineering materials. Stress–strain curves in tension and compression consisted of an initial elastic region that is nearly linear. This region was followed by yielding and a considerable nonelastic (plastic) deformation before failure occurred (Fig. 23). The nonelastic region of the stress–strain curve for the longitudinally oriented specimen reflected diffuse, irreversible microdamage created throughout the bone structure. Bone tissue that is loaded into this nonelastic region will not return to its original configuration after the load is removed. The mean value of the longitudinal elastic modulus was found to be approxi-

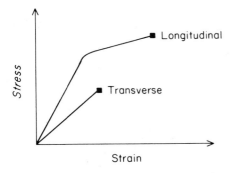

Figure 23. The influence of the direction of the applied stress on the stress–strain characteristics of bone tissue. (*From Carter and Spengler.*[1])

mately 50 percent greater than that of the transverse elastic modulus. The ultimate strength was greater in compression than in tension for specimens oriented in both the longitudinal and transverse directions. Specimens loaded in the transverse direction were significantly weaker in both tension and compression than the longitudinally oriented specimens. In addition, the transverse specimens tended to fail in a more brittle manner, with little nonelastic deformation subsequent to yielding. To establish the shear properties of cortical bone, torsional tests were conducted of square and cylindrical section specimens, which were oriented in the longitudinal direction. The results for the ultimate strength of adult femoral cortical bone as determined by Reilly and Burstein are summarized in Table 1. These results verify that the material strength of bone tissue is dependent upon the type of loading imposed, as well as upon the direction of the applied loading.

The various loading modes to which the bone specimens are subjected are associated with characteristic fracture patterns. These fracture patterns are identical to those observed with whole bone fracture. Figure 24 demonstrates the fracture patterns for longitudinally oriented bone specimens that are subjected to tensile, compressive, torsional bending and combined bending and compressive forces. Tensile specimens normally demonstrate a fracture pattern that is approximately perpendicular to the direction of applied loading. The stresses imposed on the plane of fracture are tensile stresses. Tensile loading of longitudinally oriented bone specimens, therefore, results in fracture along planes of high tensile stresses. A bone specimen that is subjected to compressive forces will generally fracture along a plane that is at an oblique angle to the direction of the applied loading. These oblique planes are subjected to significant shear stresses as the compressive forces are applied to the specimen. Compressive loading of the bone specimen, therefore, results in failure along planes of high shear stresses.

A bone specimen subjected to torsional loading demonstrates a more

TABLE 1. ULTIMATE STRENGTH OF ADULT FEMORAL CORTICAL BONE

Loading Mode	Ultimate Strength (MPa)
Longitudinal	
Tension	133
Compression	193
Shear	68
(Torsion test about longitudinal axis)	
Transverse	
Tension	51
Compression	133

Mean values from Reilly and Burstein.[7] Strain rate 0.02–0.05 per second. Age span of population 19–80 years.

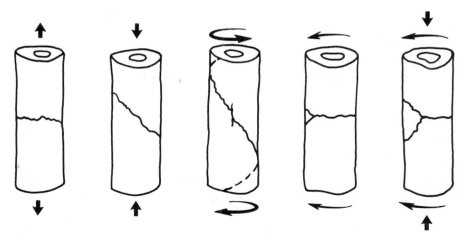

Figure 24. Fracture patterns created on cortical bone due to tensile, compressive, torsional bending and combined bending and compressive forces. (*From Carter and Spengler.*[1])

complex fracture pattern. Fracture is usually initiated at a small crack on the surface of the specimen that runs parallel to the specimen axis (on a plane of high shear stresses). After initiation of the fracture, the crack then runs in a spiral manner through the specimen following planes of high tensile stresses. The final fracture surface appears as an oblique spiral, which is characteristic of torsional fracture of bone tissue.

A bone specimen that is subjected to bending forces will be exposed to high tensile stresses on one side of the specimen and high compressive stresses on the other side. The fracture pattern that results is consistent with that observed in tensile and compressive testing of longitudinally oriented specimens. A transverse fracture surface will be present on the tensile side of the specimen, while an oblique fracture surface may be created on the compressive side. The compressive side of the specimen may contain two oblique fracture patterns, creating a loose wedge of bone as the specimen is fractured. This fracture pattern is sometimes referred to as a "butterfly" fracture. The oblique fracture lines may be accentuated in situations wherein combined compressive and bending loads are imposed, sometimes creating a large, triangular bone fragment.

The fracture patterns shown for bone specimens in Figure 24 are consistent with those observed in bone fractures clinically. Bone in vivo, however, is rarely subjected to idealized loading situations as illustrated in Figure 24. Bone fractures seen clinically are usually caused by a complex loading situation, and the resulting fracture patterns are therefore numerous. It should also be noted that very high energy (rapid strain rate) fractures result in significant comminution caused by the bifurcation and propagation of numerous fracture planes in the bone tissue.

Fatigue. The mechanical behavior of bones when exposed to a single application of high forces is of interest if one is concerned with the fracture behavior caused by a single traumatic episode. Bone tissue in vivo, however, is more commonly exposed to repetitive loading at stress levels less than those required to fracture the bone during a single traumatic episode. Repeated loading of bone in everyday activities or prolonged exercise, however, may lead to microscopic damage. If damage accumulates faster than it can be repaired by biologic processes, fatigue fracture of bone may result. In a mechanical sense, fatigue is the progressive failure of the material under cyclic or fluctuating loads. Under cyclic loading, materials may fail at load levels less than those required to cause failure with a single applied loading. Fatigue fractures may occur during prolonged exercise, such as marching or long-distance running, and are especially common in the metatarsals of young military recruits. Fatigue damage in bone due to repeated mechanical loading has also been implicated in the development of degenerative disorders, avascular necrosis, osteochondritis, senile femoral neck fractures, spondylolisthesis, pathologic fractures, and the failure of bone after orthopedic implant procedures.

Fatigue tests in the laboratory are often conducted by subjecting a number of identical, machined specimens of bone to cyclic stresses or cyclic strains of various magnitudes and noting the number of cycles to failure (fatigue life). The resulting plot of stress amplitude or strain amplitude vs cycles to failure serves to characterize the fatigue properties of the material under examination. The results of such tests have shown that bone has extremely poor fatigue resistance and that failure is a gradual process that involves the progressive accumulation of diffuse microdamage.[6,9] The tensile yield strain for bone is approximately 0.008. Fatigue loading between the strain limits of -0.003 and $+0.003$ will cause complete fatigue fracture of bone specimens after approximately 2000 cycles (Fig. 25). By comparison, fatigue loading in steel specimens to half the yield strain would never cause fatigue failure. These findings demonstrate that the range of stresses and strains that bone tissue can tolerate in vivo without major damage is quite small.

Cancellous Bone

The major physical difference between cancellous bone and cortical bone is the high degree of porosity exhibited by cancellous bone. This porosity is reflected by the apparent density, which is the mass of bone tissue present (excluding marrow) divided by the bulk volume of bone tissue (including mineralized bone and bone marrow spaces). Figure 26 demonstrates the influence of bone apparent density on the compressive stress–strain behavior of cortical and cancellous bone. The stress–strain characteristics of cancellous bone are markedly different from those of cortical bone and are similar to the compressive behavior of many porous engineering materials that are used primarily to absorb energy upon impact. The stress–strain

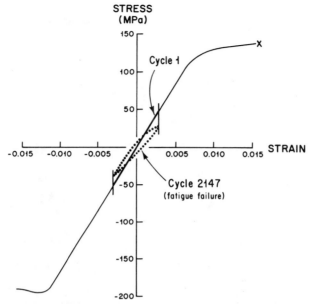

Figure 25. Schematic repesentation of bone fatigue failure superimposed on the tensile and compressive fracture stress–strain curve. A cyclic strain range of 0.006 leads to complete fracture after approximately 2000 cycles. (*From Carter et al.*[9])

for cancellous bone exhibits an initial elastic behavior followed by yield, which occurs as the trabeculae begin to fracture. Yield is followed by a long plateau region that is created as progressively more and more trabeculae fracture. The fractured trabeculae begin to fill the marrow spaces, and at a strain of approximately 0.5, most of the marrow spaces have filled with the debris of fractured trabeculae. Further loading of cancellous bone specimens after pore closure is associated with a marked increase in specimen stiffness.

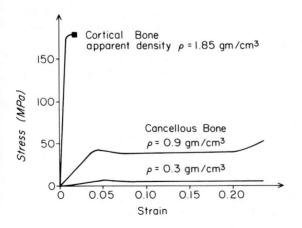

Figure 26. The influence of bone apparent density on the compressive stress–strain behavior of cortical and cancellous bone. (*From Carter and Spengler.*[1])

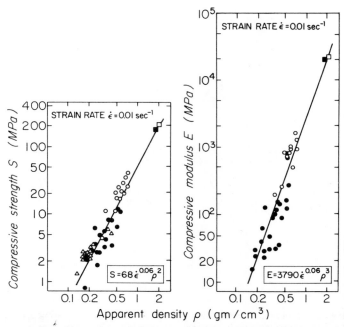

Figure 27. The influence of apparent density on the compressive strength and modulus of cancellous and cortical bone. (*From Carter and Hayes.*[10])

The strength and elastic modulus of bone tissue are markedly influenced by the apparent density of the tissue.[10] Figure 27 illustrates the influence of apparent density on the strength and elastic modulus of bone specimens tested in the laboratory. The data present on these curves represent cortical bone with an apparent density of approximately 1.8 grams per cubic centimeter, as well as numerous cancellous bone specimens with widely different apparent densities. These graphs indicate that the strength of all bone tissue in a skeleton is approximately proportional to the square of the apparent density. The elastic modulus of bone tissue is approximately proportional to the cube of the apparent density.

The stress–strain behavior of cancellous bone in tension is markedly different from that in compression.[11] Although the tensile strength and modulus of cancellous bone are similar to the compressive strength and modulus, the absorbing capacity of cancellous bone in tensile loading is markedly less. Yielding is followed by progressive fracture of trabeculae, which causes a tensile load to diminish rapidly at fairly low levels of strain. At the point of total fracture the two ends of the cancellous bone specimens separate, and the specimen can neither sustain additional loading nor absorb additional energy.

Figure 28 demonstrates the energy absorbed by cortical and cancellous bone (apparent density 0.4 grams per cubic centimeter) when the specimens

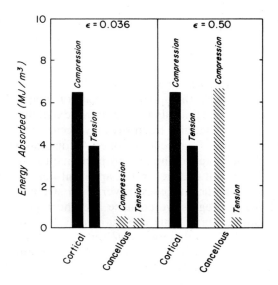

Figure 28. The energy-absorption capacity of cortical and cancellous bone in tension and compression. Apparent density of cancellous bone equals 0.4 grams per cubic centimeter. (*From Carter and Spengler.*[1])

are loaded to strain levels of 0.036 and 0.50, respectively. At a strain of 0.036 the cortical bone, both in tension and compression, is fractured and therefore can absorb no more energy with increasing deformation. The cancellous bone is also totally fractured in tension and can absorb no further energy with increasing deformation. The cancellous bone in compression, however, has failed but will continue to absorb considerable energy with increasing bone deformation. This increasing energy absorption of cancellous bone in compression is demonstrated by the bar graphs for energy absorbed at 50 percent strain. It can be seen that the energy-absorption capability of cancellous bone in compression is considerable and may even exceed the energy-absorption capacity of cortical bone.

Influence of Composition and Microstructure

Bone composition and structure may vary significantly. These variations can be characterized by assessing differences in bone tissue microstructure, porosity, mineralization, and bone matrix. Many investigations of the material behavior of bone have been based explicitly or implicitly on an examination of the influence of one or more of these parameters. However, it should be noted that in vivo these parameters seldom vary independently. It is much easier to demonstrate a correlation between a particular characteristic of bone tissue and its mechanical behavior than to positively identify a direct cause-and-effect relationship between a single parameter and the material properties of the bone tissue.

Bone mineral content is low in children but increases rapidly and is high throughout the middle adult years. After the age of 40 years, bone mass declines. Rarefaction is especially marked in postmenopausal females, since the osteogenic effect of estrogens is no longer present. The loss of bone in the elderly is common and is referred to as "physiologic osteoporosis," although the distinction between physiologic and pathologic osteoporosis may be somewhat arbitrary.

The loss of bone mass in the elderly results in a reduction of the cortical thickness of the long bones and a decrease in the apparent density of cancellous bone. The most obvious changes seen clinically in osteoporotic patients are in the spine, where multiple chronic fatigue fractures may occur. The vertebral bodies are composed primarily of cancellous bone. As discussed earlier, the strength of cancellous bone is approximately proportional to the square of the apparent bone density. Small changes in the apparent density of vertebral bodies, therefore, cause significant losses of vertebral strength and fatigue resistance. What follows is progressive collapse and shortening of the vertebral bodies. In the lumbar spine, this condition may lead to multiple collapse fractures, while in the dorsal spine, vertebral wedge fractures are created. There is an overall reduction in the height of the patient and a characteristic dorsal kyphosis. There is also a marked increase in the frequency of fractures in the peripheral skeleton of an osteoporotic patient. These fractures often occur in the hip and may lead to serious complications. The incidence of cervical or trochanteric fracture increases dramatically with age, whereas the severity of the precipitating violence generally diminishes.

The microstructural and compositional changes in cortical bone tissue associated with aging result in an overall reduction in the material properties of the bone tissue. The mechanical strength of cortical bone is greatest in the age range from 20 to 39 years. Further aging is associated with the loss of strength and stiffness as well as energy-absorption capacity. The loss of strength and stiffness is not due to a decline of the strength and stiffness of the mineralized tissue but rather to a slight increase in bone porosity. The loss of energy-absorption capacity is partly due to the embrittlement of the mineralized tissue caused by increasing collagen cross-linking and mineralization.

Disease processes can alter the mechanical response of bones by altering either bone geometry or the bone tissue material properties. Disease processes alter the material properties of bone tissue by changing the chemical composition of the bone matrix and/or the bone mineral. Bone tissue can be considered as a two-phase composite material consisting of material and collagen. Increases in collagen cross-linked density are generally associated with an increase in mineral content. The resulting bone tissue is made much stronger and stiffer not only by the increased mineral content but also by the stiffening of collagen matrix. Disease processes that tend to inhibit collagen cross-linking are often associated with a decrease in mineralization and a weakening of the bone tissue. In some disease processes, the primary deficiency leads to a direct decrease in the amount of bone mineralization, which causes significant weakening of the bone. This situation may occur in diseases such as vitamin D-deficiency rickets.

REFERENCES

1. Carter DR, Spengler DM: Biomechanics of fracture, in Sumner-Smith G (ed): *Bone in Clinical Orthopaedics*. Philadelphia, Saunders, 1982, pp 305–334.
2. Carter DR, Spengler DM: Mechanical properties and composition of cortical bone. *Clin Orth Rel Res* 1978; 135:192–217.
3. Hayes WC, Carter DR: Biomechanics of bone, in Simmons DJ, Kunin AS (eds): *Skeletal Research—An Experimental Approach*. New York, Academic Press, 1979, pp 263–300.
4. Burnstein AH, Currey J, Frankel VH, et al.: Bone strength. The effect of screw holes. *J Bone Joint Surg* 1972; 54A:1143–1156.
5. Carter DR, Hayes WC: Compact bone fatigue damage: A microscopic examination. *Clin Orthop Rel Res* 1977; 127:265–274.
6. Carter DR, Caler WE: Cycle-dependent and time-dependent bone fracture with repeated loading. *J Biomech Eng* 1983; 105:166–170.
7. Reilly DT, Burstein AH: The elastic and ultimate properties of compact bone tissue. *J Biomechan* 1975; 8:393–405.
8. Reilly DT, Burstein AH, Frankel VH: The elastic modulus for bone. *J Biomech* 1974; 7:271–275.
9. Carter DR, Caler WE, Spengler DM, et al.: Fatigue behavior of adult cortical bone—The influence of mean strain and strain range. *Acta Orthop Scand* 1981; 52:481–490.
10. Carter DR, Hayes WC: The compressive behavior of bone as a two-phase porous structure. *J Bone Joint Surg* 1977; 59:954–962.
11. Carter DR, Schwab GH, Spengler DM: Tensile fracture of cancellous bone. *Acta Orthop Scand* 1980; 51:733–741.

CHAPTER 9

Forensic Analysis of Trauma

Frank O. Raasch, Jr.

INTRODUCTION

The forensic analysis of trauma deals with the description of injuries and relates those injuries to the forces and hazards of the accident. The word *forensic* describes the application of science to law; thus, the forensic pathologist is called to examine the victims of trauma because his or her scientific findings will often find their way into legal proceedings. The pathologist's expertise lies in the interpretation of these scientific findings for the law—but it goes beyond this. The collection of data and careful description of injuries provide a basis for research into the causes, treatment, and prevention of accidents.

Trauma is an injury, and an injury is anything that brings harm, with or without physical damage. When there is physical damage, a wound is produced. Wounds can be incised, caused by a cutting edge, or lacerated, a tearing and breaking caused by blunt trauma. Additionally, blunt trauma rubbing against skin can cause an abrasion or, by breaking blood vessels, can cause a bruise.

The production of trauma depends on the magnitude of acceleration and/or deceleration and its duration.[1] If energy is discharged slowly enough to overcome the inertia of the whole tissue mass and produce a relatively uniform motion, there will be no injury. The area of impact is important. The smaller the area, the greater concentration of force and, thus, the greater the damage.

When a body is struck by a moving object, the object displaces the tissues in direction of motion. When a moving body strikes a stationary object, force tends to interrupt the forward motion of tissues. If the body is struck by a deformable object or the body part is elastic and can give way, the time of impact will be lengthened and the amount of damage will be lessened.

Thus, the localization of transmitted forces depends on the plasticity of the structures in the path of force and the manner in which the waves of energy are transmitted.

Interaction of the human, the vehicle, and the environment compose the setting for an accident.

The complex series of events can be simplified[2] by dividing an accident into three distinct collisions:

1. The vehicle vs object of impact
2. Victim vs vehicle
3. Forces transmitted into the victim producing deformation of organs

RELEVANT STATISTICS

In 1982, there were about 165,000 traumatic deaths in the United States. Since 1976, the traumatic death rate of young Americans has risen 50 percent. The National Safety Council has estimated more than 63 million dollars of earnings are lost every day from accidental trauma. The total annual cost, including medical expenses, lost wages, and indirect work losses, amounts to about 50 billion dollars.[3]

About half of the trauma-related deaths in this country involve motor vehicles.[4] The drivers of those vehicles are very often young males. Alcohol is a factor in over half of the cases. Pedestrians are usually young and inexperienced or old and frail.[5]

Natural death is not an important factor in collisions. It leads to less than 6 per 10,000 motor vehicle collisions. No major injuries are incurred.[6]

Of deaths due to trauma, 52 percent die within 1 hour, 66 percent within 6 hours, 75 percent within 24 hours, and 90 percent after 1 week.[5]

The distribution of fatal injuries is concentrated on the head and chest. Well over half of the injuries are multiple. In my experience, head injuries are the most common cause of death, followed by chest injuries, then neck and abdominal injuries. Nahum et al.[2] reported the following incidence in motorist deaths:

Head	41 percent
Chest	23 percent
Thoracoabdominal crush	15 percent
Abdomen	9 percent
Neck	6 percent
Extremities	2 percent

The problem with accumulating injury statistics is that accidents are almost never simple events and the pathologist is faced wtih a complex collection of injuries, maybe two or more of which could have caused death. Sometimes the injuries are so extensive and diverse that only the phrase "multiple injuries" can be put on the death certificate.

APPROACH

The forensic pathologist has a vital role in the team, that is analyzing and reconstructing the accident. Four basic phases to the forensic analysis of trauma are:

1. Scene
2. History
3. Autopsy
4. Toxicology

Data from all these areas are brought together, and a conclusion is reached as to the cause and manner of death.

Ideally, the pathologist should go to the scene and observe the vehicle, the vehicle's path, the site of impact(s), and the position of the victim(s) when found. This almost never happens. Instead, the pathologist relies on astute investigators who submit a concise summary of the accident to guide the pathologist in correlating the story with the findings.

Photos of the vehicle and scene are helpful. An investigator should be available to the pathologist in case additional information is needed as the autopsy progresses.

The investigator should delve into the historical background of the victim: Were there any medical problems, such as heart disease, seizure disorder, clotting disorders, or diabetes? Was the victim taking any drugs, prescribed or over-the-counter? Was the victim known to have abused drugs? Was there any evidence that the victim was depressed or under stress?

After carefully going over the investigative summary, the pathologist is ready to proceed with the autopsy. First, the body should be identified to be that of the victim listed on the investigative summary. This is usually done by noting a plastic identification band placed on the ankle. The pathologist depends on the investigative personnel as to the veracity of this identification tag.

If the victim is still unidentified, the name John or Jane Doe is listed with the case number, and the pathologist pays particular attention to the body features that may assist in identification: scars, tattoos, operations, vaccinations, hair pattern, and implanted devices, such as pacemakers that carry serial numbers that can be traced through the manufacturer. A forensic odontologist should do a complete dental chart with x-rays. A front and side identification photo should be taken. In severe facial distortion, a proficient embalmer could be called upon to restore the face to a more recognizable state.

Questions that should be going through the mind of the forensic pathologist as he or she performs the autopsy include:

1. What caused the death? What specific injury or injuries were fatal?
2. What was the manner of death? Was the death an accident, suicide, homicide, or of natural causes?
3. What are the patterns of injury?
4. What was the degree and direction of force?
5. How many and what type of impacts did the victim receive?
6. Is there any trace evidence of the vehicle on the victim, or the victim on the vehicle?
7. Was the victim alive at the time of the accident?
8. How long did the victim survive?
9. Was the victim the driver or a passenger?
10. Did any underlying disease or toxic substance cause or contribute to the accident?

The first step of the autopsy is a detailed external examination. Photographs and diagrams are helpful. If the clothing is present, it should be described with special attention to tears and amount of bleeding. Shoes may show brake or acclerator pedal marks on the soles.

External injuries should be described as to location, type (bruise, cut, laceration), appearance (pattern), and degree (depth, size). The Abbreviated Injury Scale of the American Medical Association's Committee on the Medical Aspects of Automotive Safety can help in the uniform classification of injuries.[7]

An extensive example of a medicolegal autopsy report form can be found in reference 8.

For example, very characteristic right-angled or V-shaped cutting is seen when the diced or cubical fragments of tempered glass from a car's broken side window strike the face or arms. The driver will usually incur these injuries on the left side, a passenger on the right. Various parts of the instrument panel or steering wheel may make a characteristic wound.

Pedestrian injuries are especially important to document in order to relate them to their contact with the vehicle. Leg injury locations should be measured from the heel, and the width of tissue damage also should be measured (Fig. 1). Fractures should be described and located in relation to the heel reference point. The location of these injuries may well show the height and width of the bumper or other striking point and the position of the victim in relation to the vehicle. A bumper injury lower than the suspect vehicle's bumper level would indicate that the front of the vehicle may have been dipping down in the action of braking. If the legs have been hyperextended by the impact, parallel stretch lines may be seen in the inguinal

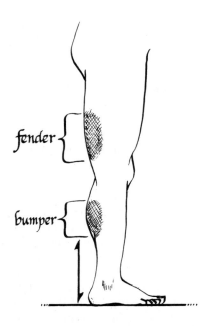

fender

bumper

Figure 1. The locations of leg injuries received by pedestrians are measured from the heel; the width of the area of tissue damage is also measured.

areas. Obese pedestrians may experience severe fat destruction and soft tissue hemorrhage, resulting in fat embolism and very severe blood loss. Usually adult pedestrians are thrown up onto the hood and then to the ground. Small children with a lower center of gravity will often be struck down ahead of the vehicle and can be run over.

The autopsy then continues with an internal examination. This should always be correlated with the external injuries. Something that continues to surprise me is the amazing resilience of the skin. Extensive internal damage may underlie skin that shows little or no injury. Incisions into suspected sites of impact in spite of little or no external injury may reward the investigator with graphic evidence of subcutaneous tissue damage and a better picture of the patterns of injury. General results of internal trauma include bleeding, laceration, contusion, and swelling.

Before opening the chest cavity, pneumothorax should be ruled out by incising the lateral thoracic wall under a pocket of water formed by the dissected-away chest wall and observing for air bubbles under pressure.

To check for air embolism, the chest plate is removed, and the pericardial sac is opened and filled with water. An incision is made into the lateral right ventricle. The presence of air bubbles indicates air of venous origin. Arterial air embolism is seen through air in the left side of the heart and air segments in various arterial systems. Caution must be used not to induce artifacts.

Now, each major organ system is examined, beginning with the central nervous system. After inspecting the reflected scalp for hemorrhages and the calvarium for fractures, the skull cap is removed. The lesions of the central nervous system can be divided into three groups according to causation:[9]

1. Primary impact lesions
2. Secondary lesions due to circulatory deficiencies
3. Indirect circulatory lesions

The primary lesions include extracerebral hemorrhages in the epidural, subdural, subarachnoid, and intraventricular spaces and intracerebral contusions and lacerations. Secondary lesions include edema and tissue necrosis either due to compression of cerebral vasculature or failure of systemic circulation. The indirect lesions are due to thrombosis or embolism.

Primary impact can injure the brain in several ways. There can be an immediate loss of consciousness or concussion. Unless it is severe, a concussion of the brain will show no significant gross findings and probably only subtle microscopic changes. A severe concussion can affect the general circulation to the brain, and small areas of bleeding can be seen in the white matter of the brain and/or the ventricular system. A concussion can be fatal. A contusion is a bruise of the brain without a direct disruption of the brain substance. Secondary hemorrhages of a contusion will destroy brain substance, but in a sharply outlined area compared to an ischemic change where adjacent tissue is softened. The most severe injury is tearing or laceration of the brain tissue.

Of special note is the mechanism of coup–contrecoup contusions, which are based on the concept that sudden stopping of a moving head results in contusion-type injuries generally greater on the side opposite the point of impact (contrecoup) (Fig. 2). The formation of coup–contrecoup injuries is quite complex and is still couched in theory. One theory holds that the brain, lagging slightly behind the suddenly stopped skull, hits against the skull opposite and is splayed into adjacent bony prominences. Another theory is based on the formation of zones of positive and negative pressure. Positive pressure occurs at the coup site, while the brain pulling away from the opposite site creates a negative pressure. Since negative pressure tends to have a more adverse effect, the injury opposite the impact is more severe.

There is one exception to the coup–contrecoup mechanism. An impact on the frontal area never produces lesions in the opposite occipital lobes.

After the brain is removed, the dura is stripped, and the base of the skull is inspected for fractures. Fractures of the base of the moving skull generally follow the line of force (Fig. 3) and are usually linear. When a great force is applied to the lower forehead or chin, a side-to-side hinge fracture through the midportion of the skull can occur. If the force is more concentrated, a depressed and often fragmented, or comminuted, fracture results. A compressed skull fractures in several areas, commonly at right angles to the direction of the compressing force.

Next comes the examination of the cervical spinal cord and spine. The

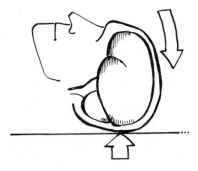

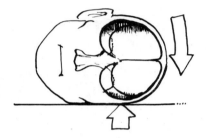

Figure 2. Contusion-type brain injuries that are generally more severe on the side opposite the point of impact are called coup–contrecoup contusions. One theory is that the brain moves away from the point of impact (short arrow) and is injured when it hits against the suddenly stopped skull on the opposite side. Another theory is that as the brain moves toward the point of impact (long arrow), this action creates a damaging negative pressure on the opposite side.

upper two cervical vertebrae are inspected through the foramen magnum to check for dislocation, ligament tears, and fracture of the odontoid process. I have found the posterior approach best for examination of the cervical cord and spine.[10] The musculature of the spine is dissected away, and the neural arches are removed to reveal the cervical cord. The cord is removed and the cervical spine is dissected away from the base of the skull and sawed away from the upper thoracic spine. After the spine is removed, the anterior and posterior ligaments are inspected, then a sagittal saw cut is made through the vertebral bodies, exposing the intervertebral disks. This approach reveals subtle disk tears and/or vertebral fractures.

An x-ray study[11] showed that a remarkable two thirds of head and neck injury cases had air in the cranial vault. This could be a source of air embolism.

In addition to the acute injuries, note should be made of any operative procedures, old injuries, and natural diseases, such as atherosclerosis, vascular malformations, epileptogenic lesions, tumors, infections, and optic nerve degeneration.

In the neck anteriorly, the tongue, hypopharynx, and laryngeal structures should be removed. The tongue should be examined for old and new bite marks that could indicate seizure disorder. Asphyxia due to compressive obstruction or aspiration should be ruled out. Neck vessel damage can obviously cause bleeding, but injury to the large veins with exposure to the air can result in air embolism.

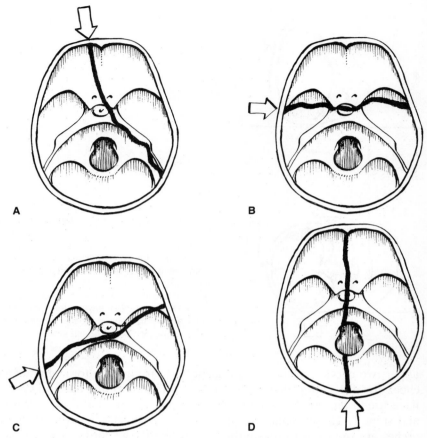

Figure 3. A–D. Fractures of the base of the skull usually follow the line of force (arrows).

The chest organs should first be inspected in situ, then removed en bloc. Amounts of blood in the cavities should be measured. The most common injury here to be aware of is a torn or ruptured aorta. This will be reflected by a bleeding into the mediastinum, followed by rupture of the blood into the thoracic cavitites, usually the left. The aortic injury is most often just distal to the arch near the attachment of the ligamentum arteriosum. The mechanism most likely is rapid acceleration or deceleration, with pulling away of the heart from the aorta relatively fixed along the thoracic spine. Cardiac ruptures occur most often in the right artrium when it is compressed while distended in diastole (Fig. 4). I have noted another cardiac lesion in a moderate number of chest injury cases: a transverse tearing of the right atrium just medial to the coronary sinus in the area of the A-V node. This may explain some arrhythmias described in cases of nonpenetrating trauma of the heart.[12,13] In rare cases, a sudden

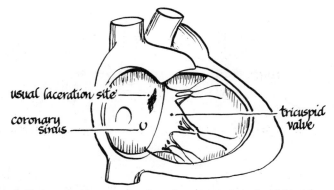

Figure 4. The most common site of cardiac rupture is in the right atrium, which occurs when the atrium is compressed while distended in diastole.

blow to the chest can suddenly stop the heart or throw it into an arrhythmia without any significant or gross microscopic findings. This is sometimes referred to as *commotio cordis*. The diagnosis depends on a good history.

The ribs can be dangerous wounding devices when fractured. They can penetrate the lung and cause hemorrhage and possibly pneumothorax. Compression of the chest alone can cause contusion and hemorrhage in the lung. A clue to compressive forces is rows of small bubbles on the lung surface. These represent a confluence of air sacs ruptured by back-pressure, the air being unable to escape from the lung quickly enough.

The extreme form of multiple rib fractures results in a flail chest, which leads soon to respiratory insufficiency.

The incidence of shock lung, or adult respiratory distress syndrome (ARDS), in trauma is debatable.[14]

Abdominal injuries are of two basic types: laceration of a solid organ and rupture of a hollow viscus. Again, the amount of blood present should be measured. The liver is the most commonly injured organ (Fig. 5) and can result in a fatal hemorrhage, but I have seen survival for several hours after massive liver damage. The spleen also can be lacerated, but this smaller organ is protected by the rib cage and is more freely movable, thus less likely to be damaged. The spleen, when lacerated or ruptured, used to be routinely removed. However, in the last few years, the spleen has been shown to have an important immunologic function, and very rapid deaths due to bacterial sepsis, especially of pneumococcal origin, have occurred months to years following splenectomy. Now surgeons try to retain at least a part of the spleen.

The rupture of an abdominal organ allows its contents to spill into the peritoneal cavity, and this leads to chemical or bacterial peritonitis, or both.

The kidneys, pancreas, and adrenals are rarely injured to the point of causing a serious problem, although hematuria is common.

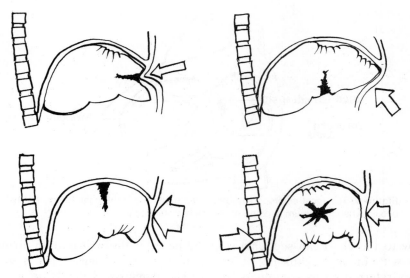

Figure 5. Common sites of laceration of the liver and the points of impact that cause them (arrows).

Seat belts are rarely associated with serious injuries. Pressure upon the abdomen, for example, can rupture the bowel or lacerate the liver, spleen, or mesentery. Backs and ribs have fractured. Some of these injuries result from a belt improperly applied across the abdomen rather than across the lap or pelvic area. I have yet to personally see a death directly due to either a lap or shoulder belt.

Injuries to the extremities cause the obvious fractures, soft tissue damage, and bleeding. Fractures of long bones sometimes release fat droplets into the systemic circulation and lead to fat embolism. When a large number of these emboli reach the brain, death can result. Thrombi can form in the injured and often immobilized extremity. These can break off and lead to a rapidly fatal pulmonary embolism.

Inactivity brought about by injuries can also reduce the pulmonary exchange, setting up a nidus for infection and so-called hypostatic pneumonia.

Late deaths due to trauma are usually caused by infections and/or multiple organ failure. Risk factors for this development include: shock, head injury, peritoneal contamination and malnutrition.[3]

The microscopic analysis of tissue is important in selected cases. This is especially true in the evaluation of underlying disease, such as infection, cancer, chronic lung obstruction, cirrhosis of the liver, and/or fatty liver and recent myocardial infarction.[15]

Bacteriologic studies are valuable in the study of delayed deaths due to infection.

Finally, the forensic pathologist is responsible for obtaining samples for toxicologic analysis and later interpreting the results in consultation with the toxicologist. The blood and the urine are the basic fluids analyzed, but some circumstances require more extensive or special study.

For example, the blood clot of a recent subdural or epidural hematoma will reflect the alcohol level of blood when the bleeding occurred. Thus, a significantly lower alcohol level in the blood would indicate that the hematoma formed sometime earlier. When little blood is available, the vitreous fluid from the eye can be analyzed.

The blood should be routinely analyzed for alcohol. Carbon monoxide and drugs should be looked for as the history indicates.

The urine can be screened for the commonly abused drugs, including metabolites of marijuana.

Other organs and fluids from which samples can be taken in more involved cases include liver, lung, kidney, brain, heart, and bile.

Newer methods of diagnosis in the forensic autopsy are mostly in the field of postmortem radiology. These findings lead the pathologist to subtle injuries and to areas he or she may not usually examine. Angiography of the various vascular beds, especially the head and neck, and computerized axial tomography (CAT) scanning are two studies that can be helpful. In the case of the CAT scan, the brain should be cut in the plane of the scans.

A recent study has shown that early myocardial ischemia can be demonstrated with electron-probe microanalysis of the cellular sodium–potassium ratio.[15]

PREVENTION

A fundamental principle in reduction of injury is prevention of ejection and the second collision with the hard interior of the vehicle.

Information from the autopsy should be fed back to the clinicians, researchers, and car designers. Recurring patterns of injury and death should alert these personnel to look for changes in treatment and design.

Trauma centers have been set up in some metropolitan areas in response to injury and death statistics.[3] It has been shown that patients treated in hospitals without a center had significantly higher death rates. In Orange County, California, after the trauma centers were formed, the preventable death category dropped from 73 percent to 4 percent.

Once the injury has occurred, prompt and proper treatment can reduce both morbidity and mortality. With trauma centers, 20 percent of patients described as "dead on arrival" can be resuscitated and will recover without permanent neurologic damage.[3] The pathologist can provide important information as to the effectiveness of treatment and diagnosis.

A recent report by the American Medical Association Council on Scientific Affairs[4] makes a long list of recommendations to reduce the number and degree of automobile injuries. These include use of passive and active crash protection, improvement of car design, and continued studies of drug and behavioral problems related to accidents.

CONCLUSION

Forensic analysis of trauma leads to a better understanding of the mechanics of injury and the body's response to it. The forensic pathologist forms a vital link between the accident victim and those seeking to prevent accidents and reduce mortality after the accident.

REFERENCES

1. Gikas PW: Mechanisms of injury in automobile crashes. *Clin Neurosurg* 1972; 19:175.
2. Nahum A, Lasky I, Noguchi T: Automobile-accident injuries and accident pathology, in Brinkhous KM (ed): *Accident Pathology: Proceedings of an International Conference.* Washington, DC, US Government Printing Office, 1968, pp 14–25.
3. Trunkey D: Trauma. *Sci Am* 1983; 249:28.
4. Council on Scientific Affairs: Automobile injuries: components, trends, prevention. *JAMA* 1983; 249:3216.
5. Tonge JI, O'Reilly MJ, Davison A, et al.: Traffic crash fatalities, injury patterns and other factors. *Med J Aust* July 1972; 2:5.
6. Baker S, Spitz W: An evaluation of the hazard created by natural death at the wheel. *N Engl J Med* 1970; 283:405.
7. American Medical Association Committee on Medical Aspects of Automotive Safety: Rating the severity of tissue damage. I. The abbreviated scale. *JAMA* 1971; 215:277–280.
8. A Panel Report: Medicolegal autopsy report, in Brinkhous KM, (ed): *Accident Pathology: Proceedings of an International Conference.* Washington, DC, US Government Printing Office, 1968, pp 217–249.
9. Lindenberg R: The postmortem examination of brain injuries, in Brinkhous KM (ed): *Accident Pathology: Proceedings of an Internal Conference.* Washington, DC, US Government Printing Office, 1968, pp 144–153.
10. Horsch JD, Schneider D, Kroell CK, Raasch F: Response of belt-restrained subjects in simulated lateral impact, in *Proceedings of Twenty-third Annual Stapp Conference.* Warrendale, Pa, Publishing Society of Automotive Engineers, 1979, pp 69–103.
11. Alker GJ, Young S, Leslie EV, et al.: Postmortem radiology of head and neck injuries in fatal traffic accidents. *Radiology* 1975; 114:611.
12. Louven B, et al.: Nonpenetrating trauma to heart. *Deutsch Med Wochenschr* 1972; 97:1627–1631 (abstract in *JAMA*).
13. Sutherland G, Calvin J, Driedger A, et al.: Anatomic and cardiopulmonary responses to trauma with associated blunt chest injury. *J Trauma* 1981; 21:1.
14. Pietra GG, Ruttner JR, Wust W, Glinz W: The lung after trauma and shock—

fine structure of the alveolar–capillary barrier in 23 autopsies. *J Trauma* 1982; 21:454.

15. Singh S, Abraham J, Raasch F, et al.: Diagnosis of early human myocardial ischemic damage with electron-probe microanalysis. *Am J Forensic Med and Pathol* 1983; 4:85.

CHAPTER 10

Clinical Biomechanics of Sports Injuries

Cyril B. Frank, Savio L-Y Woo

INTRODUCTION

With increasing participation in sports at all levels of competition and increasing leisure commitment to more loosely defined sporting activities, such as jogging, cycling, weight lifting, and so on, there has been a rapid increase in the need to better understand sports-related trauma, the diverse group of so-called sports injuries.

Our purpose in this chapter is to progressively subdivide the subject of sports injuries into more quantitative parts, beginning with the large and confusing clinical spectrum of injury possibilities and leading toward their definition on both joint and tissue levels. Toward this goal, sports injuries will first be separated according to mechanical considerations into two major groups. These groups will then be broken down into structural and mechanical terms so that the common clinical syndromes can be expressed more simply on the basis of tissues involved. Finally, we will introduce mechanical concepts of injury, diagnosis, treatment, and prevention of sports injuries, both in the clinic and in the laboratory. In this way, we hope to provide current insights into some useful applications of biomechanics in better explaining and treating both common and complex sports injuries.

Since the emphasis of this paper is on a combined clinical and mechanical relevance, the reader may wish to consult more detailed clinical[1-3] or mechanical[4,5] references for further elucidation of the material discussed. Further reading and research in this area is encouraged, as well, to promote a much needed critical approach to this interesting, complex, and expanding new field.

BACKGROUND

Incidence of Sports Injuries

With the difficulty in defining sports[6] and, therefore, sports injuries, and given the likelihood that the majority of sports injuries are never reported or documented,[7] it is difficult to accurately assess their incidence and impact in terms of trauma care. It has been written[8] that in each year about 1 in every 10 people in the United States will seek care for some acute musculoskeletal condition. Athletic activities may account for a surprising proportion of these conditions, since it has been estimated[9] that more than 20 million "weekend athletes" are injured in the United States each year. Even if only a fraction of these injuries are significant (requiring either hospitalization or some period away from work), it can be seen that the total socioeconomic costs of sports injuries are substantial.[8,10] The personal costs of sports trauma, in terms of dollars, changes in lifestyle, and potential chronic disabilities (e.g., arthritis) are also of serious concern, providing compelling combined reasons to study these problems.

Distribution of Sports Injuries

There are many ways of describing the distribution of sports injuries. The most common ways are by sport, by anatomic location, and by type of injury (or general diagnosis). If statistics are combined from two well-documented injury surveys,[11,12] reasonable estimates of sports injuries according to these categories can be made. In this way, the sports yielding the highest number of injuries (Table 1), the parts of the body most commonly injured (Table 2), and the most common diagnoses made (Table 3) can be provided, giving some indications of which areas deserve the greatest attention in this review.

TABLE 1. DISTRIBUTION OF INJURIES BY SPORT (APPROXIMATE)

Sport	Percent
North American football*/soccer†	20.0
Skiing	19.8
Rugby	15.8
Judo*	11.1
Wrestling	9.7
Gymnastics	8.1
Swimming/diving	4.6
Skating/hockey	4.5

*North American statistics.
†United Kingdom statistics.
All other entries are combined United Kingdom and North American Statistics.

TABLE 2. DISTRIBUTION OF SIGNIFICANT SPORTS INJURIES BY LOCATIONS*

Location	Percent
Knee	20.4
Ankle	12.1
Elbow/shoulder	11.6
Hand/wrist	7.8
Head/neck	6.2
Hip	5.0

*Note that the majority of remaining injuries are muscular, occurring "between joints."

From these statistics, we can see that the majority of athletic injuries have apparently occurred in relatively high energy sports (football, skiing) and most commonly involve the soft tissues of the lower extremities. From a mechanical point of view, there are good reasons for these patterns. Before discussing these reasons, however, there are some important qualifying considerations regarding the apparent nature of these distributions of sports injuries that should be mentioned. Although we are forced to generalize about the causes and effects of sports trauma, we should be aware that there are several basic factors that have a significant bearing on these relationships.

Special Considerations

Level of Competition. The level of competition of a sport clearly has some influence on the number, distribution, and seriousness of resulting injuries.[13] While the statistics in Table 1 are probably not surprising, conforming to the usual impressions of athletic injuries (e.g., knee injuries in football players), they really only represent the distributions of injuries in a fairly select, relatively competitive portion of the population. Less aggressive competition with less body contact (at lower energies) would almost certainly affect the types and degrees of injuries and patterns described. The previously mentioned weekend athlete would probably have a different injury profile than these numbers would suggest, with different areas of anatomic and mechanical relevance.

Sports Specificity. Certain sports are clearly predisposed to causing acute injuries through their extreme physical demands, high velocities, rules about physical contact, and so on. These are the sports that are commonly

TABLE 3. PATHOLOGIC DISTRIBUTION OF INJURIES

Diagnosis	Percent
Sprains/strains	33.1
Contusions	24.6
Fractures/dislocations	19.3

surveyed, therefore, to document their trauma. However, there are many types of injuries that are more chronic in nature and probably more common; secondary to more endurance physical demands during conditioning. Long distance running, for example, has its own patterns of injury[14] secondary to its unusual mechanical demands (Table 4). While the more dynamic sports and their injuries (e.g., football knee injuries) receive more attention, a greater number of casual participants in conditioning sports may have a greater incidence and prevalence of injuries. For that reason, our classification of sports injuries should include a major category for these chronic conditions, a classification scheme that would be less sport-specific.

In addition, of course, it is important to note that most sports are relatively joint-specific (e.g., tennis elbow, jumper's knee). While the joints that are stressed and the stresses themselves may be different for each sport, there are nonetheless certain mechanical principles and tissue limitations that they share. Our classification, therefore, intends to define injuries in terms of these tissues and their properties, presuming that these tissues and their limits are at least similar for all synovial joints.

Age. Most sports injury statistics represent a restricted proportion of the overall participating population in terms of age, usually in the range of 18–30 years. This is, of course, the group that includes most professional sports with their highly competitive athletes and their injuries. It should be pointed out, however, that injuries in this group are somewhat age-specific[15] in addition to sport-specific because of certain biologic factors.

Skeletally immature participants, by virtue of lower body masses, different body mechanics, and, most importantly, different tissue properties,[13,15,16] have different injury patterns and severities as compared with adults. Due to unusual bone properties, including exceptional weakness of growth plates (Fig. 1), decreased bending stiffness, and increased deformation to failure, fracture patterns in immature athletes are unique. Certain soft tissue injuries of the joints (e.g., torn ligaments) are also more unusual in younger athletes due to a similar combination of structural and mechanical differences. Growth plates often fail before the soft tissues in skeletally immature individuals.

It is quite likely that the age of maximum performance of an athlete, in terms of physiologic capability, falls within the narrow age range that was

TABLE 4. DISTRIBUTION OF RUNNING INJURIES

Location	Percent
Knee	30
Achilles' tendon	20
Shin splints/stress fractures	15
Plantar fasciitis	10

From Brody.[14]

Copy

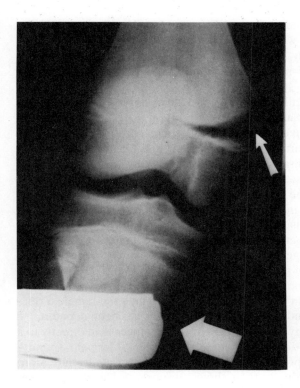

Figure 1. Valgus stress x-ray of juvenile knee injury. Injury to growth plate (small arrow) is demonstrated by laterally directed force on tibia (large arrow) and is age-specific.

previously mentioned. Growth is completed, and there is a maximum opportunity to condition and train for the performance of maximum mechanical advantage during their athletic activity. The material properties of the athlete's tissues are probably also at a peak shortly after skeletal maturity, giving maximum protection against injury but encouraging a stressing of these tissues to their limits.

Middle-aged or elderly sports may be mechanically predisposed to certain other injuries because of different levels of competition and also because of changes in body mass (and its distribution), differences in technique, and all-important changes in tissue properties (e.g., resorption of bone, breakdown of joint cartilage).

Age is therefore an extremely important variable in the distribution and cost of sports injuries[17] that should not be ignored.

Sex. A considerable amount has been written about the significant differences between males and females in terms of respective injury patterns and their causes.[18–20] Nearly twice as many males suffer serious musculoskeletal injuries at all ages than do females[13,18,19,20,21] as a result of a combination of cultural, structural, and mechanical factors.

Cultural factors have contributed to female avoidance of certain high-energy and contact sports and have therefore prevented a certain number of related injuries. As more women are now participating in these sports, however, there will probably be a parallel increase in such trauma.

There are a number of musculoskeletal differences between males and females that contribute to structural predilections for certain sports and injuries. On the basis of smaller body masses, for example, females are less inclined to compete in contact sports where body momentum is an advantage. Similarly, due to relative decreases in muscle mass, females are unable to generate the same momentum for activities, such as tackling or checking, putting them at a severe disadvantage when competing against the larger male athlete. These factors, combined with having shorter limbs (shorter lever arms), tend to limit the velocities of the female extremities (in throwing, jumping), providing an intrinsic protective device as well as a discouraging factor for using the joints in these ways.

The female athlete is better suited to certain sports activities than the male. The fact that females generally have shorter legs and a wider pelvis, in addition to a lower center of gravity, makes them ideally suited for activities requiring narrower stance and balancing activities, such as gymnastics. It has been argued that increases in subcutaneous fat stores give the female a relative advantage in terms of endurance activities, but this has been debated.[20] It is less controversial that many of these structural differences lead directly to a different distribution of injuries in the female athlete. Shoulder pain, spondylolysis, scoliosis, vertebral apophysitis, bunions, and stress fractures are more common in women,[18] partly due to their choice of sports and levels of competition but also due to these unique body mechanics.

Chondromalacia of the patella (degeneration of the patellar cartilage) is the best example of a structural difference in the female leading to a clinical condition. The higher incidence of genu valgum (knock knees) secondary to having a wider pelvis in females leads to a relatively higher incidence of abnormal tracking of the patella in the femoral groove. A chronic tendency for lateral patellar displacement is one factor that leads to a higher incidence of chondromalacia in the female—a significant cause of disability and perhaps an important factor in discouraging many women from activities involving running.[18]

It is likely that there are certain differences in the mechanical properties of various tissues in females vs males as well. Hormonal differences may contribute to differences in bone metabolism or its responsiveness and may account for the higher incidences of stress fractures in women. Changes in ligament properties secondary to hormonal shifts may also be significant and may allow some of the greater joint flexibility that is usually noted in the female athlete. Other mechanical differences (in the tissues) may also be found. However, these are probably not the most important factors affecting the conditions described.

Quite clearly, a combination of congenital and acquired biologic *and* mechanical factors control the sexual differences in sports trauma.

TYPES OF INJURIES

Definitions

Sports injuries can be classified into two major groups based on a scale of severity.[22,23] These two groups have been called "microtrauma" and "macrotrauma" and differ mainly by forces applied, the quality of tissues being stressed (Fig. 2), and the timing or degree of biologic responses to them. The most severe injuries are secondary to the greatest forces, with tissue properties being exceeded by such a great extent that preinjury tissue quality is not a great factor. With decreasing amounts of force, however, whether the tissue is normal or abnormal (e.g., previously injured, inflamed, or pathologic in some other way) has a lot to do with its ease of injury and subsequent responses.

Signs and Symptoms

Biologic responsiveness to varying degrees of trauma can be either acute or chronic, to a large extent contributing to the definition of the injury and also to its diagnosis. Macrotrauma presents with acute signs and symptoms of injury, with objective evidence of tissue disruption. Microtrauma, on the other hand, has a more insidious onset, more chronic symptoms, and less objective proof of specific tissue trauma. There is clearly a spectrum of injuries in between these extremes, based on various degrees of macroscopic and microscopic contributions to the clinical presentation. More specific descriptions, with examples, will be helpful in better defining this spectrum

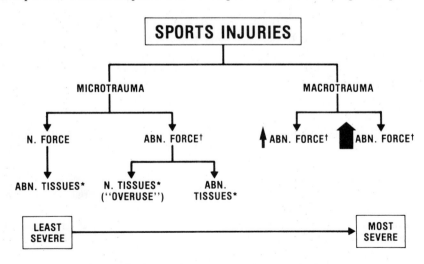

† ABN. FORCE = Abnormal force (magnitude, direction or duration)
* TISSUES = Bone, Cartilage, Muscle, Ligament, Tendon, Fascia

Figure 2. Classification of sports injuries by extent of damage, force, and tissue quality.

and in understanding their respective mechanisms. It is important to realize, however, that biologic influences on mechanical factors can be enormous across this entire spectrum of trauma.

Microtrauma

Microtrauma describes a variety of tissue damage that occurs on a relatively microscopic scale. It may be an actual primary microscopic disruption of the tissue (tearing, cracking, stretching), or it may simply be a secondary mechanical change in the tissues due to the inflammatory process, such as those secondary to increased local pressure, inadequate blood supply, or impaired venous drainage of an area or a tissue. Normal tissues will behave with normal mechanical responses (stress–strain, time, and history-dependent viscoelastic properties), while abnormal tissues will not. No tissue is exempt from having mechanical fatigue limits. However, the range of their responsiveness is considerable and linked to both their mechanical and biologic history.

Forces that cause microtrauma are generally of small magnitude but fall over a wide range, depending on the quality of the tissues involved. Normal activities of daily living may be the beginning of the microtrauma force spectrum, stimulating the gradual normal turnover of microscopically injured tissues over time. More strenuous activities, placing more stress on the tissues, may accelerate these turnover processes. Beyond a certain point, however (which almost certainly varies with each tissue), these stresses become pathologic. Tissues no longer respond by regeneration, and a symptomatic injury takes place.

These processes and limits are not yet well defined for each of the tissues. Since all of the tissues are anisotropic (having different mechanical properties in different stress directions), the permutations and combinations of force, direction, duration, recovery, and other factors are almost impossible to determine in vivo for each of them and nearly as difficult in vitro. (Some of their characteristics under limited circumstances are described separately in a section to follow.)

Certain in vivo examples of microtrauma, in which these limits have obviously been exceeded, can be found clinically, beginning at the least severe end of the sports injury spectrum.

Clinical Examples. Jogging produces by far the best and most common example of microtrauma in clinical practice. The range of forces and tissues affected by jogging are immense, providing a good idea of the potential diversity of this portion of the injury spectrum.

By definition, normal forces (magnitude, direction, and duration) should not injure normal tissues. Abnormal tissues (previously injured, inflamed, or pathologic), however, may be more sensitive to such forces, being prone to acute injury or having a lower than normal tolerance for recovery. Jogging of normal force may exceed the fatigue limits of uncondi-

TYPICAL "MICROTRAUMA"

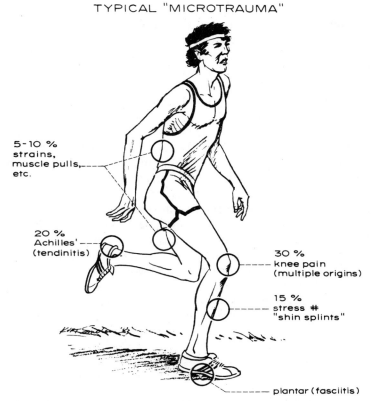

5-10 %
strains,
muscle pulls,
etc.

20 %
Achilles'
(tendinitis)

30 %
knee pain
(multiple origins)

15 %
stress #
"shin splints"

plantar (fasciitis)

PHOTOCOPY

Figure 3. Typical areas of apparent microdamage from repeated cyclic loading. Excessive forces can lead to both symptoms and signs in these areas, even in conditioned runners.

tioned (therefore, abnormal) tissues, producing any number of areas of microdamage with secondary symptoms.[14] (Fig. 3).

Moving up the scale of injury severity, jogging may also exceed normal tissue limits by being excessive in terms of force. Abnormal magnitudes and/or durations of force may exceed the endurance and recovery limits of the normal tissues and cause either their disruption or irritation (inflammation) followed by a secondary disruption. These are the so-called overuse syndromes, where recurrent cyclic loading in some way traumatizes the tissues. The mechanism of this trauma, whether directly mechanical, by fatigue, or secondary to impairment of poorly defined biologic recovery processes, is unknown. Probably the best example of an overuse syndrome is the stress fracture, occurring most frequently in athletes in the otherwise normal tibia,[24] but also occuring in the metatarsals, talus, fibula, ulna, and other bones. Stress fractures have been most frequently reported in runners but also occur in ballet, football, baseball, basketball, and a variety of other sports and activities[25] involving repeated tissue loading. Stress fractures are at the link between macrotrauma and microtrauma, where microscopic injury may suddenly become macroscopic.

These same excessive forces on abnormal tissues (by the above definition) will clearly cause even more severe damage and symptoms. A stress fracture into a bone cyst, for example, may occur sooner and be more extensive than one occuring in normal bone.

Mechanisms. The mechanism of a stress fracture is through cyclic fatigue of cortical bones, as reported by Carter et al. and by Hayes,[26–29] with subsequent crack propagation and failure. The fatigue behavior of cortical bone can be described in vitro, with the approximate number of cycles to failure at each of three levels of force (walking, mild exercise, rigorous exercise) being shown (Fig. 4). It can be seen from that data that the equivalent of many miles of very rigorous exercise might be needed to fatigue a cortical bone. However, these cyclic strain values can fall within the limits of a legitimate athletic training program. As noted above, any underlying abnormality of the bone (e.g., cyst, cortical defect) would seriously decrease these fatigue limits and predispose to much lower fracture stresses.

Soft tissues may demonstrate an analogous fatigue behavior with proportionately higher strain tolerance and cyclic resistance than bone. These limits, however, have not been well defined for the various tissues in their

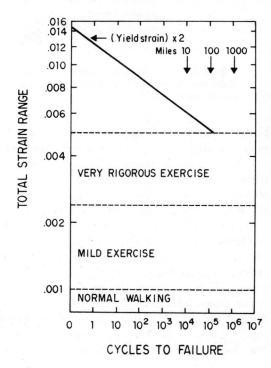

Figure 4. Cyclic fatigue of cortical bone in vitro with approximate limits where stress fracturing would occur. *(After Carter and Hayes.[26–29])*

various biologic environments. As composite materials of less rigid nature, soft tissues probably fatigue more slowly than bones, but this property is difficult to quantitate due to viscoelastic predominance. Because of difficulty testing soft tissues in vitro over time (they probably deteriorate over the length of time needed to demonstrate fatigue behavior), the group of soft tissue microtrauma conditions is by far the most difficult to mechanically document and quantitate.

Chronic abnormal forces will result in chronic microtrauma and, therefore, chronic biologic responses. Soft tissue calcification is often a chronic response to bleeding or soft tissue disruption and can be a cause of irritation as well as a sign of mechanical injury. Articular cartilage may be unable to mount a sufficient biologic response or repair to chronic trauma and may therefore be doomed to deteriorate. Cumulative mechanical trauma with subsequent biologic degradation and failure of regeneration is one of the classic explanations of the high incidence of degenerative arthritis in athletes.

Macrotrauma

Clinical Examples. Macrotrauma can be subdivided by severity of tissue disruption in direct proportion to the forces involved. Forces greatly exceeding ultimate properties of the tissues have an explosive effect, damaging much greater amounts of tissues over much broader areas. Smaller macrotrauma forces will also disrupt tissue but in more isolated areas and more predictably simple patterns.

Based on our previously mentioned distributions of sports injuries, the most common examples of macrotrauma are those involving the knee. The most common injuries of the knee are to the ligaments and the cartilage, with bony disruptions occurring less often. Cartilage injuries can be subdivided into fibrocartilage (menisci) or hyaline cartilage (joint surfaces), each with its own signs, symptoms, and complications.

Mechanisms. Understanding the mechanisms of macrotrauma to the structures and tissues of the knee requires a simple understanding of their normal functions. Contoured articular surfaces of diarthrodial joints, such as the knee, are normally covered by glassy hyaline cartilage, forming a low friction gliding surface. This cartilage distributes some of the joint loads to subchondral bone and serves mechanical as well as immunologic protective functions. The fibrocartilagenous menisci serve as molds to accommodate the shapes of the femur and tibia, forming a pocket of protection and guidance for the femur and transferring a considerable portion of joint loads.[30] The ligaments of the knee serve as static check reins, resisting tensile forces in various planes with particular fiber orientations having individual functions (Fig. 5). The anterior cruciate ligament mainly resists anterior translation of the tibia on the femur. The posterior cruciate resists posterior tibial translation. The medial collateral ligament resists lateral

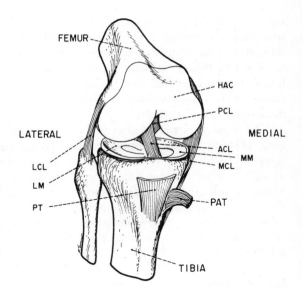

Figure 5. Schematic representation of normal knee with major stress carrying structures labeled. HAC, hyaline articular cartilage; ACL, anterior cruciate ligament; MCL, medial collateral ligament; MM, medial meniscus; PAT, pes anserine tendons; PT, patellar tendon; LM, lateral meniscus; LCL, lateral collateral ligament; PCL, posterior cruciate ligament.

(valgus) bending forces, and the lateral collateral ligament resists medially directed tibial bending (varus) forces.

The mechanism of a simple medial collateral ligament tear (the most common ligamentous injury in the athlete's knee) is a strong lateral force directed medially across the surface of the joint, causing tensile failure of the MCL (and other medial structures). Depending on the force and direction (bending, rotation, translation), other structures will also fail in tension on the medial side of the joint, while the lateral side of the joint may be compressed (Fig. 6). The combination of MCL tearing with disruption of the medial meniscus (to which the deep portion of the MCL is normally attached) and a tear of the anterior cruciate ligament (from a continued separation of the medial side, hyperextension of the knee, or internal tibial rotation) is one of the most common, complex athletic knee injuries. Other combinations of knee ligament injuries can also occur and are specific for the magnitude and directions of the forces involved.

Commonly described cartilage tears of the knee refer to failure in substance of either the medial or the lateral meniscus (fibrocartilage) in one of several planes. Although it is possible for a normal meniscus to fail catastrophically, many believe that meniscal injuries are almost always secondary to acute forces of intermediate magnitudes, failing relatively abnormal (previously injured) fibrocartilage.[31] If this is the case, attrition of the menisci and microscopic tears may easily be extended by slightly greater than normal joint loads, particularly if the menisci happen to become caught between the bones. The medial meniscus, being less mobile than the lateral meniscus, is more prone to being trapped in this fashion and is failed by compressive forces that are distributed throughout its ringlike collagen

VALGUS STRESS

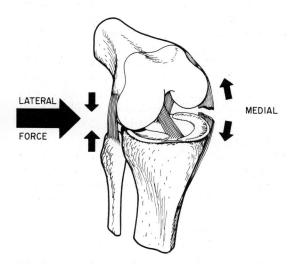

Figure 6. Valgus stress injury to knee. A medially directed force at the level of the knee will cause medial tensile failure and lateral compressive failure in addition to any shearing or rotational components. Injury patterns are specific for force, direction, and magnitude.

structure.[32] The compressed meniscus tries to expand to dissipate these forces but ultimately tears (in tension) along the weak spots between fiber planes in its structure.[31] Characteristic patterns of tearing with subsequent mechanical interference with joint function by blocking of normal knee motion (locking) and joint reaction (swelling) complete the clinical picture of meniscal damage.

Macrotrauma to hyaline cartilage usually involves a fracture of subchondral bone, although these fractures may be age related. Loose articular fragments can be single or multiple, depending on the forces and directions involved. Compressive and shear injuries to the articular surfaces are the common modes of injury. Due to their deep location (in the center of the joint) and their frequent association with other more obvious superficial injuries, joint surface injuries and fractures are frequently missed entirely or misdiagnosed. Since joint contours and gliding surfaces are critical to both acute and chronic joint function (stability and bending), even a minor macrotraumatic change may be fatal to the life of the joint.

MECHANICAL CONSIDERATIONS

Injury

There are some basic mechanical considerations in describing and explaining sports injuries as a whole.

Forces and Strains. The forces generated in sports are enormous. Even in individual activities, the forces generated on the joint surfaces are surprisingly large. The force generated in the quadriceps tendon during the kicking of a football, for example, has been estimated to be more than 3 times body weight.[33] The vertical force on the ground during the high jump takeoff has been estimated to be greater than 350 kg, a force that must clearly be transmitted as reactive forces through the joints of the lower extremities.[34] The patellofemoral reaction force has been measured during various activities and was found to be approximately 3.3 times body weight during normal stair climbing and as much as 7.6 times body weight during a deep knee bend.[35] Simply extending the knee from 90 degrees of flexion with a 9 kg weight on the leg created patellofemoral forces in excess of 1.4 times body weight, possibly explaining the high incidence of retropatellar symptoms during certain types of leg-strengthening exercise.[33] From these figures it can readily be appreciated that during even relatively simple noncontact athletic activities compressive forces on the joint surfaces may be tremendous, requiring considerable force distribution, dissipation, and reaction to prevent injury to the tissues.

Forces during contact sports will be many times greater than forces during these simple activities. The amount of kinetic energy possessed by a 70 kg football player traveling at between 5 and 10 meters/second, for example ($KE = \frac{1}{2} mv^2$), is up to 3500 joules. Using the equation of impulse and momentum:

$$F\Delta t = m\Delta v,$$

it can be seen that if that individual is hit and stopped over a short distance and over a short period of time (high Δv and small Δt, e.g., rapid deceleration of over 10g), forces of more than 10 times his body weight may be generated and absorbed over the tissue area that is contacted. Assuming a rigid body situation, if that amount of force were transmitted to the cross-sectional area of one knee (whose foot was fixed on the ground), virtually all of the tissues of that knee would be failed by stresses several orders of magnitude greater that their calculated ultimate values (Table 5). [36–51]

Fortunately for the athlete, however, that type of situation is rare. Force dissipation is usually over larger tissue areas and greater distances (high Δt, low Δv) than this extreme example, so that peak stresses in the tissues are substantially reduced. Shock absorption through bending of the joints, dynamic muscular forces,[52] padded equipment, compliant playing surfaces, and shoes that do not rigidly fix the extremities to the ground are critical factors in reducing these high-energy types of injuries.

Of major significance to the amounts and types of musculoskeletal damage, as well, is the rate at which the tissues are loaded. Cortical bone, for example, is sensitive to increase in loading rates by a significant increase in its modulus of elasticity and ultimate strength. Ligaments, on the other hand, are not quite as responsive to similar changes in loading rate, with only slight improvement in failure capabilities at high speeds (Fig. 7). These adaptations are of critical importance when it is noted that typical in vivo

TABLE 5. PROPERTIES OF VARIOUS MUSCULOSKELETAL TISSUES*

Tissues (In Order of ↑ Strength)	Ultimate Stress (mPa)	Ultimate Strain (%)
Muscle[†]		
Noncontracted	0.1–0.3	40–60
Ligament[†]		
Elastic (nuchae)	1–2	30–125
Bone (cancellous)	1–2	0.1
Tension	1.5–2	0.03–0.6
Compression		
Fascia[†]	15	15–17
Cartilage (hyaline)		
Tension	1–40	10–100
Compression	7–23	3–17
Shear	6	—
Cartilage (fibrocartilage)		
Tension	10–50	10–20
Compression	20	30
Tendon[†]	40–100	10–17
Collateral ligaments[†]		
Nonelastic	60–100	5–14
Bone (cortical)		
Tension	90–170	0.7–5
Compression	100–280	1–2.4
Shear	50–100	—

*All values are approximate and based only on specific test conditions.[36–51]
[†]Tension only.

strain rates at the knee (e.g., tensile strain of the medial collateral ligament in a skier rapidly decelerating from 10 meters/second) may be anywhere from 50 percent/second to 4200 percent/second.[53] Estimated ultimate properties, as in Table 5, taken in vitro at relatively slow strain rates (fractions of these values) are therefore really underestimates of in vivo stress capacities. In vitro values are similarly also probably overestimates of true ultimate strains, since decreased deformation to failure occurs at very fast in vivo rates.

Structural vs Mechanical Properties. In subdividing sports injuries, we have noted that there is generally a higher incidence of soft tissue injuries than bony injuries in both micro- and macro-types of trauma. This is not surprising from a biomechanical point of view for both structural and mechanical reasons.

Structural Behavior. In an anatomic sense certain joints (and their soft tissues) are more predisposed to injuries than others. The knee, for example, with its concentration of movers (muscle–tendon units) on the anterior and posterior portions of the joint, leaves the medial and lateral sides relatively unprotected. Forces in these planes, therefore, are more likely to

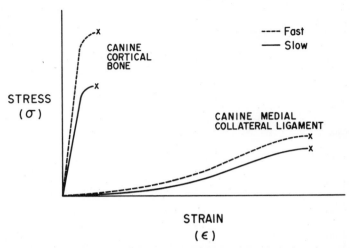

Figure 7. Schematic representation of mechanical properties of canine cortical bone and canine medial collateral ligament tested at two strain rates. Bone is more sensitive to increased loading rate with a greater increase in peak stiffness and stress. The ligament absorbs more energy to failure than the bone and is slightly strain rate-sensitive. At very fast loading rates, ligaments will fail first.

strain the static stabilizing structures (ligaments) in these areas, rather than being able to rely on the rigid or energy-absorbing nature of the contracted joint musculature. During slow loading situations, it is likely that muscle feedback is fast enough to react to abnormal joint bending, either stabilizing it by rigid contraction or simply moving it to a more stable position. Under faster loading situations, however, it is doubtful that the muscle proprioceptive system gives fast enough feedback to adequately protect the joint.[52] In these situations, partial contraction of muscles against large forces may lead to either muscle injury or injury to those other stabilizing structures.

There are at least two other structural reasons for the high incidence of soft tissue failures at the knee. The static stabilizers of the knee are oriented to permit motions in one plane and resist most others. Abnormal joint forces, therefore, put these structures on stretch very early during displacement. This fact, combined with the mechanical advantage given to abnormal bending forces at the knee (inertial forces of the trunk over a long lever arm) and the multiplication of strain rates over the width of the joint,[53] means that the peripherally lying static stabilizers of the knee not only get stretched first but get stretched faster and with more force than most other soft tissues in the body.

Mechanical Behavior. From a mechanical point of view the soft tissues are ideally suited to their functions. Ligaments, for example, through both organizational and chemical characteristics are well suited to absorb energy over relatively large amounts of deformation (Fig. 7). Their allowance of great deformation clearly protects the bones from reaching ultimate failure deformations (strains) in most loading situations. Their stiffness is critical, allowing some joint displacement but with progressively increasing resistance.

Ligaments or their insertions at the same time often represent the weak mechanical link in the joint linkage system. Their ultimate failure values in in vitro test conditions are inferior to those of cortical bone. They are even more disadvantaged at higher loading rates (closer to in vivo conditions) on the basis of their relative lack of sensitivity to increasing strains. This is well demonstrated by comparing the responses of bone and ligaments to comparable increases in loading rates (Fig. 7). Bone responds with marked increases in stiffness. Ligaments also respond, but with much lower percentage increments. In other words, in rapid, high-stress loading situations—to which certain ligaments, *particularly* those of the knee, are anatomically and functionally predisposed—there is a much higher chance of ligament rather than bony failure.[53]

From the standpoint of examining the mechanical properties of other tissues, as well as comparing them to the properties of cortical bone and ligaments, Table 5 has been organized in approximate order of increasing ultimate stresses. With the exception of ligamentum nuchae (an extremely elastic ligament in the neck) and cancellous bone (which has the benefit of cortical protection), there is a relative correlation between mechanical properties and predilections for clinical injury of these tissues. In other words, there is an increasing stress capacity and decreasing injury propensity from top to bottom in that table. It must quickly be pointed out, however, that these numbers represent relatively static values and do not consider the important dynamic, cyclic, and viscoelastic properties of the tissues or the biologic variables that may affect them.

Research Objectives. In order to better define the spectrum of sports injuries, a combination of mechanical and biologic considerations are needed. Examination of both structural and mechanical properties at simulated in vivo loading rates of the various tissues with various simulated afflictions (e.g., partial tears, inflammation) will be a research goal for the future. Better definitions of their dynamic, cyclic, and viscoelastic behaviors, functional interdependence, and healing characteristics are also clear objectives for combined clinical and research studies.

Diagnosis

There are some important mechanical considerations in the diagnosis of sports injuries in addition to understanding their etiologies. In general terms, for example, the physical examination of a sports injury attempts, in a controlled way, to duplicate its causative mechanism and thereby reproduces either its symptoms (if chronic) or its signs (if acute). Much lower forces than those of the injury are, of course, used to prevent further dam-

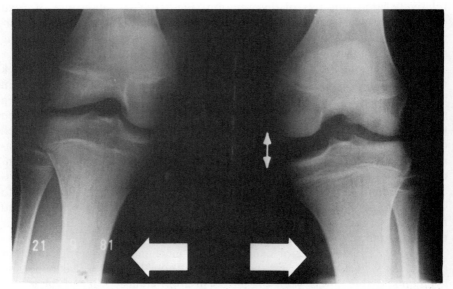

Figure 8. Valgus stress x-ray demonstrating medial collateral ligament injury by medial joint widening (arrows). Large arrows demonstrate laterally directed forces on tibias. This is a gentle reproduction of the mechanism of injury.

age. Patellar compression (to reproduce the pain of chondromalacia), simple palpation (to reveal an area inflamed by chronic pressure, e.g., bursitis), and joint stress testing (Fig. 8) are all really only gentle mechanical reproductions of the mechanisms of tissue injury. Some awareness of the structural and mechanical implications of each test for each injury is therefore important in reaching a proper diagnosis.

Treatment and Prevention

Clinical Considerations. The clinical treatment of sports injuries depends on an understanding of its anatomy,[54] its mechanism, and its subsequent pathology,[55,56] as well as an understanding of its repair mechanics.[50] Treatment is based on restoration of normal anatomy and kinematics in as short a time as possible (limited by the biologic processes), with a considerable emphasis on prevention of reinjury during the rehabilitation process. Immobilization, while important in minimizing damage and possibly decreasing symptomology in the early phase of recovery from both macrotrauma and microtrauma, has its own detrimental effects.[57] Early mobilization and muscle rehabilitation, therefore, depend on a mechanical understanding of both tissue tolerances and joint stresses.

In general, there is no substitute for normal structure and function after any type of injury, but this is particularly true in sports injury because of the likelihood of early and repeated stresses during recovery. Protection in the form of casts, braces, molds, and other protective equipment can be useful,

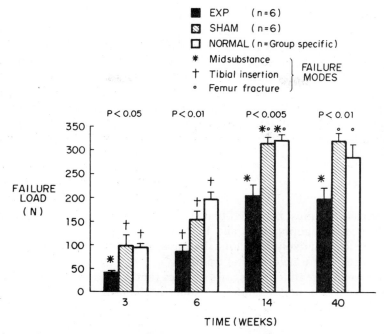

Figure 9. Failure properties of rabbit bone–medial collateral ligament–bone complexes. Experimental ligaments (midsubstance injuries without repair or immobilization) are consistently inferior to normal and sham-operated complexes. Modes of failure also change with age, demonstrating the importance of maturity to structural mechanics. (*From Frank et al.*[50])

but it is not encouraged as a replacement for normal anatomy or conditioning. Proper joint conditioning and selective muscle strengthening for the demands of the individual (age, sex, level of competition) and the sport are probably the best ways to treat and prevent the majority of sports injuries.

It is beyond the scope of this chapter to discuss the implications of mechanical principles to clinical treatments of sports injuries other than to emphasize their essential importance *in combination with* biologic facts.

Research Considerations. Sports injuries are similar to all musculoskeletal injuries in requiring better mechanical definition. We have recently shown, for example, that the structural and mechanical properties of an untreated medial collateral ligament scar are grossly abnormal for long periods of time (Fig. 9), reaching only about 60 percent of normal structural and about 30 percent of normal mechanical capabilities.[50] Optimization of this process through repair, exercise, motion, drugs, and so on will require careful mechanical quantitation. Although exercise has been shown to improve the capabilities of some normal tissues, it may have serious negative effects as well (such as prolonging healing, inducing laxity).

Patterns of injury, as implied in our introduction, may be level, sport, age, and sex specific. However, after the injury occurs, its response and recovery become more a function of both the individual and his/her tissues than of the original cause.[56] For that reason, we advocate that future sports injury research should address the same topics that we have introduced: understanding injury mechanics, diagnosis, and treatment (or prevention) *on the tissue level* and in biologically relevant circumstances. In that way, we will derive a better understanding of sports injury "principles" rather than "empiricals."

SUMMARY

In conclusion, it must be stated that sports injuries are in some ways unique, affecting certain populations and in certain definite patterns, but in other ways they are relatively common, sharing a spectrum of severities with many other causes of musculoskeletal trauma.

Improving treatment of sports injuries (and all musculoskeletal injuries) will require a firm understanding of their mechanisms, diagnosis, treatment, and prevention. Defining and classifying the clinical syndromes and injury patterns in biomechanical terms and understanding the problems at the individual (kinematic), joint (structural), and tissue (mechanical) levels will be the optimal way to improve end results.

Combining mechanical and clinical research information will minimize trial and error and establish a scientific basis for the treatment of sports trauma.

ACKNOWLEDGMENTS

The authors gratefully acknowledge the financial support of the Alberta Heritage Foundation for Medical Research, the Veterans Administration Medical Center, and the facilities of the University of California, San Diego. We would also like to thank Dr. D. Daniel for the use of his clinical material in this manuscript.

REFERENCES

1. Banks HH (ed): Symposium on sports injuries. *Orthop Clin North Am* 1980; ll:685.
2. Banks HH (ed): Major sports injuries: Tribute to Thomas B. Quigley, M.D. *Clin Orthop* 1982; 164:2.
3. Radin EL, Simon SR, Rose RM, et al.: *Practical Biomechanics for the Orthopaedic Surgeon*. New York, Wiley, 1979.

4. Frankel VH, Burstein AH: *Orthopaedic Biomechanics.* Philadelphia, Lea & Febiger, 1979.
5. Fung YC: *Biomechanics: Mechanical Properties of Living Tissues.* New York, Springer-Verlag, 1981.
6. Gove PB (ed): *Webster's Third New International Dictionary.* Springfield, Mass, GC Merriam Co, 1976, p 2206.
7. Haddon W Jr: Principles in research on the effect of sports on health. *JAMA* 1966; 197:885.
8. Kelsey JL, Pastides H, Bisbee GE, et al.: *Musculo-Skeletal Disorders: Their Frequency of Occurrence and Their Impact on the Population of the United States.* New York, Prodist, 1978, pp 2–3.
9. Mellin M: The sports medicine breakthrough. *San Diego Magazine,* August 1983, p 111.
10. Kelsey JL, White AA, Pastides H, et al.: The impact of musculoskeletal disorders in the population of the United States. *J Bone Joint Surg* 1979; 61A:959.
11. MacIntosh DL, Skrien T, Shephard RJ: Athletic injuries at the University of Toronto. *Med Sci Sports* 1971; 3:195.
12. Muckle DS: *Injuries in Sport.* Chicago, Year Book, 1978, pp 4–10.
13. Zaricznyj B, Shatluck LJM, Mast TA, et al.: Sports related injuries in school-aged children. *Am J Sports Med* 1980; 8:318.
14. Brody DM: *Running Injuries.* CIBA Pharmaceutical Co., 1980; No. 32.
15. Wilkins KE: The uniqueness of the young athlete: Musculoskeletal injuries. *Am J Sports Med* 1980; 8:377.
16. Torzilli PA, Takebe K, Burstein AN, et al.: The material properties of immature bone. *J Biomech Eng* 1982; 104:12.
17. Pritchett JW: High cost of high school football injuries. *Am J Sports Med* 1980; 8:197.
18. Hunter LY, Andrews JR, Clancy WG, et al.: Common orthopaedic problems of female athletes. *American Academy of Orthopaedic Surgeons Instruc Course Lect* 1982; 31:126.
19. Whiteside P: Men's and women's injuries in comparable sports. *Phys Sports Med* 1980; 8:130.
20. Protzman R: Physiologic performance of women compared to men. *Am J Sports Med* 1979: 7:191.
21. Crompton BA, Tubbs N: A survey of sports injuries in Birmingham. *Br J Sports Med* 1977; 11:12.
22. Davies GJ, Wallace LA, Malone T: Mechanisms of selected knee injuries. *Phys Ther* 1980; 60:1590.
23. Travers PR: Sports injuries. *Physiotherapy* 1980; 66:215.
24. Belkin SC: Stress fractures in athletes. *Orthop Clin North Am* 1980; 11:735.
25. Protzman RR, Griffis CG: Stress fractures in men and women undergoing military training. *J Bone Joint Surg* 1977; 59A:825.
26. Carter DR, Hayes WC: Compact bone fatigue damage. A microscopic examination. *Clin Orthop* 1977; 127:265.
27. Carter DR, Spengler DM: Mechanical properties and composition of cortical bone. *Clin Orthop* 1978; 135:192.
28. Carter DR, Caler WE, Spengler DM, et al.: Fatigue behavior of adult cortical bone: The influence of mean strain and strain range. *Acta Orthop Scand* 1981; 52:481.

29. Hayes WC: Response of bone tissue to loading. Orthopaedic Biomechanics Course. White Plains, NY, June 1982.
30. Shrive NG, O'Connor JJ, Goodfellow JW: Load bearing in the knee joint. *Clin Orthop* 1978; 131:279.
31. Mueller W: *The Knee: Form, Function and Ligament Reconstruction.* Berlin, Springer-Verlag, 1983, pp 102-103.
32. Wagner HJ: Die Kollagen faserarchit ektur der Menisken des menschlichen Kniegelenkes. *Z Mikrosk Anat Forsch* 1976; 90:302.
33. Frankel VH, Hang YS: Recent advances in the biomechanics of sports injuries. *Acta Orthop Scand* 1975; 46:484.
34. Gombac R: The mechanics of take-off in high jump. *Med Sport Biomechan II* 1971; 6:232.
35. Reilly DT, Martens M: Experimental analysis of the quadriceps muscle force and patello-femoral joint reaction force for various activities. *Acta Orthop Scand* 1971; 43:126.
36. Yamada H: Evans FG (ed): *Strength of Biological Materials.* Baltimore, Williams & Wilkins, 1970, pp 19–105.
37. Freeman MAR: *Adult Articular Cartilage.* Oxford, Pitman & Sons, 1973, pp 174-219.
38. Mow VC, Schoonbeck: Comparison of compressive properties and water, uronic acid and ash contents of bovine nasal, bovine articular and human articular cartilages. *Trans Orthop Res Soc* 1982; 7:209.
39. Myers ER, Mow VC: Biomechanics of cartilage and response to biomechanical stimuli, in Hall BK (ed): *Cartilage.* New York, Academic Press, 1981.
40. Reilly DT, Burstein AH: The mechanical properties of cortical bone. *J Bone Joint Surg* 1974; 56A:1001.
41. Fung YC: *Biomechanics: Mechanical Properties of Living Tissues.* New York, Springer-Verlag, 1981, p 385.
42. Elftman H: Biomechanics of muscle. *J Bone Joint Surg* 1966; 48A:363.
43. Barfred T: Experimental rupture of the Achille's tendon. *Acta Orthop Scand* 1971; 42:406.
44. McMaster PE: Tendon and muscle ruptures. *J Bone Joint Surg* 1933; 15:705.
45. Elliott DH: The biomechanical properties of tendon in relation to muscular strength. *Ann Phys Med* 1967; 9:1.
46. Benedict JV, Walker LB, Harris EH; Stress–strain characteristics and tensile strength of unembalmed human tendon. *J Biomech* 1968; l:53.
47. Abrahams M. Mechanical behavior of tendon in vitro: A preliminary report. *Med Biol Eng* 1967; 5:433.
48. Woo S L-Y: Mechanical properties of tendons and ligaments. I. Quasi-static and nonlinear viscoelastic properties. *Biorheology* 1982; 19:385.
49. Woo S L-Y, Gomez MA, Woo YK, Akeson WH: Mechanical properties of tendons and ligaments. II. The relationships of immobilization and exercise on tissue remodeling. *Biorheology* 1982; 19:397.
50. Frank C, Woo S L-Y, Amiel D, et al.: Medial collateral ligament healing. *Am J Sports Med* 1983; 11:379.
51. Seering WP, Piziali RL, Nagel DA: The effect of strain rate on the mechanical properties of human ligaments. *Trans Orthop Res Soc* 1978; 4:241.
52. Pope MH, Johnson RJ, Brown DW, et al.: The role of the musculature in injuries to the medial collateral ligament. *J Bone Joint Surg* 1979; 61A:398.

53. Crowninshield RD, Pope MH: The strength and failure characteristics of rat medial collateral ligaments. *J Trauma* 1976; 16:99.
54. Warren LF, Marshall JL, Girgis F: The prime static stabilizer of the medial side of the knee. *J Bone Joint Surg* 1974; 56A:665.
55. Krause WR, Pope MH, Johnson RJ, et al.: Mechanical changes in the knee after meniscectomy. *J Bone Joint Surg* 1976; 58A:599.
56. Noyes FR, Mooar PLA, Matthews DS, et al.: The symptomatic anterior cruciate deficient knee. Part I: The long-term functional disability in athletically active individuals. *J Bone Joint Surg* 1983; 65A:154.
57. Woo S L-Y, Kuei SC, Gomez MA, et al.: The effect of immobilization and exercise on the strength characteristics of bone–medial collateral ligament–bone complex. *Am Soc Mech Eng* 1979; 32:62.

CHAPTER 11

Radiography and Other Imaging Modalities in the Diagnosis of Trauma

Jose Guerra, Jr.

INTRODUCTION

Radiography plays an important role in the diagnosis of trauma. Most of the required information is provided by routine plain radiographs. There are occasions, however, when special techniques are necessary to detect and/or to better characterize the trauma. Normal plain radiographs do not preclude an injury in the patient when there is a strong clinical suspicion for significant trauma.

Special radiographic techniques and studies, as well as other imaging modalities, are available and should be utilized depending upon the clinical problem. Indications for the use of some of these techniques may be unfamiliar to clinicians. The radiologist should be able to suggest the appropriateness of the special studies that may elucidate the clinical problem when the need arises.

RADIOGRAPHY

Routine Studies

The number and type of projections obtained in the routine study of a traumatized region may vary from one hospital to another. Certainly, the larger the number of radiographs and the more varied the type of projections obtained, the smaller the incidence of missed trauma. This reasoning, however, does not automatically dictate that every routine study include the greatest number of radiographs possible. The patient benefits if the examination is tailored to the type of trauma he sustained.

For example, if the patient received trauma to the great toe, the ordering of a routine foot series is inappropriate. Whenever possible, clinical history should be given to indicate the region of interest. In this instance, the appropriate examination to request is anteroposterior, oblique, and lateral radiographs of the great toe.

The examination of the patient with a cervical spine injury is a more serious example. If the injury is minor and there is no loss of consciousness or neurologic deficit, anteroposterior, lateral, and open-mouth views of the cervical spine will usually suffice. If the injury is severe with loss of consiousness and/or neurological deficit, a more complete examination is called for. This would include obliques, the pillar view, and possibly dynamic studies, such as flexion, extension, and distraction views in the lateral projection. Therefore, the appropriate study will include more radiographs, depending upon the degree of injury incurred and the level of suspicion that important trauma has occurred. One can appreciate that some clinical and radiographic judgment must be exercised. Proper communication between the clinician and the radiologist will optimize the study for the patient.

A single anteroposterior or lateral view of the traumatized region should not be requested or accepted as proper screening to exclude trauma. This is true for such regions as the cervical spine in particular and articulations in general. Even in the case of long bones, there are innumerable examples of fractures being visualized in only one projection and not the other, and this is particularly true in children.

Finally, it is important to reiterate that there is no substitute for good rapport between the clinician and the radiologist. The radiologist can best serve the clinician and the patient if he is aware of the type of trauma that occurred and of the clinician's suspicion for occult trauma. The radiologist can then suggest to the clinician the appropriate study to properly evaluate the patient for his particular injury.

Dynamic Studies
The term "dynamic studies" generally refers to radiographs obtained while the involved region is stressed or is at the extremes of range of motion. Admittedly, the impression derived from these studies can also be obtained with fluoroscopy. For the sake of discussion, however, we will use the above definition for dynamic studies, and fluoroscopy will be reserved for any plain film examination that is concurrently visualized utilizing the image intensifier and the video screen.

Stress views are included in this category and are helpful in diagnosing ligamentous injuries. They are indicated when there is a clinical question of instability or disruption of the joint or when the routine radiographs are suggestive of this.[1] The detachment of a ligament may be equivalent to a fracture insofar as the stability of the joint is concerned.

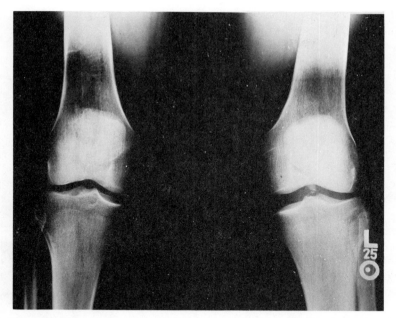

Figure 1. Valgus stress view of both knees. A restraining strap has been placed around the thighs and equal valgus stress exerted on the knees. The medial joint space of the left knee opens excessively, indicating that the medial collateral ligament has been torn. The osseous fragment in the intercondylar notch is attached to the avulsed anterior cruciate ligament.

Stress films of the joint must be obtained without delay after an injury because swelling and muscle spasm will mask the presence of a torn ligament.[2] Pain and guarding may be so severe that stress films can be obtained only with general anesthesia. A stress comparison view of the opposite uninjured side is essential for proper interpretation. The joints most frequently examined with stress views are the acromioclavicular, the knee, and the ankle (Fig. 1).

The cervical spine is probably the most important region examined by dynamic studies. These would include films obtained in flexion and extension, rotation to either side, lateral bending to either side, and distraction (Fig. 2). Since neurologic injury is either present or highly suspected, the clinician is generally not eager to move the patient's neck. Only patients that are fully conscious and cooperative undergo the studies (with the exception of the distraction examination), and they themselves determine how far the spine may be taken in the type of motion examined. The clinician and radiologist may help the patient and provide both mental and physical support, but it is the patient who determines how far the neck motion can be taken. The distraction study is usually performed by the traumatologist in one of two ways: (1) pulling uniformly on both mandibles in the cephalad

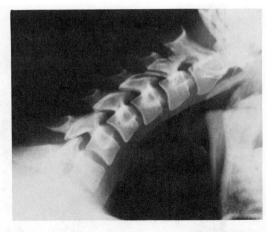

A

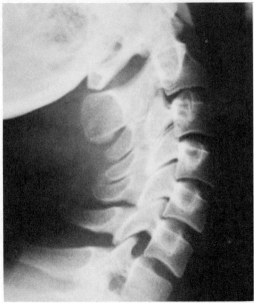

Figure 2. Normal flexion (**A**) and extension (**B**) views of the cervical spine. Note the degree of widening and narrowing that occurs with movement at the interspinous spaces, the disk spaces, and the apophyseal joints. The movement is gradual and coordinated without disproportionate changes.

B

direction or (2) placing the patient in tongs hooked up to a pulley system and adding weights until the diagnosis is made or the study is discontinued.

The reason for studying the traumatized patient in a dynamic fashion is obvious. Many injuries cannot be documented unless the proper motion or stress, usually the same one that caused the injury, is placed on the traumatized region.

Many injuries to the cervical spine would go undiagnosed if it were not for the appropriate radiographic examination. It is not uncommon to have a

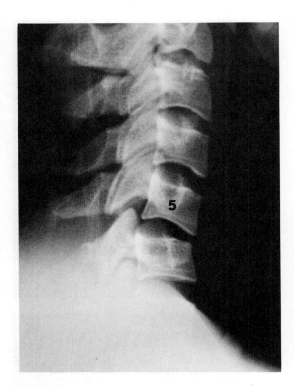

Figure 3. Limited flexion view in patient with a C5-6 hyperflexion injury. Note the disproportionate narrowing at the disk space and widening of the interspinous space and apophyseal joints, indicating that the soft tissues at this level have been disrupted. The supraspinous and interspinous ligaments, as well as the apophyseal capsules, have been torn by distraction, and the C5-6 disk has been partially crushed.

normal static cervical spine examination in a patient with severe neurologic deficit. Tearing of the supporting soft tissues has occurred and may only be documented with flexion and extension radiographs in the lateral projection (Fig. 3). In the unconscious or sedated patient, distraction of the cervical spine is a satisfactory means of examination. The spinal cord has a certain amount of give and is not damaged by mild distraction. Yet, distraction may allow the demonstration of the injury site (Fig. 4).

Fluoroscopy
All of the so-called dynamic studies can be monitored fluoroscopically to see the real-time pathologic process at the site of injury and to prevent a catastrophic event from happening in a patient who is already severely compromised. In addition, fluoroscopy can be used to obtain the exact projection necessary for diagnosis or comparison.

Another important use for fluoroscopy is the examination of joints, particularly in those patients who describe a clicking or snapping with movement. Some of these patients have intra-articular osteochondral bodies that may be seen to become interposed between the major bones of the articulation and then snap out of the way as the patient completes the arc of motion. Arthrography, of course, is necessary to make the diagnosis if the

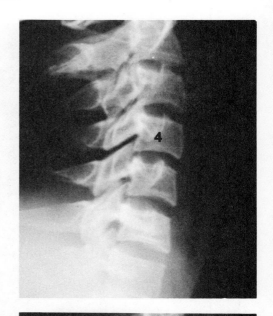

A

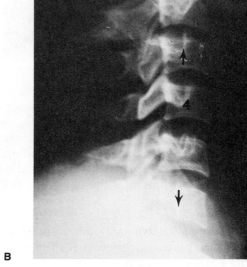

Figure 4. Lateral cervical spine radiographs of a patient with quadriplegia from a motor vehicle accident. **A.** The neutral and supine lateral radiograph demonstrates retrotracheal soft tissue swelling without evidence of a fracture. The C4-5 apophyseal joint spaces are unusually prominent. **B.** The distraction lateral radiograph using 40 pounds of weight demonstrates widening of the apophyseal joints and the C4-5 disk space, indicating through-and-through tearing of soft tissues.   **B**

suspected intra-articular body is radiolucent. This will be discussed in a later section.

Tomography

Conventional tomography still remains an important tool in detecting and characterizing trauma.[3-5] In some respects, computed tomography (CT) has replaced or supplemented conventional tomography; this is particularly true in the spine. Although computed tomography with reconstruction can adequately characterize trauma in appendicular joints, some radiologists and clinicians still favor conventional tomography because it gives a more suitable anatomic representation.

The knee is one of those joints in which conventional tomography may be quite helpful. This is particularly true in cases of tibial plateau fractures and avulsion of one or both of the major intercondylar eminences.[6] Since the surface of both tibial plateaus is not perfectly flat, a large part of it will not be tangential to the x-ray beam in the anteroposterior, lateral, and oblique radiographs of the knee. A central depressed or separated fragment may, therefore, not be seen on plain films. Tomography is useful here in determining how much of the plateau is fractured and whether or not the fragments are depressed (Fig. 5).

Before the advent of computerized tomography, conventional tomography was the preferred method of choice for evaluating the traumatized spine. The anteroposterior and lateral tomograms give an anatomic rendition of the spine, which is familiar to both clinicians and radiologists. (The CT scan with high-grade reconstruction can provide adequate coronal and sagittal sections and it excels in demonstrating osseous fragments in the spinal canal.) Transversely oriented fractures in the axial plane, such as the fracture through the base of the odontoid and the chance (seatbelt) fracture, however, may be better demonstrated by conventional tomography than by CT.

Arthrography

Most arthrography today is performed on the knee, but essentially every joint can be examined using this technique.[7–10] With regard to trauma, the knee is examined for meniscal tears, chondral or osteochondral fractures, and cruciate ligament injuries (Fig. 6). Arthrography can be performed using contrast media alone or contrast media with air; the latter is usually preferred and is known as the double-contrast arthrogram.

Arthrography can be used in other joints, such as the elbow and shoulder. Trauma may result in an osteochondral fracture that may persist and actually grow in size in the joint, or it may be resorbed. Arthrography in conjunction with tomography can be used to demonstrate both the fracture defect and the fracture fragment (Fig. 7).

RADIOISOTOPE BONE SCANNING

Bone scanning is useful in several clinical situations. It is particularly applicable in evaluating the battered child and the osteopenic individual for occult fractures. It may detect subtle trauma when the radiographic examination is negative or equivocal and there exists a strong clinical suspicion for

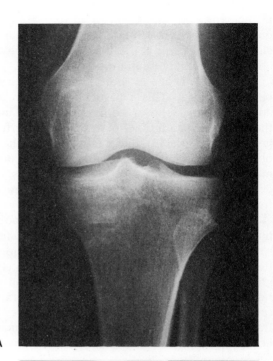

A

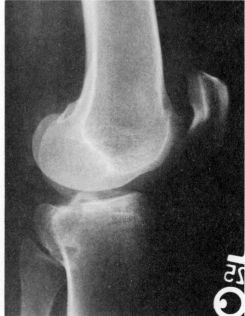

Figure 5. Patient is status post-trauma with knee pain. **A.** The anteroposterior view demonstrates bone fragments suspicious for a proximal tibial fracture. The articular surface is not disrupted in this view. **B.** The lateral view indicates that there is a plateau fracture, but the number of fragments and the degree of depression are not accurately determined. (*cont.*)

B

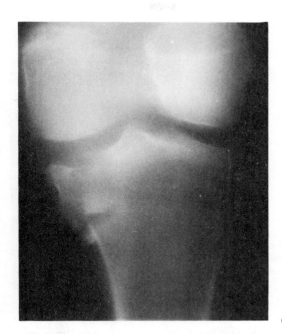

C

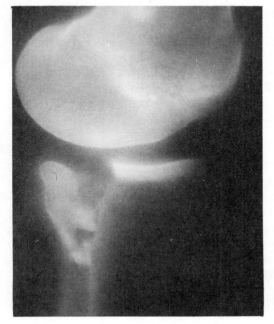

D

Figure 5. (*cont.*) **C.** The antero-posterior tomogram better delineates the fracture. **D.** The lateral tomogram best demonstrates the articular disruption, the number of fragments, and the separation and depression of the main and minor fragments.

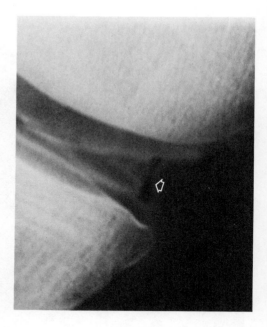

Figure 6. Double-contrast arthrogram of the knee. A vertical tear filled with air is seen in the peripheral portion of the posterior horn of the medial meniscus (arrow).

a fracture.[11] In this respect, the hip, spine, carpal scaphoid, and stress fractures are regions and situations in which bone scanning has proved most useful.[12]

Bone scanning is performed with technetium-99m-labeled phosphate or diphosphonate compounds. The bone will demonstrate increased nuclide activity at the site of fracture in 80 percent of patients within 24 hours and in 95 percent of patients within 72 hours.[11] Nearly all patients with acute fractures in whom scans become positive after 24 hours are over 65 years old. False negative scans are infrequent and occur in patients with poor nutrition or old age.[11,12]

There are three phases of fracture healing, as demonstrated by bone scanning.[11] In the first stage, diffuse activity surrounds the fracture site and the actual fracture line may be visible. This stage lasts 3 to 4 weeks. The second stage lasts approximately 8 to 12 weeks and demonstrates well-defined and intense nuclide activity at the fracture site. In the third stage, the activity gradually diminishes until the scan returns to normal. The bone scan reverts to normal sooner with rib fractures than with vertebral body or long bone fractures. The bone scan in approximately 90 percent of fractures of ribs, vertebrae, and extremities returns to normal in 2 years, and in 95 percent of the fractures it is normal in 3 years.

Bone scanning has proven itself useful in the evaluation of the possibly battered child. It not only can detect recently induced fractures but may also demonstrate fractures in various stages of healing. This sort of image in the

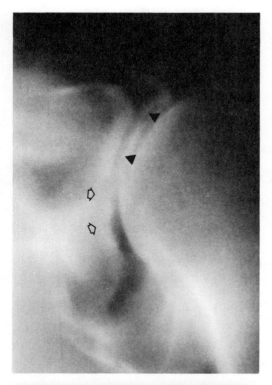

A

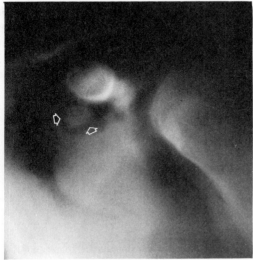

Figure 7. Double-contrast antero-posterior arthrotomograms of the shoulder in patient with history of previous anterior dislocation. (Plain film radiographs were normal.) **A.** Note loss of articular cartilage and subchondral bone from the anteroinferior lip of the glenoid (arrows). The superior aspect of the glenoid is covered by cartilage (arrowheads). **B.** One intra-articular fragment (arrows) is noted in the subscapularis bursa, which communicates with the glenohumeral joint.

B

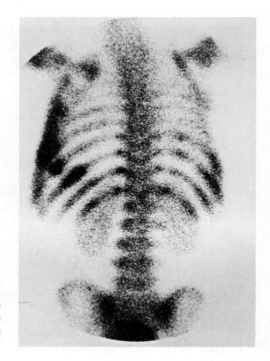

Figure 8. Bone scan of the thorax in a battered child with negative chest and rib radiographs. Different degrees of nuclide activity in the ribs indicate that multiple fractures have occurred at different times and are in variable stages of healing.

proper clinical setting is virtually pathognomonic of the battered child syndrome (Fig. 8).

Plain radiographs may fail to detect fractures in old and osteopenic individuals. This is particularly true of fractures involving the thoracolumbar spine, sacrum, and the femoral neck. As previously mentioned, the bone scan may not be positive for 2 or 3 days after the injury. Even so, bone scanning may provide more convincing evidence that a fracture has occurred (Fig. 9).

Stress fractures may go undetected radiographically for up to 2 weeks after the onset of symptoms but are demonstrated by bone scan within 1 to 3 days.[13] Many regions can develop a stress fracture, but one of the most important areas remains the femoral neck. Old and osteopenic patients, as well as those patients taking steroid medications, have an increased susceptibility to developing femoral neck stress fractures (Fig. 10). This is impor-

Figure 9. Old and osteopenic patient who has suffered a recent fall and has sacral pain. **A.** The plain anteroposterior radiograph of the sacrum demonstrates osteopenia but no fractures. Lateral radiograph and frontal and lateral tomograms (not shown) were equivocal at best. **B.** The bone scan of the sacrum demonstrates moderately intense nuclide activity in a slightly arcuate distribution characteristic of a sacral fracture.

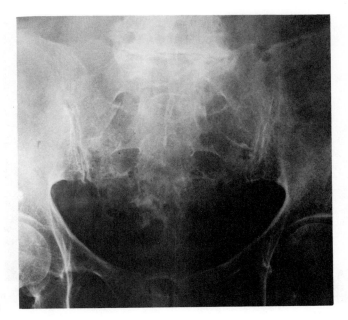

A

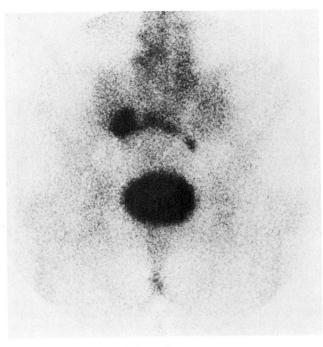

B

Figure 9.

218

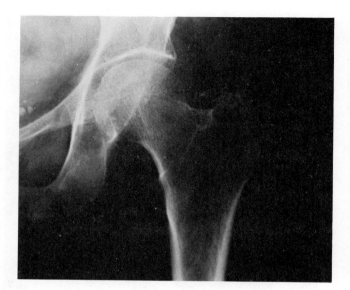

A

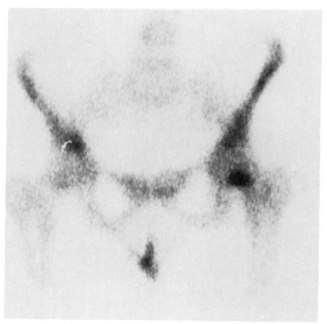

B

Figure 10. Elderly female with rheumatoid arthritis taking steroid medication with new left hip pain. The differential diagnosis with this history would be stress fracture versus steroid-induced osteonecrosis. **A.** Plain anteroposterior radiograph of the hip demonstrates osteopenia, but no other abnormalities are detected. **B.** The bone scan performed 1 day later demonstrates focal activity in the medial femoral head–neck junction consistent with a stress fracture. *(cont.)*

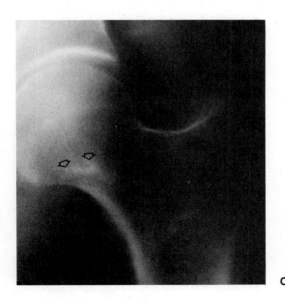

Figure 10. (*cont.*) **C.** An antero-posterior tomogram obtained 5 days after the bone scan demonstrates a linear band of sclerosis in the typical distribution for a stress fracture.

C

tant to recognize, for if activity continues, the fracture may become complete with displacement. The patient may then face a complicated course, with possible surgery and avascular necrosis of the femoral head.

COMPUTED TOMOGRAPHY

Computed tomography provides excellent visualization of special relationships in the transverse plane, and modern CT scanners have sagittal and coronal plane reconstruction capability. Regions, such as the hips, pelvis, spine, and shoulders, which previously were difficult to examine, are now exquisitely visualized. Depending upon the type of fracture, the CT scan should be performed to examine these regions in conjunction with, or in place of, conventional tomography. Fractures that are small, nondisplaced, and in the transverse plane are difficult to detect on axial CT scans and may require reconstruction. A fracture through the base of the odontoid is such an example. (On the other hand, it is precisely this type of fracture that is best visualized by conventional tomography).

The CT scan provides the best overall assessment of the traumatized spine.[14,20] Not only are the fractures easily seen, but also paraspinal bleeding and bony encroachment on the spinal canal are noted (Fig. 11). Epidural bleeding and direct trauma to the spinal cord may at times be visualized, particularly if small amounts of metrizamide are introduced into the subarachnoid space. The CT scan is also useful in patients who have developed an acute radiculopathy (Fig. 12).

The examination of pelvic and acetabular fractures has been revolution-

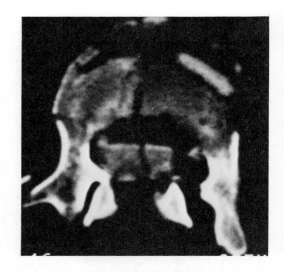

Figure 11. The patient is status post-motor vehicle accident and near complete paraplegia. Plain films (not shown) demonstrated a severe fracture of the spine, but spinal canal encroachment could not be assessed. The CT scan exquisitely demonstrates the degree of spinal canal encroachment by the retropulsed vertebral body fragment. The facet joints are disrupted, indicating instability.

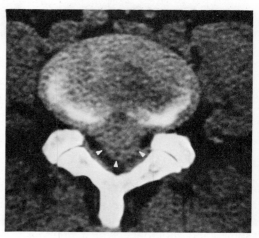

A

Figure 12. The patient suffered a sports injury, with an acute right L-5 radiculopathy. Lumbar spine series (not shown) was within normal limits. **A.** The axial CT scan demonstrates a large herniated nucleus pulposus (arrowheads) compressing the thecal sac near the origin of the L-5 roots. **B.** The midline sagittal reconstruction demonstrates the magnitude of the L4-5 herniated nucleus pulposus (arrows).

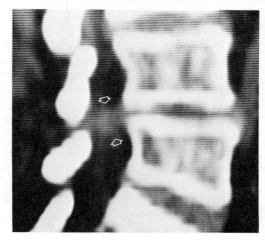

B

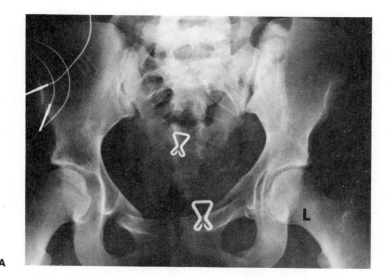

A

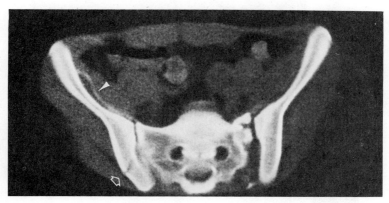

B

Figure 13. Patient with a lateral compression type pelvic fracture as a result of an auto accident. **A.** The anteroposterior radiograph demonstrates fractures of all four pubic rami and diastasis of the left sacroiliac joint. A fracture in the left sacroiliac joint region is not seen. **B.** The CT scan demonstrates the compression type fracture involving the anterior aspect of the sacrum with disruption of the posterior sacroiliac joint ligament complex. An unsuspected fracture of the right iliac bone (arrow) is noted and is also the result of the compression type fracture. Periosteal new bone (arrowhead) is present, as this scan was performed 3 weeks after the injury.

ized by the CT scan.[21–23] Many fractures that previously went undetected are now demonstrated. This is particularly true of compressive type fractures involving the anterior aspect of the sacroiliac joint region (Fig. 13). The CT scan is the best examination to assess the position of the fractures, the amount of hemorrhage, and the integrity of the pelvic ligaments.

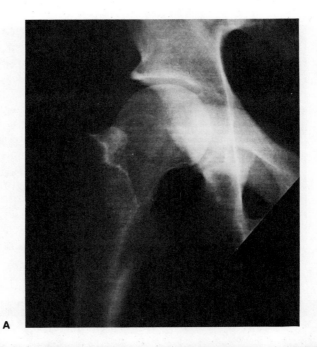

A

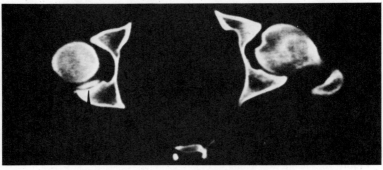

B

Figure 14. Patient after right hip dislocation and intraoperative reduction. **A.** The postoperative anteroposterior radiograph demonstrates residual widening of the hip joint. A fracture in this region is not seen. **B.** The axial CT scan demonstrates a large intra-articular fragment (arrowhead) along the inner margin of the posterior acetabulum that is preventing normal reduction of the hip joint.

Intra-articular osteochondral fracture fragments are exquisitely demonstrated by the CT scan. In the past, these fracture fragments have been notoriously difficult to detect radiographically. This is particularly true for the hip joint. If fragments are left within the hip joint, early changes of posttraumatic osteoarthritis may occur. The patient with trauma in the hip region should have a CT scan to exclude an intra-articular fracture fragment (Fig. 14).

ANGIOGRAPHY

Angiography is used to evaluate acute and late sequelae of vascular injuries. Immediate manifestations of vascular damage are hemorrhage, thrombosis,[24] and embolism, and late sequelae of vascular insufficiency arise from false aneurysms and arteriovenous fistulae.[25]

The most frequently injured vessels lie close to a bone and are held in a relatively fixed position by fascial or muscular attachments. The femoral artery in the adductor canal, the entire course of the popliteal artery, the brachial artery adjacent to the elbow and in its proximal third adjacent to the humeral shaft, the subclavian artery adjacent to the distal third of the clavicle, and the branches of the hypogastric artery in association with pelvic fractures are the arteries most frequently at risk.

REFERENCES

1. Lauge-Hansen N: Ligamentous ankle fractures: Diagnosis and treatment. *Acta Chir Scand* 1949; 97:544.
2. Hughston JC: Acute knee injuries in athletes. *Clin Orthop* 1962; 23:114.
3. Bokstrom I: Principles of vertebral tomography. *Acta Radiol* 1953; 103 (Suppl):1.
4. Maravilla KR, Cooper PR, Sklar FH: The influence of thin-section tomography on the treatment of cervical spine injuries. *Radiology* 1978; 127:131.
5. Norman A: The use of tomography in the diagnosis of skeletal disorders. *Clin Orthop* 1975; 107:139.
6. Elstrom J, Pankovich AM, Sasson H, Rodriguez J: The use of tomography in the assessment of fractures of the tibial plateau. *J Bone Joint Surg* 1976; 58:551.
7. Ala-Ketola L, Puranen J, Kolvisto E, Puupera M: Arthrography in the diagnosis of ligament injuries and classification of ankle injuries. *Radiology* 1977; 125:63.
8. Mink JH, Richardson A, Grant TT: Evaluation of glenoid labrum by double-contrast shoulder arthrography. *Am J Roentgenol* 1979; 133:883.
9. Pavlov H, Ghelman B, Warren RF: Double contrast arthrography of the elbow. *Radiology* 1979; 130:87.
10. Resnick D: Arthrography in the evaluation of arthritic disorders of the wrist. *Radiology* 1974; 113:331.
11. Matin P: The appearance of bone scans following fractures, including intermediate and long-term studies. *J Nucl Med* 1979; 20:1227.
12. Rosenthall L, Hill RO, Chuang S: Observations on the use of ^{99m}Tc-phosphate imaging in peripheral bone trauma. *Radiology* 1976; 119:637.
13. Geslien GE, Thrall JH, Espinosa JL, Older RA: Early detection of stress fractures using ^{99m}Tc-polyphosphate. *Radiology* 1976; 121:683.
14. Roub L, Drayer B: Spinal computed tomography: Limitations and applications. *Am J Roentgenol* 1979; 133:267.
15. Brant-Zawadzki M, Miller EM, Federle MP: CT in the evaluation of spine trauma. *Am J Roentgenol* 1981; 136:369–375.
16. O'Callaghan JP, Ullrich CG, Yuan HA, Kieffer SA: CT of facet distraction in

flexion injuries of the thoracolumbar spine: The "naked" facet. *Am J Roentgenol* 1980; 134:563–568.

17. Kilcoyne RF, Mack LA, King HA, et al.: Thoracolumbar spine injuries associated with vertical plunges: Reappraisal with computed tomography. *Radiology* 1983; 146:137–140.

18. Brown BM, Brant-Zawadski M, Cann CE: Dynamic CT scanning of spinal column trauma. *Am J Roentgenol* 1982; 139:1177–1181.

19. Brant-Zawadski M, Jeffrey RB Jr, Minagi H, Pitts LH: High resolution CT of thoracolumbar fractures. *Am J Roentgenol* 1982; 138:699–704.

20. Steppe R, Bellemans M, Boven F, et al.: The value of computed tomography scanning in elusive fractures of the cervical spine. *Skeletal Radiol* 1981; 6:175–178.

21. Sauser DD, Billimoria PE, Rouse GA, Mudge K: CT evaluation of hip trauma. *Am J Roentgenol* 1980; 135:269–2

22. Shirkhoda A, Brashear HR, Staab EV: Computed tomography of acetabular fractures. *Radiology* 1980; 134:683–688.

23. Walker RH, Burton DS: Computerized tomography in assessment of acetabular fractures. *J Trauma* 1982; 22:227–234.

24. Staple TM: Vascular radiological procedures in orthopedic surgery. *Clin Orthop* 1975; 107:48.

25. Bassett FH, Silver D: Arterial injury associated with fractures. *Arch Surg* 1976; 92:13.

CHAPTER 12

Biomechanics of Head Injury:
Clinical Aspects

John M. Seelig, Lawrence F. Marshall

INTRODUCTION

Trauma is the leading cause of death for Americans between the ages of 1 and 44 years of age.[1] The overwhelming predominance of victims is young males who have been involved in some type of motor vehicle accident. Injury to the brain is present at autopsy in nearly 75 percent of all road traffic deaths.[2] There are approximately 60,000 deaths per year in the United States that can be directly attributable to traumatic brain injury, while an additional 40,000 Americans are left severely disabled from their injuries and face enormous difficulties in social reintegration.[3]

In addition to those who suffer severe brain injury, it is increasingly obvious that a substantial number of patients admitted to hospitals in the United States each year with mild head injuries (400,000) suffer long-term consequences of their injury. While no more than 10 percent of these individuals are chronically disabled, a large percentage are unable to work for the first several months following what has previously been considered to be a trivial injury.[4] The identification of this large population has resulted in a remarkable escalation of interest in the pathophysiologic substrate of brain injury.

Thus, it can be seen that following head trauma, brain injuries occur in a spectrum ranging from mild to severe. Mild injuries are those that occur when a person is only temporarily dazed, and the most severe injuries involve extensive cortical and brain stem damage, leading to immediate death.

In this country alone, more than one million people each year require medical attention for various degrees of pathologic damage to the brain.[5] Because of the ubiquitous nature of brain injury and because of the tremendous heterogeneity of the disease, a high level of sophistication is now required to separate out the various factors that are responsible for the ultimate outcome in patients suffering head injury. Head injury is not a unitary phenomenon. Any modern approach to the biomechanics of head injury must concentrate on the pathophysiologic disturbances to the skull and its contents.

The most common traumatic brain injuries are due either to penetrating wounds or impact injuries that result in differential movement of the skull, meninges, and brain. In this chapter, the biomechanics of nonmissile head injuries are discussed.

THE PATHOLOGIC PROCESS

In patients suffering severe head injury, damage to the brain from impact usually results in multiple macroscopic and microscopic injuries. These injuries can be characterized as either resulting from direct impact, or as those injuries that occur later as secondary insults to an already-injured brain. These secondary insults are a product of the dynamic pathologic process that is head injury.

Although certain types of head injury may exist in isolation, more frequently combinations of injuries to the skull and brain occur. For example, epidural hematoma, once thought to be a relatively benign lesion, is frequently associated with intracranial contusion. Such contusions should be seen as evidence of a more widespread and diffuse injury.[6] The concept that impact injury results in diffuse changes within the brain parenchyma, some of which are reversible and some of which are not, is a well-established concept in the pathologic literature. It is a new one for the neurosurgeon, however. The tendency has been for clinicians who care for patients suffering acute brain injury to focus on those lesions that result in a neurologic deficit and/or that can be immediately visualized with computerized tomographic (CT) scanning.

The results of neuropsychometric testing demonstrate, however, that patients with acute focal lesions often have cognitive dysfunction indicative of a more widespread or diffuse process. The outcome in such individuals often is not a product of the focal lesion but rather of the degree of diffuse damage. Thus, damage to the brain from impact can occur from direct injury or can be indirect (secondary) and may cause focal and/or diffuse brain damage. Although one type of brain injury may exist in isolation, various combinations of injuries to the skull and intracranial contents frequently occur. Brain injury, therefore, should be viewed as a composite process with outcome dependent on exactly how each component contributes to the overall level of functioning.

DIRECT BRAIN INJURY

Direct types of injury to the head are manifested by skull fractures, shearing, tearing, lacerations, contusions, and/or hemorrhage. Skull fractures may be simple, compound, or comminuted. The amount of brain damage that occurs is dependent upon the degree of depression and the location of the fracture. Fractures are most harmful when underlying structures are lacerated, i.e., cranial nerves, blood vessels, and/or brain parenchyma. When a simple fracture occurs in the temporal bone, an epidural hematoma can develop from the laceration of the middle meningeal artery, as shown in Figure 1. On rare occasions, especially when multiple skull fractures are present, hemorrhage from the bone can cause deadly epidural hematomas or, in infants, fatal cephalohematomas. More than 75 percent of intracranial hematomas are associated with a skull fracture.[7]

It has been shown, using a large number of cadaver heads with the scalp and intracranial contents intact, that for a linear skull fracture to occur, a force of 450 to 750 inch-pounds must be generated in a striking blow to the

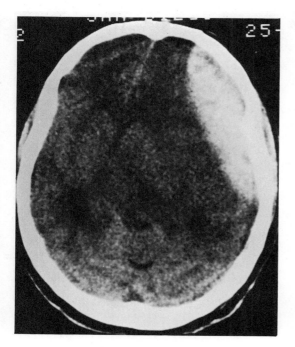

Figure 1. Epidural hematoma from middle meningeal arterial tear.

head.[8] The presence of a linear fracture is a good indication that such a force has been applied to the head; but it is not a good index of the probability of physical disruption of neural tissue. Fatal intracranial damage does not necessarily occur in the presence of a skull fracture, and, conversely, the presence of a skull fracture is not necessary for there to be severe intracranial damage.

Angular acceleration of the intracranial contents along the irregularities of the skull results in direct contusions of the brain. Contusions classically occur at the crests of gyri and are predominantly hemorrhagic during the acute postinjury period. Typically, these contusions can be demonstrated at the orbital surfaces of the frontal lobe, at the temporal tips, and at the sharp edges of the lesser wing of the sphenoid bone or within the tentorium cerebelli (Fig. 2). As the brain rebounds off these fixed bony structures, contracoup cortical contusions can also develop. Such contusions are mainly impact related and are significantly more common and severe in patients with fractures of the skull.[9]

Direct neuropathologic damage may be reversible at the cellular and subcellular level, but in more severe cases, laceration, shearing, tearing, and hemorrhage tend to occur at locations where they are irreversible. Acute subdural hematomas classically develop from lacerations of bridging cortical veins (Fig. 3). Tearing of a pial arteriole can also lead to the formation of an acute subdural hematoma.

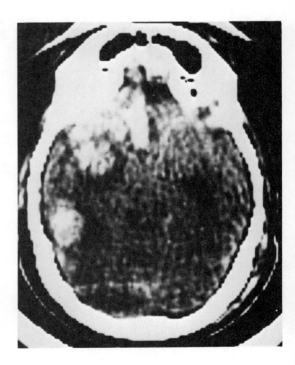

Figure 2. Typical temporaparietal contusion as a result of angular acceleration of the brain along the irregularities of the skull. Note the surrounding edema.

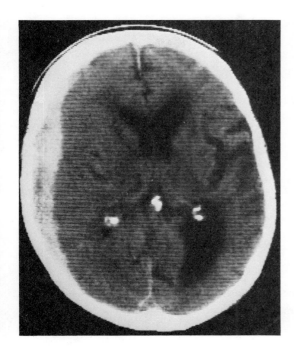

Figure 3. Acute subdural hematoma from a laceration of bridging veins.

Severe deformation of the brain rarely leads to intraventricular hemorrhage but not infrequently leads to diffuse axonal shearing, [10–15] deep intraparenchymal hemorrhage, or primary brain stem damage.[16–25] Diffuse axonal injury (DAI) was first clearly defined in 1956 by Strich[10] on the basis of characteristic lesions in the corpus callosum, the brain stem, and the white matter. These diffuse abnormalities in the white matter are manifested by axonal retraction balls, microglial stars, and long tract degeneration.[16]

DAI appears to be a form of primary brain damage. Some investigators, however, have contended that this degeneration of the white matter is secondary to hypoxic or ischemic brain damage resulting from an intracranial expanding lesion and/or from cerebral edema.[26,27]

There are three distinct pathologic features of DAI that argue in favor of this process being a form of primary brain damage.[11] The tear in the corpus callosum is the feature most easily seen without the aid of the microscope (Fig. 4). This tear usually extends over an anteroposterior distance of several centimeters and may involve the interventricular septum. Hemorrhage is characteristically a hallmark of this lesion in patients surviving longer.

Tearing of the dorsolateral quadrant of the rostral brain stem in the region of the superior cerebellar peduncle is the second distinct pathologic feature of DAI. Again, the presence of hemorrhage may appear on gross examination, but in some cases, only microscopic changes characterized by the presence of hemosiderin or a cystic scar can be identified.

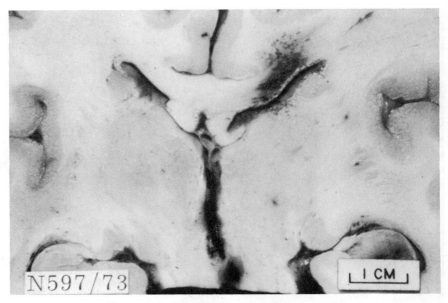

Figure 4. Corpus callosum tear in tumor material.

The third feature of DAI is only recognizable by microscopic examination of the brain. With appropriate silver impregnation techniques, numerous axonal retraction balls can be seen in the parasagittal white matter of the cerebral hemispheres and in the midbrain or rostral pons.

Isolated primary brain stem injury, though it is rare among survivors, is commonly found in fatal cases. The midbrain and pontomedullary junction are quite vulnerable to trauma.[12,16,17,19-25] The basic lesions found in victims of acute injury are transections, tears, or hemorrhages that range from visibly gross to those only demonstrable with the aid of a microscope. Tomlinson[23] has pointed out that failure to fully appreciate this has resulted in an underestimation of the true incidence of primary brain stem injury. Jellinger described them in 43.5 percent of autopsy cases,[24] Rosenblum et al. in 88 percent,[19] and Mayer in 100 percent of cases.[25]

The majority of the traumatic lesions within the midbrain were seen in the colliculi, adjacent to and including the roof of the aqueduct, in the dorsolateral quadrant of the rostral stem including the lateral mesencephalic sulcus, and along the lateral margins of the peduncles, as well as dorsal to or within the substantia nigra.

Hyperextension injury can produce tears or complete transection at the pontomedullary junction and is often associated with ring fractures around the foramen magnum. Atlanto-occipital and atlantoaxial fractures and ligamentous tears at these junctions are commonly found in this type of injury. Clinically, these lesions result in death on impact or shortly thereafter.

Rarely do patients survive this injury. Two patients in Britt's series who survived were quadriplegic with facial paralysis and communicated with residual eye and eyelid movements.[16]

INDIRECT (SECONDARY) BRAIN INSULTS

Recognition that secondary insults to the already-injured brain play a major role in the ultimate outcome of patients suffering severe head injury has increased over the past decade. The introduction of advanced emergency medical services in many regions of the United States resulted in the first attempts to study the influence of early evacuation on ultimate outcome. In our region, the incidence of uncontrollable intracranial hypertension in severely head-injured patients fell from 25 percent to approximately 12 percent.[28] There was no change in acute hospital care management during the period studied, but there had been a substantial change in initial roadside management. Thus, there appears to be a link between the early treatment of hypoxia and shock and the late development of intracranial hypertension. Harr and his colleagues at the University of Texas at Houston showed that there was a strong correlation between the presence or absence of severe hypoxia ($PaO_2 \leq 59$ mm Hg) and the later development of severe intracranial hypertension (ICP >25 mm Hg) in patients in whom blood gases were sampled at the scene and prior to road or aircraft evacuation.[29] Coupled with our findings, this observation provides strong evidence that prehospital care substantially influences secondary insults to the brain and further focuses our attention on their role in exacerbating primary brain injury.

Cellular damage from metabolic derangements can occur as a result of hypercapnia, hypoxia, acidosis, hyperthermia, and hyponatremia and give the clinician a very small margin of error from which to work in the care of patients suffering acute brain injury. Even relatively trivial episodes of hypoxia or hypercapnia may result in catastrophic deterioration in such patients.

Major indirect or secondary brain insults following impact injuries can result from compression of the brain by a slowly expanding mass lesion, delayed intracerebral contusion, cellular damage from metabolic derangements, compromise of cerebral perfusion pressure (CPP), or infection. Any one or a combination of these secondary insults can be more devastating to the brain than the initial impact injury.

An expanding mass lesion, whether from hematoma or edema formation, can compress vital structures directly or indirectly by herniation. Delayed intracerebral contusions have been reported to occur in most series of severe head injury at a rate of 5 to 10 percent.[30-32] Cooper et al., however, reported a 33 percent incidence in their series, with a 79 percent mortality rate.[33]

Delayed intracranial contusions usually appear within the first 48 hours of injury but can be delayed for as long as 5 to 10 days. They tend to occur primarily in the frontal and temporal lobes and are less likely to occur in the thalamus or basal ganglia. Factors associated with delayed intracerebral contusions include the following: prior surgical procedure(s), hypoxia, acidosis, multiple systemic injury, coagulopathy, and severity of injury. These associated factors have led to four schools of thought regarding the etiology of delayed contusion: (1) The release of the tamponade effect after removal of an intracranial hematoma. Interestingly, two thirds of delayed contusion hematomas are contralateral to the surgically removed hematoma. (2) Weakness in the cerebral arteriolar wall. This weakness may be a product of endothelial cell damage secondary to tissue acidosis[34] or actually a sheared blood vessel. (3) Growing areas of contusion not appreciated on the initial CT scan, especially at the base of the frontal lobes or anterior temporal poles, are frequently seen. The mechanism of their production is not well understood. In part, mass effect is secondary to late edema, as shown in Figure 2, but in some cases the hematoma also enlarges, as seen in Figures 5A and 5B. (4) Coagulopathy following structural brain injury. Kaufman and co-workers have reported a 75 percent incidence of abnormal coagulation studies in patients with Glasgow Coma Scale scores of 8 or less.[35] While a clinical picture of overt disseminated intravascular coagulopathy (DIC) is rare, more modest changes in clotting factors may play a significant role in increased hematoma size and in the occurrence of delayed hematomas.

Brain swelling and edema can be attributed both to loss of vasomotor tone and to increased permeability of cerebral capillaries. Under normal circumstances, the brain is capable of regulating cerebral blood flow within very narrow limits. Normal flow is usually 50ml/100grams/minute, with cerebral metabolic rate being coupled to flow. Following brain injury this autoregulatory phenomenon is often lost, and we speak of either a loss of vasomotor tone or autoregulation. As the majority of patients suffering substantial brain injury are hypertensive following their injuries, it can be seen readily that this loss can result in increased intravascular volume and in a subsequent increase in intracranial pressure (ICP).

There is recent evidence that increased permeability of cerebral capillaries accompanies acute brain injury. This permeability appears to be particularly important in the development of edema and in the increase in brain water content adjacent to contusions, while it plays a lesser a role in diffuse swelling and in unilateral hemispheric swelling.[11] Brain swelling following removal of an acute subdural hematoma, for example, appears to be the product of vasomotor paralysis, i.e., the loss of autoregulation.

In children, Bruce and his colleagues have described a very specific syndrome in which diffuse bilateral hemispheric swelling occurs, resulting in almost complete obliteration of the ventricular system and the presence of very high ICP.[36] Cerebral blood flow studies have demonstrated diffuse hyperemia in these children, indicating that most of the increase in brain

A

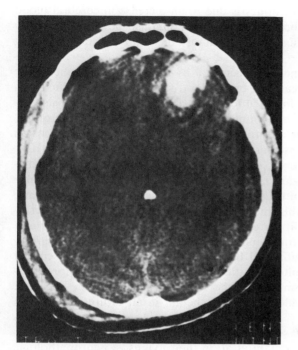

B

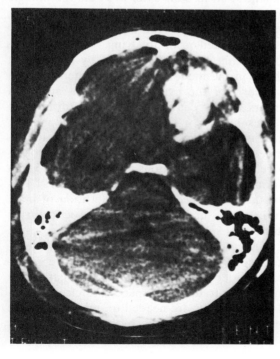

Figure 5. A. Initial CT demonstrating right frontal intracerebral hematoma. **B.** Follow-up CT scan showing enlargement of right frontal intracerebral hematoma.

bulk in this particular type of injury is due to an increase in cerebral blood volume. Such increases in brain blood volume can also be exacerbated by the presence of tissue acidosis, which almost inevitably occurs following severe head injury.

Hyperthermia, which increases cerebral as well as systemic metabolic rate, may also result in increases in brain blood volume. Figure 6 illustrates the vicious cycle that can be initiated following acute head injury and also describes, in schematic form, the dilemma that faces the treating neurosurgeon.

In the face of a loss of autoregulation, systemic arterial hypertension can increase cerebral intravascular volume, resulting in an increase in ICP. Systemic hypotension, however, causes a decrease in perfusion and provision of substrate to the brain and can result in brain tissue acidosis, cerebrovasal dilatation, and, ultimately, in cellular ischemia. The dilemma, therefore, is to provide within the clinical setting an advantageous physiologic milieu that will result in the optimal environment for patient recovery.

In experimental head injury models, the greater the severity of injury the more likely autoregulation will be lost and the greater the subsequent effects of derangement in autoregulation will be.[37] One can therefore envision the occurrence of circumstances where CPP, which would be adequate in the uninjured brain, would result in ischemia in an injured brain with impaired autoregulation.

The high frequency of secondary ischemia following head injury has been well documented by Graham et al. in their extensive neuropathologic studies.[38] Reilly, working with the group in Glasgow, has demonstrated that the incidence of secondary ischemic injuries to the brain was remarkably high (90 percent) in patients who talked prior to dying.[39] These studies indicate that secondary brain ischemia, frequently caused by intracranial hypertension, is a major problem in the modern day management of such patients. This lends strong support for the extremely vigorous approach many clinical head injury centers in this country have taken toward the treatment of intracranial hypertension.

What remains to be defined is what level of elevation in ICP is tolerable. Recent experience at our center has demonstrated that there are occasional patients in whom relatively modest elevations of ICP (20 to 25 mm Hg) result in transtentorial herniation and brain stem compression.[40] This has led to the development of the new concept that the location of mass effect in patients suffering brain injury may be as important a factor as overall pressure within the intracranial space.

Our observations strongly suggest that multiple compartmental pressures may exist within the intracranial space. Previous workers have generally resisted such a concept except under circumstances where herniation has occurred and the supratentorial and infratentorial spaces are clearly separated. Laboratory and clinical investigations are needed to further delineate the mechanisms by which secondary insults to the brain occur and how differential mass effects can exacerbate the primary brain injury.

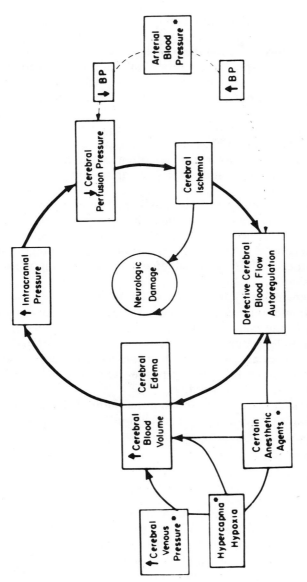

Figure 6. The complex interplay of factors that occurs following injury is shown.

DIAGNOSIS

The advent of CT scanning has revolutionized the early diagnosis of intra-cranial pathology following traumatic brain injury. Routine sequential CT scanning permits the serial evaluation of primary and secondary neuropatho-logic changes, such as brain swelling, delayed intracerebral hematomas, and hydrocephalus. Diffuse axonal shearing has been recognized on CT scanning by eccentric hemorrhage in the corpus callosum, diffuse cerebral swelling, and hemorrhage in the region of the third ventricle, cerebral white matter, and brain stem.[41-44] Continuous ICP monitoring is a good indicator of the expansion of mass lesions, changes in brain shift, abnormalities in cerebral hemodynamics, and systemic metabolic derangements (i.e., hypoxia, hypo-natremia, and hyperthermia). Patients dying of head injury without raised ICP usually have diffuse axonal injury often complicated by hypoxic brain damage, CNS infection, other systemic complications, or a combination of these.

In comatose patients in whom muscle relaxants are used for controlled hyperventilation, the clinical assessment is of limited value. Multimodality-evoked potentials of the visual, auditory, and sematosensory responses are an excellent means of assessing the electrophysiologic integrity of the vari-ous sensory tracts.[45-47] Evoked potentials aid in localizing major brain and brain stem lesions.

Rosenblum et al.[19] have correlated neuroanatomic and pathologic changes in the midbrain of head-injured victims with early evoked potential abnormalities in the midbrain or pons. Thirty-eight of forty-four patients in the series had midbrain or pontine abnormalities demonstrated by multimo-dality-evoked potential criteria. All had characteristic and appropriately lo-cated pontine or midbrain lesions at postmortem. The surviving six patients were severely disabled or vegetative.

Refinement of these diagnostic tools has improved our understanding of the biomechanics of head injury and allowed for careful scrunity of treat-ment modalities.

TREATMENT

The critical factor determining outcome in patients sustaining nonmissile head injury is the severity of damage to the brain. The consequences of head injury, both direct and secondary, must be fully appreciated before effective treatment protocols can be derived.

Table 1 summarizes the present methods of management in our institu-tion for patients who have sustained severe head injury. It must be recalled that the basic principles of treatment in patients with severe brain injury include preventing secondary insults, particularly intracranial hypertension.

CPP, which can be roughly calculated by subtracting the intracranial

TABLE 1. MANAGEMENT OF SEVERELY HEAD-INJURED ADULT PATIENTS WITH INTRACRANIAL HYPERTENSION

1. Head up 30 degrees and in neutral plane
2. Controlled ventilation to a PCO_2 of 25–30 mm Hg with adequate paralysis and sedation
3. Maintain PO_2 > 70 mm Hg
4. Maintain SAP between 100 and 160 mm Hg systolic
5. Maintain normothermia
6. Use prophylactic anticonvulsants
7. Maintain fluid balance with 0.5 normal saline

For ICP Control

8. Ventricular drainage if possible
9. Dexamethasone 20 mg IV every 6 hours
10. Mannitol 0.25 gm–1 gm/kg as needed
11. Avoid anesthetics and other drugs that are cerebral vasodilators

pressure from the systemic arterial pressure (CPP = SAP − ICP), serves as a crude guide for judging the adequacy of substrate supply. As it is relatively simple to calculate and easily available, CPP can often be a useful tool. However, because of the heterogeneity in perfusion to the injured brain, one must be careful to err on the side of caution when concluding that cerebral perfusion is adequate when the CPP exceeds 65 to 70 mm Hg. Futhermore, it should be emphasized that clinical management must be instined early; otherwise the outcome will often be poor or nil because valuable time has been lost in returning the patient to an adequate physiologic milieu.

Approximately 35 to 40 percent of patients with severe brain injuries will require surgical intervention for removal of intracranial hematomas or for debridement. To avoid the devastating consequences of herniation, uncontrolled brain swelling, and refactory elevations in ICP, the evacuation of mass lesions as promptly as possible from the time of initial impact[48,49] or before neurologic deterioration occurs[6] is essential.

Rapid triage to a qualified and prepared medical facility, effective and efficient evaluation, and accurate diagnosis are also mandatory.

Once the patient reaches such a facility, the principal objectives of clinical assessment are to determine which insults are treatable and which complications are preventable. Numerous articles have been written elucidating the management of patients with head injury and anticipating the various direct and indirect consequences of head injury.[50,51] The rules of resuscitation for any traumatized patient apply in head injury as well. Of utmost importance is that patients be transported without delay to an optimal care hospital where CT scanning and comprehensive neurosurgical care are immediately available. Refractory shock and/or obstruction of the airway should be the only indications for transport to a local receiving hospital.

A key factor influencing patient outcome is the immediate availability of neurosurgical facilities with provision for the continued monitoring and treatment of elevated ICP.

Besides surgical removal of intracranial hematomas, most therapies are aimed at control of ICP. Treatment of increased ICP with sedation, mechanical hyperventilation, CSF drainage, diuretics, mannitol, and barbiturates is usually effective if instituted prior to irreversible secondary ischemic damage.

PREVENTION

Brain stem damage is often a secondary injury, as Mitchell and Adams have pointed out.[21] In many instances, these secondary injuries can be avoided by paying prompt attention to the principles previously alluded to here. It is important to recognize that the brain stem is susceptible to direct biomechanical injury engendered at the moment of impact. As a matter of priority, the neurosurgeon must pay attention to any means of preventing such impacts and should become involved in the development of improved safety equipment for both automobiles and motorcycles.

Preventive measures, such as helmets for the motorcyclist and restraining devices for those in automobiles, are absolutely essential if a substantial reduction in mortality in brain injury is to occur. Two thirds of the deaths from motor vehicle accidents in the United States occur prior to patients reaching the hospital, indicating that these patients are unsalvageable from the moment of impact in almost every instance. The motorcyclist runs five times the risk of having a fatal accident per mile traveled as does a person traveling in a car or truck. The total lack of protection for most motorcyclists places them at substantially increased risk of death or severe head injury in any accident.[52] In Kansas, 93 percent of severely injured motorcyclists without a helmet had fatal head injuries compared to a 33 percent fatality rate for those with helmets. Watson et al. estimated that the repeal of helmet laws in 1975 resulted in a 40 percent increase in motorcycle-related fatalities.[53] The Department of Transportation, in a summary of the relationship between head injury and the wearing of a helmet, noted twice as many head injuries and three to four times as many fatalities in the nonhelmeted group.[54]

For those traveling in automobiles, any device preventing the passenger from striking his or her head on the interior surface of a vehicle should reduce head injury deaths. In 28,000 automobile accidents reported by Bohlin, no fatalities were noted where the vehicle was traveling less than 100 kilometers per hour and upper torso restraints were used.[55] A compulsory seat belt law was passed in Australia at the end of 1971.[56] In 1972, a 25 percent drop in automobile deaths occurred when only 75 percent of the passengers complied with the new law. From these data it is clear that head injuries are decreased by the use of upper torso restraints.

IMPACT ON THE NATION

The emphasis in this chapter has been on the biomechanics of brain injury and clinical assessment of the effects of severe head injury. Neurotraumatologists have become increasingly interested in more moderate or mild injuries because of the pioneering work of Gronwall et al. in New Zealand[57,58] and the more recent studies of Rimel et al. at the University of Virginia.[59,60] These investigators have demonstrated the remarkably high incidence of social and intellectual dysfunction in patients suffering mild brain injury.

Many of the patients studied by Gronwall et al. were discharged from the accident ward without having been admitted. In the study of Rimel et al., minor head injury patients were defined on the basis of a normal Glasgow Coma Scale (GCS) score within 48 hours of injury and a normal CT scan. Half of these patients had a substantial diminution in function (at a level of 2 SD or more below national norms) in such areas as short-term memory, abstract learning, and information processing. In an uncompleted study of the data from the Comprehensive CNS Trauma Centers in the United States, similar findings have been observed in three independent communities.

These data strongly indicate that the consequences of minor head injury are substantial. If one extrapolates the findings of our Comprehensive Center and the Rimel study, approximately 135,000 of the 400,000 patients hospitalized with minor head injury will be out of work for 3 months following injury. In the Gronwall experience,[58] approximately 10 percent of this entire population will be chronically disabled and will never return to their formerly productive roles in society.

What is fascinating about the observations of our New Zealand colleagues is the fact that students who suffered trivial head injuries years prior to testing demonstrated a substantial deterioration in cognitive function when placed in a simulated high-altitude environment.[61] These data indicate that minor head injury results in irreversible brain injury that can be elicited by mild to moderate hypoxic stress. This is the first clinical study to demonstrate that the axonal injuries seen in animals concussed for less than 10 minutes in the Philadelphia Head Accelerating Device appear to be analogous to minor head injury in man. Whether these axonal injuries are the cause of the later deterioration that appears under circumstances of hypoxic stress is not known. They may simply be markers of much more subtle organelle or neurochemical changes in the brain.

Nevertheless, the concept that relatively minor injuries may result in long-term sequelae for patients suffering brain injuries has enormous implications for society, as well as for individuals suffering such injuries. It is clear that the neurosurgeon cannot and will not be able to cope with the majority of these patients. Thus, it is incumbent upon us both to prevent what appear to be relatively minor brain insults and to develop mechanisms to return these people to a useful role in society as soon as possible after such brain insults occur. As we come to better understand the biomechanics of brain injury and the steps we may take to prevent it, we should not lose sight of the fact that a large majority of the patients suffering cranial trauma in the United States and elsewhere sustain modest injuries. The resulting neurologic deficits may have overwhelming social consequences during the first 2 to 3 months post-injury for many and, for some, for the rest of their lives.

The focus of clinical care on the severely brain injured was appropriate when acute care delivery systems and intensive care were in their relatively early stages of development. Consolidation of our gains in these areas has been rapid in many regions, and it is now appropriate that the whole spectrum of brain injury gain the attention of bioengineers, neurosurgeons, and the political and legislative bodies.

Billions of dollars of income and productivity are lost each year as an economic by-product of the occurrence of traumatic brain injury. In 1977, Americans spent 2,120,000 days in the hospital because of head injuries (400,000 patients hospitalized for on average of 5.3 days) at a cost of $3 billion.[62] This loss could be greatly curtailed if our knowledge of the biomechanics of brain injury and the use of our resources to treat and prevent head trauma were fully applied. The facts show that use of such simple preventive measures as the implementation of mandatory seat restraints could go farther in reducing the toll of head injury than any major advance in intensive care.

REFERENCES

1. National Center for Health Statistics. Monthly vital statistics report. *Annual Report: Final Mortality Statistics,* 1978; vol 29, no 6, suppl 2. (DHHS publication no. (PHS)80-1120). Washington, DC, US Government Printing Office, 1980; 1–39.
2. Kihlberg JK: Head injury in automobile accidents, in Careness WE, Walker EA (eds): *Head Injury.* Philadelphia, Lippincott, 1966; pp 27–36.
3. Kalsbeck WD, McLaurin RL, Harris BSH, et al.: The national head and spinal cord injury survey. *J Neurosurg* 1980; 53:519–531.
4. Marshall LF, Ruff R: Disability of minor head injury: Unrecognized morbidity. Unpublished observations.
5. Caveness WF: Incidence of craniocerebral trauma in the United States with trend from 1970–1975, in Thompson RA, Green JR (eds): *Advances in Neurology. Complication of Central Nervous System Trauma.* New York, Raven Press, 1979; vol 22, p 1.

6. Seelig JM, Marshall LF, Toutant SM, et al.: Traumatic acute epidural hematoma—Unrecognized lethality in comatose patients. *Neurosurgery*, accepted for publication, 1983.
7. Jennett B, Teasdale G: *Management of Head Injury*. Philadelphia, Davis, 1981; p 1.
8. Gurdjian EG, Lissner HR, Hodgson VR, et al.: Mechanisms at head injury. *Clin Neurosurg* 1966; 12:112–128.
9. Adams JH, Scott G, Parker LS, et al.: The contusion index: A quantitative approach to cerebral contusions of head injury. *Neuropathol Appl Neurobiol* 1980; 6:319–324.
10. Strich SJ: Diffuse degeneration of cerebral white matter in severe dementia following head injury. *J Neurol Neurosurg Psychiatry* 1956; 19:163–185.
11. Adams JH, Gennarelli TA, Graham DI: Brain damage in non-missile head injury: Observations in man and subhuman primates, in Smith WT, Cavanaugh JB (eds): *Recent Advances in Neuropathology*. Edinburgh, Churchill Livingstone, 1982; pp 165–190.
12. Strich SJ: Shearing of nerve fibres as a cause of brain damage due to head injury. *Lancet* 1961; 2:443–448.
13. Adams JH, Graham DI, Murray LS, et al.: Diffuse axonal injury due to nonmissile head injury in humans: An analysis of 45 cases. *Ann Neurol* 1982; 12:557–563.
14. Adams JH, Graham DI, Scott G, et al.: Brain damage in fatal non-missile head injury. *J Clin Pathol* 1980; 33:1132–1145.
15. Gennarelli TA, Thibault LE, Adams JH, et al.: Diffuse axonal injury and traumatic coma in the primate. *Ann Neurol* 1982; 12:564–574.
16. Britt RH, Hernick MK, Mason RT, et al.: Traumatic lesions of the pontomedullary junction. *Neurosurgery*. 1980; 6:623–631.
17. Adams JH, Mitchell DE, Graham DI, et al.: Diffuse brain damage of immediate impact type: Its relationship to "primary brain-stem damage" in head injury. *Brain* 1977; 100:489–502.
18. Budzilovich GH: On pathogenesis of primary lesions in blunt head trauma with special reference to the brain stem injuries, in McLaurin RL (ed): *Head Injuries: Second Chicago Symposium on Neural Trauma*. New York, Grune & Stratton, 1976; pp 39–43.
19. Rosenblum WI, Greenberg RP, Seelig JM, et al.: Midbrain lesions: Frequent and significant prognostic feature in closed head injury. *Neurosurgery* 1981; 9(6):613–620.
20. Lindenberg R: Significance of the tentorium in head injuries from blunt forces. *Clin Neurosurg* 1964; 12:129–142.
21. Mitchell DE, Adams JH: Primary focal impact damage to the brainstem in blunt head injuries: Does it exist? *Lancet* 1973; 2:215–218.
22. Crompton MR: Brainstem lesions due to closed head injury. *Lancet* 1971; 1:669–673.
23. Tomlinson BE: Brain stem lesions after head injury. *J Clin Pathol* 1970; 4(Suppl):154–165.
24. Jellinger K: Haufigkeit and Pathogenese zentraler Hirnlasionen mach stumpfer Gewalteinwirkung auf den Schadel. *Wien Z Nervenheildk* 1967; 25:223–249.
25. Mayer E Th: Zentrale Hirnschaden nach Einwirkung stumpfer Gewalt auf den Schadel. *Arch Psychiatr Neurol* 1967; 210:238–262.

26. Adams RD, Discussion IN, Walker AE, et al. (eds): *The Late Effects of Head Injury*. Springfield, Ill., Thomas, 1969; p 524.
27. Jellinger K: Pathology and pathogenesis of apallic syndromes following closed head injuries, in One GD, Gerstewbrand F, Lucking CH, et al. (eds): *The Apallic Syndrome*. Berlin, Springer-Verlag, 1977; p 88.
28. Marshall LF, Bowers SA: Medical management of intracranial pressure, in Cooper PR (ed): *Head Injury*. Baltimore, Williams & Wilkins, 1982.
29. Harr FL, Phillips S, Huchton JI, et al.: The incidence and significance of early hypoxemia in head injury patients. *Trans Am Assoc Neurosurg* 1981.
30. Diaz FG, Yock DH Jr, Larson D, et al.: Early diagnosis of delayed posttraumatic intracerebral hematomas. *J Neurosurg* 1979; 50:217–223.
31. Jamieson KG, Yelland JDN: Traumatic intracerebral hematoma: Report of 63 surgically treated cases. *J Neurosurg* 1972; 37:528–532.
32. Gudeman SK, Kishore PRS, Miller JD, et al.: The genesis and significance of delayed traumatic intracerebral hematoma. *Neurosurgery* 1979; 5:309–313.
33. Cooper PR, Maravilla K, Moody S, et al.: Serial computerized tomographic scanning and the prognosis of severe injury. *Neurosurgery* 1979; 5:566–569.
34. Povlishock JT, Becker DP, Sullivan HG, et al: Vascular permeability alterations to horseradish peroxidase in experimental brain injury. *Brain Res* 1978; 153:223–239.
35. Kaufman MH, Moake JL, Olson JD, et al.: Delayed and recurrent intracranial hematomas related to disseminated intravascular clotting and fibrinolysis in head injury. *Neurosurgery* 1980; 7:445–449.
36. Bruce DA, Alavi A, Bilaniuk L, et al.: Diffuse cerebral swelling following head injuries in children: The syndrome of "malignant brain edema." *J Neurosurg* 1981; 54:170–178.
37. Seelig JM, Lewelt W, Jenkins JD, et al.: Autoregulation of CBF to changes in arterial and intracranial pressure after experimental head injury. *Intracranial Press V*. Berlin, Springer-Verlag, 1983; pp 487–489.
38. Graham DI, Adams JH, Doyle D: Ischaemic brain damage in fatal non-missile head injuries. *J Neurol Sci* 1978; 39:213–234.
39. Reilly PL, Graham DI, Adams JH, et al.: Patients with head injury who talk and die. *Lancet* 1975; 2:375–377.
40. Marshall LF, Barba D, Toole BM, et al.: The oval pupil: Clinical significance and relationship to intracranial hypertension. *J Neurosurg* 1983; 58:566–568.
41. Bruce DA, Raphaely RC, Goldberg AI, et al.: Pathophysiology, treatment and outcome following severe head injury in children. *Child Brain* 1979; 5:174–191.
42. Zimmerman RA, Larissa TB, Gennarelli T: Computed tomography of shearing injuries of the cerebral white matter. *Radiology* 1978; 127:393–396.
43. Zimmerman RA, Bilaniuk LT, Bruce DA, et al.: Computed tomography of pediatric head traumas. Acute general cerebral swelling. *J Radiol* 1978; 126:403–408.
44. Cooper PR, Maravilla K, Kirkpatrick J, et al.: Traumatically induced brain stem hemorrhage and the computerized tomographic scan: Clinical, pathological, and experimental observations. *Neurosurgery* 1979; 4:115–124.
45. Greenberg RP, Becker DP, Miller JD, et al.: Evaluation of brain function in severe human head trauma with multimodality evoked potentials: Part 2. Localization of brain dysfunction and correlation with post-traumatic neurological conditions. *J Neurosurg* 1977; 47:163–177.

46. Greenberg RP, Mayer DJ, Becker DP, et al.: Evaluation of brain function in severe human head trauma with multimodality evoked potentials: Part I. Evoked brain-injury potentials, methods, and analysis. *J Neurosurg* 1977; 47:150–162.
47. Greenberg RP, Newlon PG, Hyatt MS, et al.: Prognostic implications of early multimodality evoked potentials in severe head injury patients: A prospective study. *J Neurosurg* 1981; 55: 227–236.
48. Becker DP, Miller JD, Ward JD, et al.: The outcome from severe head injury with early diagnosis and intensive management. *J Neurosurg* 1977; 47:291–302.
49. Seelig JM, Becker DP, Miller JD, et al.: Traumatic acute subdural hematoma: Major mortality reduction in comatose patients treated within four hours. *N Eng J Med* 1981; 304:1511–1518.
50. Miller JD, Sweet RC, Narayan R, et al.: Early insults to the injured brain. *JAMA* 1978; 240:439.
51. Miller JD, Becker DP, Ward JD, et al.: Significance of intracranial hypertension in severe head injury. *J Neurosurg* 1977; 47:503–516.
52. National Safety Council: *Motorcycle Facts.* Chicago, National Safety Council, 1978.
53. Watson GS, Zador PL, Wilks A: The repeal of helmet use laws and increased motorcyclist mortality in the United States, 1975–1978. *Am J Public Health* 1980; 70:579–585.
54. US Department of Transportation; National Highway Traffic Safety Administration: The effect of motorcycle helmet usage on head injuries, and the effect of usage laws on helmet wearing rates. A preliminary report. Washington, DC, US Government Printing Office, 1979.
55. Bohlin NI: A statistical analysis of 28,000 accident cases with emphasis on occupant restraint value, in *Proceedings of the Eleventh Stapp Car Crash Conference.* New York, Society of Automotive Engineers, 1967, pp 455–478.
56. Henderson M, Wood R: Compulsory wearing of seat belts in New South Wales, Australia. An evaluation of its effect on vehicle occupant deaths in the first year. *Med J Aust* 1973; 2:797–801.
57. Gronwall D, Sampson H: *The Psychological Effects of Concussion.* Auckland, Auckland University Press, 1974.
58. Gronwall D, Wrightson P: Delayed recovery of intellectual functions after minor head injury. *Lancet* 1975; 2:955–997.
59. Rimel RW, Giordani B, Barth JT, et al.: Disability caused by minor head injury. *Neurosurgery* 1981; 9:221–228.
60. Barth JT, Macciocchi SW, Giornani B, et al.: Neuropsychological sequelae of minor head injury. *Neurosurgery* 1983; 13:529–533.
61. Ewing R, McCarthy D, Gronwall D, et al.: Persisting effects of minor head injury observable during hypoxic stress. *Clin Neuropsychol* 1980; 2:147–155.
62. US Department of Health, Education, and Welfare; Public Health Service Office of Health Research, Statistics and Technology. Utilization of short-stay hospitals. Annual Summary of the United States, 1977, series 13, no 41, Hyattsville, Md., National Center for Health Statistics, 1979.

CHAPTER 13

Biomechanics of Head Injury:
Experimental Aspects

Ayub K. Ommaya

INTRODUCTION

The experimental study of head injury mechanisms is unique because it examines the effects of damage to a part of the body containing the organ that controls all other parts, as well as itself and the flow of information into and out of the body as a whole via its sensory receptors and motor effectors. The highest level of brain function constitutes the state of consciousness that may be defined as a state of self-reflective and intentional actions, one measure of which is the capacity to receive, record, and recall new information over time (memory). Experimental and theoretical studies of head injury use a combination of biomechanical and physiopathologic techniques not only to prevent (or minimize) the effects of obvious and visible damage, such as skull fractures and brain contusions, but, most importantly, to prevent the loss or disturbance of consciousness. In general, such disturbances of consciousness are produced either directly (and dynamically) via the mechanisms of reversible or irreversible concussive brain injury on neural elements or indirectly (and statically) via either the relatively slower volumetric distortions and pressure gradients induced by skull fracture, intracranial bleeding, and cerebral edema or by the onset of delayed tissue destruction secondary to vascular and metabolic factors (including hypoxia and ischemia), many of which are as yet unknown but potentially reversible. This latter aspect, important as it is for the final outcome, will not be reviewed in this chapter. The primacy of the traumatic disturbances of consciousness are clearly recognized in most current classifications of head injury types and in most systems of grading head injury severity.

Any listing of the types (and lesions) of head injuries, such as is given in Table 1, shows only their anatomic loci and, while suggesting some of their physiopathologic effects, does not clearly indicate their effects on consciousness. It is important, therefore, to relate each injury type to its mechanism and to show how the functional effects follow, i.e., to connect the trauma input biomechanics to the physiopathologic response and its contribution to the outcome for the patient. The two most commonly used measurement tools for estimating head injury severity are the Glasgow Coma Scale and the Abbreviated Injury Scale. The first is used primarily by clinicians to record the intensity of coma and, by repeated use, can also show its duration. The second is a simple estimate of the threat to life caused by the injury at an early stage of the trauma and is used primarily by motor vehicle accident investigators. It is useful, therefore, to attempt to bridge the gap between these two disparate methods of grading head injury severity, which I will do later in this chapter.

TABLE 1. TYPES OF HEAD INJURY LESIONS

I. Scalp
  A. Bruise (leakage of blood from a vessel into adjacent tissue)
  B. Abrasion (traumatic removal of some outer layers of scalp)
  C. Laceration (cutting injury, tearing of scalp)
  D. Avulsion (extreme laceration causing peeling of whole scalp)

II. Skull
  A. Suture separation (diastasis, more often in younger skulls)
  B. Indentation (e.g., ping-pong fracture, usually in younger skulls)
  C. Linear fracture (may occur at points remote from impact location due to tensile stresses generated by sudden return of skull to its original shape after deformation)
  D. Depressed fracture (may be accompanied by perforation, fragmentation, or comminution of skull)
  E. Crushed skull (massive comminution, usually due to extreme static loading)

II. Extracerebral bleeding (focal or diffuse)
  A. Subarachnoid hemorrhage
  B. Epidural hematoma (with skull fracture in 90% of cases)
  C. Subdural hematoma (with skull fracture in 50% of acute cases—usually due to torn bridging veins)

IV. Brain tissue damage (neural and/or vascular)
  A. Brain concussive injuries (grades I through VI, with grade III including classic cerebral concussion and associated with increasing intensity and distribution of DAI—see Table 2)
  B. Brain contusions (bruises of brain tissues located at any site, e.g., cortical, intracerebral, brain stem)
  C. Intracerebral hematoma (visible intracerebral blood clots)
  D. Cerebral laceration (visible tearing of brain tissues)

CURRENT CONCEPTS OF HEAD INJURY MECHANISMS

We have previously presented a paradigm for head injury mechanisms (Fig. 1).[1] This paradigm will be used to discuss the current concepts of how the various types of head injuries listed in Table 1 are produced and will also serve to summarize the current hypotheses seeking to integrate knowledge in this field. Modern concepts owe much to the work of Holbourn, Gurdjian, Symonds, and Sabina Strich, although two of the most fundamental facts about the nature of head injuries are very ancient. Thus, Al-Razi (Rhazes) and Abu Ibn Sina (Avicenna) on the basis of clinical observations alone were probably the first to recognize and distinguish between the causation and effects of skull fracture, brain wounds, and cerebral concussion in the tenth century.[2] Holbourn was probably the first physicist who, on the basis of physical and theoretical modeling, was able to present a rational and testable hypothesis for the mechanics of head injuries.[3] Gurdjian, working closely with Lissner and Evans, pioneered the experimental study of skull fractures and brain injuries using linear accelerometers. Their work, later developed by others, led to the Wayne State Curve and via the work of Gadd to the current head injury criterion (HIC).[4-7] Strich made the original observations in 1961 on diffuse axonal injury (DAI) in severely head-injured patients, a lesion she postulated to be a significant factor in the pathogenesis of symptoms and signs in the final outcome after head injuries of lesser severity.[8] This prediction was supported by Oppenheimer, who in 1968 found histologic indicators of such tissue damage in patients with minor concussive head injuries who died of other causes.[9] The classic review of theoretical, clinical, experimental, and pathologic data on concussive brain injuries by Sir Charles Symonds in 1962 was the first statement of hypothesis that has proved to be very fruitful in generating new experiments, namely that the term "cerebral concussion" or "concussive brain injuries":

> should not be confined to cases in which there is immediate loss of consciousness with rapid and complete recovery, but should include the many cases in which the initial symptoms are the same but with subsequent long continued disturbances of consciousness often followed by residual symptoms . . . . concussion in the above sense depends upon the diffuse injury to nerve cells and fibers sustained at the moment of the accident. The effects of this injury may or may not be reversible.[10]

Referring to our paradigm for head injury mechanisms (Fig. 1) and to the list of head injury types (Table 1), we can now elaborate the current concepts for such mechanisms. Applied loads to the head may be either static (occuring at time durations exceeding 200 msec) or dynamic (time durations less than 200 msec and usually in the range of 5 to 50 msec). Static loads causing head injury occur infrequently, e.g., when the jack of an automobile fails and allows the car to crush the head of the victim trapped underneath. A key observation in such patients is that in spite of severe

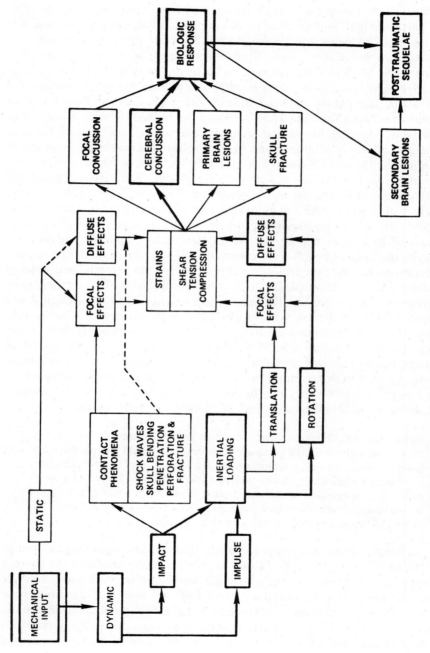

Figure 1. A paradigm for head injury biomechanics. (*From Ommaya and Gennarelli.*[1])

skull fractures and brain contusions with spinal fluid leakage through ears and nose, disturbance or loss of consciousness is seldom seen and the disassociation between the severe focal injuries and the absence of coma is striking.[11] This is strong supportive evidence for the concept that concussive brain injuries (Table 1, IVA) require some mechanism for which focal injuries to the brain and its adnexae (including skull fractures) are not necessary. Impacts to the head, e.g., in falls, automotive accidents, and assaults, are by definition dynamic events that include *two* components: the contact phenomena and inertial loading of the head and its contents. It should be noted that while contact phenomena can only be produced by direct impacts, inertial loading can be produced also by indirect impacts or impulsive loading of the head, for example, during the violent flexion–extension movements of the head after a rear end car collision. Of course inertial loading is always a highly significant component of all direct impacts when the head is free to move, and the rate of onset and/or duration of such loading pulses may be critical factors in the genesis of the types of brain injuries that result. Again with reference to our paradigm, all the items in categories 1, 2, and 3, and IVB, C, and D in Table 1, can be produced as focal injuries induced by contact phenomena alone. Although shock waves are listed as one of the contact phenomena, thus raising the possibility of diffuse brain injuries caused by this mechanism, it has been argued by Holbourn and others that the extremely short transit times for such waves (in microseconds) are exceeded by an order of magnitude by the impact durations and, therefore, should be ignored.

Figure 1 also indicates that inertial loading results in two types of head motion: translation and rotation. Holbourn's original hypothesis predicted that because of the incompressible nature of brain tissue, translation would be harmless and only rotation could initiate the diffuse tensile and shear strains required to produce the diffuse effects required to generate the concussive brain injuries.[3] Our experimental test of this prediction refutes the notion of the innocuous nature of translation but does support the critical but not exclusive role of rotation in causing diffuse effects and cerebral concussive injuries, as will be described below.[1,12]

Other concepts that have been advanced for the causation of concussive brain injuries include the idea of intracranial pressure gradients causing a primary and direct (or focal) effect on the brain stem and also by cervical cord stretch, an effect on the lower brain stem, or spinomedullary junction.[4,13] While there is little doubt that a traumatic focal lesion in the brain stem, particularly in the rostral midbrain zone, would certainly cause severe and usually prolonged disturbances of consciousness, it is much more difficult to explain how such a mechanism as a *primary* event could explain at least two sets of clinical and pathologic data. First, there is the well-established one-way dissociation between what may be called the *amnestic* and the *paralytic* phenomena of concussive brain injuries. Thus, the amnestic or memory functions appear much more fragile than the more robust sensorimotor control functions of the brain. In lesser degrees of head injury, it is possible to see amnestic disturbance without the victim falling down unconscious, although when such a knockout does occur, amnesia is always present. Second, when careful autopsy studies of severely head-injured patients are made, it is not possible to find primary brain stem damage *without* extensive associated damage elsewhere in the brain.[14] It is for this reason that we introduced the *centripetal theory of cerebral concussion and concussive brain injuries,* which invokes the geometric structural and material properties of the cranium and its contents, to state that the diffuse effects of the rotational component of inertial loading are produced by a centripetal progression of strains from the outer surfaces to the core of the brain (coinciding with the midbrain and basal diencephalon). These strains are exaggerated and enhanced at all regions where abrupt changes in material and structural properties occur, e.g., in relation to dural partitions, bony protuberances (e.g., the sphenoid wing), and vascular tethers.[1] This hypothesis explains most of the available clinical pathologic and experimental data, including the phenomenon of one-way dissociation of amnestic and paralytic phenomena in concussive brain injuries. We have stated this hypothesis as follows: Concussive brain injuries include cerebral concussion and constitute a graded set of clinical syndromes following head injury wherein increasing severity of disturbance in level and content of consciousness is caused by mechanically induced strains affecting the brain in a centripetal sequence of disruptive effects on function and structure. These effects begin at the surface of the brain and extend inward to the diencephalic–mesencephalic core at the most severe coma-producing levels of trauma.

Our proposed classification of six grades of increasing severity of cerebral concussion thus produced is shown in Table 2, where a correlation with AIS level, pathologic data, and outcome at 1 month is also proposed. Because of the variable material properties, inhomogeneities, and anisotropy of the neural tissues, it is not proper to assume that strain distributions from the cortex through subcortical gray and white matter will necessarily correlate with visible tissue tears. Indeed, without a clear understanding of the precise relative fragility of axons, synaptic clefts, and neuronal bodies with which one could assess the three-dimensional distribution of all known structural lesions (including DAI), a definitive test of this hypothesis cannot be achieved. The classic case of cerebral concussion is thus understood as being in the middle of a range of concussive syndromes of lesser and greater severity. On the one hand may be found less severe cases where memory disturbances occur without loss of motor control, and consciousness is only partially impaired (grades I to II). In such cases, we suggest that significant strains did not reach the reticular activating system in the rostral brain stem. On the other hand are the more severe cases in coma with greater degrees of diffuse irreversible damage. When such diffuse impairment reaches a

TABLE 2. AIS—CONCUSSIVE BRAIN INJURY GRADING CORRELATION

AIS Level	Concussive Brain Injury Grade	Clinical Descriptions	Pathologic Description (13)(15a)	Outcome (1 Month)
1	I	Confusion without amnesia (ding, stunned)	Not known; CT scans usually normal; skull fractures and intra-cranial bleeding uncommon	Normal except for PCS* and occasional vascular complications
2	II	Amnesia without coma (type A, slow onset; type B, rapid onset)		
3	III	Coma < 6 hours (includes classic cerebral concussion, minor and moderate head injuries)	Increasing intensity and distribution of diffuse axonal injuries and/or intracranial	
4	IV	Coma 6–24 hours (severe head injuries)	bleeding (e.g., acute subdural clots) and other visible brain	Morbidity increasing to 35%+
5	V	Coma > 24 hours (severe head injuries)	lesions (or tissue tears); CT scans usually abnormal; skull	and mortality to 50%+
6	VI	Coma → death within 24 hours (fatal head Injuries)	fracture incidence 20–50%	

*Postconcussive syndrome.

critical amount and therefore when the rostral brain stem shows visible structural damage, the worst subgroup of the grade V case may develop, and the patient never emerges from a vigilant coma state. This type of result is aptly described by the term "persistent vegetative state." Immediate death due to severe concussion may also occur (grade VI). Our hypothesis leads to three critical predictions: (1) When the level of trauma is severe enough to produce what is described as traumatic unconsciousness, the extent of simultaneous primary disruption of functions in the brain is more severe in cortical and subcortical structures, particularly in the critically vulnerable frontotemporal zones, than in the rostral brain stem. (2) It follows that because the mesencephalon appears to be the last of the vulnerable zones to be affected by trauma, primary damage to the rostral brain stem will not occur in isolation in the vast majority of head injuries that are associated with acceleration or deceleration trauma. A truly primary lesion of this part of the brain stem found at postmortem should be always in association with diffuse damage to the brain. If a patient with a lower grade of cerebral concussion dies from other causes, we predict that isolated primary rostral brain stem lesions will not be found. (3) Although confusion and disturbances of memory can occur without loss of consciousness, the reverse should never be seen, i.e., every case of head injury with a grade IV cerebral concussion must have an associated period of traumatic amnesia, the mesencephalon being less vulnerable than the temporal lobes and limbic system.

Traumatic unconsciousness develops after head injury of the usual accelerative type, when the centripetally directed, mechanically induced disconnection isolates the well-protected mesencephalic–diencephalic core from the overlying cerebral mantle. This disconnection is primarily functional, although in more severe cases it is reinforced by structural disconnection, e.g., in the white matter, as shown by the work of Strich and others.[8,9,15–17] Thus a condition akin to the diaschisis of spinal shock is induced. Monakow's "permanent diaschisis" would, therefore, be analogous to structural reinforcement of the functional disconnections by such lesions as the Strich lesions.[18] The possible mechanism I would propose for such a diaschisis is a modification of that described by Sherrington in 1910,[19] when in his careful analysis of spinal shock he concluded that it was due to "the number and character of the descending nerve paths through which the lesion breaks." Sherrington concluded, "the condition of the spinal reflex arcs in spinal shock appears to resemble a general spinal fatigue rather than an inhibition . . . suggests a loosening of nexus between the links of the neuron chain comprising the arc; a defect of transmission at the synapse."[19] That this conception of the mechanism of diaschisis is useful is supported also by Kempsinsky's observations[20] that electrical activity in regions remote from a cerebral lesion will be depressed if such regions are in neuronal contact with the damaged zone. Our modification of Sherrington's concept is that the irreversibility of the diaschisis is directly proportional to the extent of structural transection of axons and persisting abnormalities of membrane potential maintenance associated with synaptic disorganization. In this way, it becomes possible to conceive how trauma could induce the graded syndromes of cerebral concussion and also to consider spinal shock as being functionally similar to grade I–II concussion.[21] As a result of an extensive series of elegant denervation experiments (posterior root and cord transections) on cat spinal cord and histologic observations on the synaptic organization of the motor neuron surface, Illis suggested that the depression in nerve cell activity after trauma was due to profound alteration of the synaptic zone.[22] Thus, denervation of a critical number of boutons termineaux led not only to degeneration of those boutons but also to disorganization of the entire synaptic zone. With time, only the boutons with intact fibers reorganized and recreated a partial mosaic pattern of bouton distribution that was seen in the stage of functional reorganization (after 9 to 15 days). It is interesting to speculate whether such synaptic clearing at the motor neural surface after only partial denervation could be the structural analog of the diaschisis type of phenomenon seen in concussion of the brain or spinal cord. Even when axons are not transected, a sufficiently severe mechanical loading can produce a sufficiently prolonged functional paralysis so as to induce the connecting synaptic zone to undergo the above described disorganization of bouton patterns. In experiments done by Thibault and Gennarelli in our laboratory it has been shown that dynamic mechanical loading of the frog sciatic nerve can modulate the compound action potential between reversible decrements of amplitude to complete abolition of the waveform, which in some severe loadings resulted in a complete inability of the nerve to conduct even though it appeared to be histologically intact (at light microscope levels).

It is suggested, therefore, that in trying to understand neural disintegration after trauma it would be most profitable to study the organization of he synaptic zone and transmembrane potential states in order to elucidate the structural and functional correlates of the syndromes of cerebral concussion. Correlations of such changes with the resultant sequelae of neural trauma should then provide useful insight into more rational management.

CURRENT RESEARCH APPROACHES AND RESULTS

Our definitive experimental test of Holbourn's hypothesis for the relative contributions of the translational and rotational components of inertial loading clearly showed that *both* contribute to injurious effects on the brain.[1] These experiments utilized an inertial loading apparatus that eliminates contact phenomena, and they also clearly indicated that pure translation did not produce diffuse damage although the focal lesions seen were dissimilar from those induced by the contact phenomena as seen in previous experiments.[12] In particular, some of the contusions seen under pure translation loads suggested a cavitation mechanism. It was only when rotation was added to translation that all the diffuse damage effects were noted. Thus, cerebral concussion, bilateral diffuse subdural, petechial, and subarachnoid hemorrhages were only seen after the rotational component was introduced (Figs. 2 and 3). The continuation of this work by my colleagues Gennarelli and Thibault in Philadelphia, in collaboration with the neuropathologists Adams and Graham in Glasgow, has resulted in further documentation of the significance of diffuse axonal injuries (DAI) in closed head injuries in Rhesus monkeys subjected to inertial loading with a rotational component.[15] The Glasgow group has clearly demonstrated the similarity of these lesions to those seen in autopsy specimens from human victims of closed head injuries.[16] In these studies, Gennarelli et al. have further defined the kinematics of such rotational effects and demonstrated that the extent of axonal injury, duration of coma, and outcome of this type of experimental head injury correlate better with coronal rather than sagittal plane rotations.[15]

In our experimental testing of Holbourn's hypothesis, we also were concerned about the role of contact phenomena. Holbourn implied that the response of the brain within the skull could be modeled by a single degree of freedom springmass system. Thus, short-duration impacts would produce injury proportional to the rotational velocity induced by the blow, while long-duration impacts would display a dependence on the head's rotational acceleration for injury. We were able to confirm the rotational velocity dependence of short-duration pulses producing cerebral concussion in both direct impact head injury and whiplash trauma in Rhesus monkeys. The main facet of Holbourn's hypothesis was, however, *not* proven. Thus, if head rotation were indeed the crucial brain injury mechanism for diffuse injuries, cerebral concussion should be produced at an identical threshold for rotational velocity of the head irrespective of how the head rotation was induced, directly or indirectly, and the local effects of impact should have no influence on the threshold for cerebral concussion. Our experimental data did not support this prediction and indeed showed that about twice the rotational velocity was required to produce cerebral concussion when the animal experienced indirect impact (whiplash).[12] This suggested a significant contribution to brain injury by one or more of three possible factors: a direct contribution of the contact phenomena, a summation of the rotational input with translational effects, or a significant difference in the rate of onset of the acceleration pulse under the two loading conditions.[1,12]

One of the major problems in extrapolating experimental data from subhuman primate and other animal species to humans is the question of interspecies scaling. We have published one suggestion of how this may be done for rotational velocities and accelerations on the basis of experimental

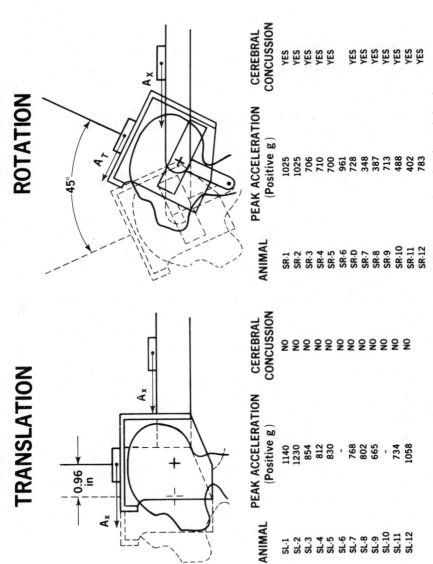

TRANSLATION

ANIMAL	PEAK ACCELERATION (Positive g)	CEREBRAL CONCUSSION
SL-1	1140	NO
SL-2	1230	NO
SL-3	854	NO
SL-4	812	NO
SL-5	830	NO
SL-6	-	NO
SL-7	768	NO
SL-8	802	NO
SL-9	665	NO
SL-10	-	NO
SL-11	734	NO
SL-12	1058	NO

ROTATION

ANIMAL	PEAK ACCELERATION (Positive g)	CEREBRAL CONCUSSION
SR-1	1025	YES
SR-2	1025	YES
SR-3	706	YES
SR-4	710	YES
SR-5	700	YES
SR-6	961	YES
SR-D	728	YES
SR-7	348	YES
SR-8	387	YES
SR-9	713	YES
SR-10	488	YES
SR-11	402	YES
SR-12	783	YES

Figure 2. Diagrammatic display of subhuman primate kinematics and results of translational versus rotational (+ translation) acceleration at equivalent inputs. Note absence of concussion under condition of pure translation.

Key

- Subdural blood
- ...... Subarachnoid blood
- Cortical contusion
- ▲▲▲▲ Petechial hemorrhage

TRANSLATED

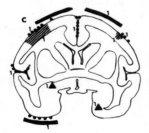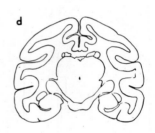

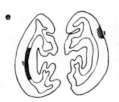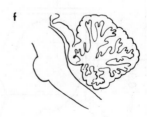

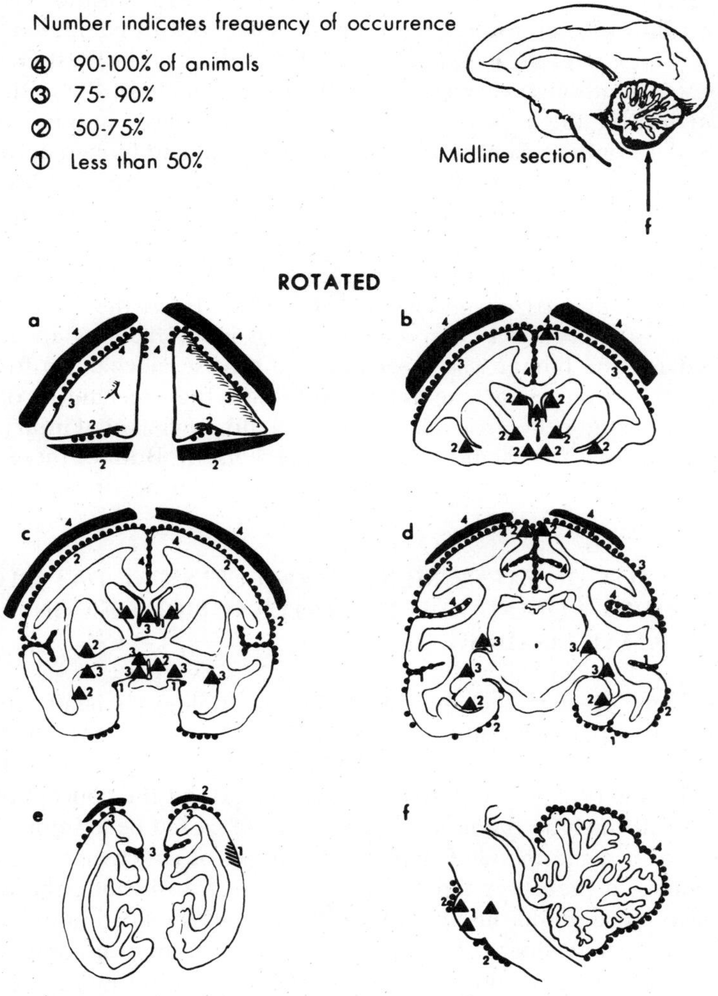

Figure 3. Cross-sections of squirrel monkey brains obtained 24 hours after head injury. Note legends in top part of diagram and markedly diffuse nature of lesions when rotational acceleration is added.

data in three subhuman primate species of gradually increasing brain mass.[23] In an attempt to bypass the problems of scaling, a number of workers have attempted to use cadavers as models for obtaining head injury biomechanical data. Injury responses have been measured in terms of intracranial pressure measurements and by noting the extent of extravasation of pigments perfused into the cerebrovascular system. The work of Nahum et al. in San Diego,[24] Tarriere et al. in France,[25] and Schmidt et al. in West Germany[26] is to be noted. Although very useful for the development of theoretical models of head injury, the lack of physiopatholigic response cautions one to use cadaver data only in close correlation with animal and human accident data, where CT scan data on fatal as well as surviving cases are most useful.[7,25,26,27]

The contributions of contact phenomena and inertial loading to head injury mechanisms has thus been identified separately on the basis of these observations in subhuman primate species. What is not clearly known at this time can be summarized under two categories. First, the quantitative biomechanics of the impact components and their interactions are not yet established for the human case. Second, the precise mechanism for the concussive brain injuries at the microlevel, i.e., the mechanism of cerebral concussion at the cellular level, is unknown although axonal injury is a good candidate for one key element of such a mechanism. Both of these issues are addressed in the final section on suggested areas for new research.

APPLICATION OF EXPERIMENTAL AND CLINICAL DATA TO HEAD INJURY CODING AND THE INTERACTION OF CONTACT AND INERTIAL LOADING

Current data on the mechanics and physiopathology of head injuries continue to support the findings of Symonds' 1962 report, which has provided us with a valuable heuristic in understanding the mechanisms of head injuries. He was probably the first to explicitly extend the concept of cerebral concussion to include the more severe forms of closed head injury.[10] Indeed, we have also recommended that the definition of concussive brain injuries be extended in the other directions as well, i.e., to include the less severe (but much more numerous) minor and subconcussive head injuries.[1] Table 2 shows our current recommendation for such a system of grading concussive brain injuries. This has been designed to conform with most of the current clinical and neuropathologic data, as well as to correspond numerically with the AIS scale levels. The pioneering neuropsychologic outcome studies of Gronwall, Wrightson, and others have shown very clearly that the postconcussive sequelae of even the most minor head injuries can be of major significance in terms of ability to return to work and competence in previous occupations.[28,29] These data fit very well with our classification of concussive brain injuries (Table 2) as well as with the neuropathologic observations of Oppenheimer and others on neuropathologic lesions present in mild to moderate head injuries.[9,30] More recent work confirming and extending the observations on the disabilities of the victims of minor head injury has been reported by Rimel and her colleagues at Charlottesville.[31]

The correlation of lesion types with the mechanics in head injuries is an ongoing task. Thus, the lesions shown in Table 1 have for the most part been well described in qualitative terms. Review of the available data on so-called DAI (diffuse axonal injuries) and non-DAI cases of fatal head injuries was carried out in order to propose the following hypothesis, which seeks to explain how the relative contributions of contact phenomena and inertial loading components favor or reduce the production of DAI.

Conditions Favoring DAI Type of Case

1. Distributed loading (impact over large area of head)
2. Soft impact (longer duration, > 10 msec)
3. Partition of kinematics to favor more rotation and less translation
4. Contact phenomena negligible

Conditions Favoring Non-DAI Type of Case

1. Focused loading (impact over small area of head)
2. Hard impact (shorter duration, < 10 msec)
3. Partition of kinematics to favor less rotation and more translation
4. Contact phenomena prominent

It should be emphasized that DAI cases form a minority of fatal head injury cases, and thus this lesion is only one of the possible array of structural alterations at the cellular level after trauma.

HEAD INJURY SEVERITY LEVELS, INJURY CRITERIA, AND TOLERANCE LEVELS

The European Economic Council Biomechanics program has recently produced a consensus report defining these terms as follows:

Injury Level or Injury Severity Level. This term denotes the magnitude of changes in terms of physiologic changes and/or structural failure that occurs in a living body as a consequence of mechanical violence. The AIS scale is widely used for this purpose, but other scales have also been proposed.

Injury Criterion, Injury Criteria. This term denotes a physical parameter that correlates well with the injury severity of the body region under consideration. Currently, the HIC is the accepted criterion for head injury, but the precision of its correlation is being questioned.[7] This term is often used in other ways, thus causing much confusion in the literature.

Tolerance Level. This term denotes the magnitude of loading of the living body or body part that produces a specific type of injury and injury severity level. When used, this term must be specified by defining the following aspects: the physical parameter expressing the magnitude of loading, the type of living body (animal or human, sex and age), the type of body part, what kind of injury, and what injury severity level is being connsidered. Extrapolations of tolerance levels that do not follow these procedures are not recommended.

By inference, an injury criterion may represent a legal specification and as such may be used to form part of a standard that protective devices or crash environments must satisfy in order to permit their recommendation and distribution to the general public. However, in such cases, the standard or criterion must not be confused with the tolerance level for the reasons given above. Moreover, it is almost inevitable that the specific prescription required for such a standard requires the use of an arbitrary mechanical system, having only a distant and partial correlation with its human counterpart. Such test systems properly permit only comparative evaluation of the relative merits of injury-mitigating components, e.g., helmets, but do not permit their use to duplicate a degree of trauma that would be experienced by a human under corresponding circumstances—nor should they be used to predict human tolerance levels. The following discussion on head injury criteria is based on an earlier report by Goldsmith and Ommaya.[32]

Injury Criteria for the Head

As stated, the terms "injury criteria" and "tolerance levels" should not be confused, and the distinction made in this report as suggested by the EEC Biomechanics program is recommended. Clearly, a distinction should be made for the skull and brain and possibly also for the vascular components. The stress and strain fracture characteristics of the skull can be quantified by a mean value and a standard deviation, which may be only slightly smaller than the 50 percentile magnitude but which is preferable to the specification of a breaking load or energy. On the other hand, brain failure cannot be so conveniently delineated because (1) physiologic dysfunction, including death, can occur at levels well below that producing mechanical disruption of neural tissues, and (2) the tissues comprising the brain and blood vessels are so complex and inhomogeneous that neither functional nor structural failure limits for either the entire system or for specified regions have been adequately established. In consequence, an injury level for the brain is normally specified in terms of the magnitude or history of some mechanical parameter considered to be a major indicator of cerebral trauma. Peak and average linear and angular accelerations, their duration and rate of onset, intracranial pressure, volume changes of the skull, and force and energy applied to the cranium have been utilized for this purpose, but it is not yet conclusively proven that any of these parameters are optimal for trauma correlation. Whether kinematic parameters can be employed at all for this purpose has been questioned, but a recent interesting paper by Ono et al. concludes that it is possible to model head injury assessments from head motions derived from animal and cadaver data.[33] Current practice divides the brain injury generation mechanisms as those resulting from linear or angular motion, respectively.

Linear Acceleration Limits
The first of these demarcations is the Wayne State Tolerance Curve (WSTC), in which it is claimed that the dividing line represents the onset of concussion. This tolerance curve is based on the hypothesis that the dominant head injury mechanism is linear acceleration. Its initial enunciation was based on six experiments on embalmed cadavers striking rigid surfaces at the forehead in the A/P direction in the duration range from 1 to 6 msec. The results were correlated with concussive effects generated in animals and were later supplemented by additional experiments on primates and cadavers and employment of long-duration acceleration tolerance information from human volunteers.

Within a few years, a determined effort had been made to represent the WSTC in analytic form. The resulting expression for a severity index (SI) is:

$$SI = \int [a\,(t)]^{2.5} dt \qquad (1)$$

This is a straight line on a logarithmic plot and did not quite match the curved transposition of the WSTC, but the difference was ignored. A tolerance level of SI $\leq$ 1000 was stipulated as acceptable.

Serious objections to the use of this criterion were raised on the basis of the arbitrary value of the exponent and inconsistencies due to failure to distinguish between the approximation to tolerance data, the scaling of severity, and the definition of effective acceleration. It was suggested that the latter term might be replaced by a time-averaged, weighted acceleration, although such a representation was still inadequate in view of insufficient biomechanical validation data and lack of a specific severity designation. However, NHTSA mandated the use of a modified version of this idea, expressed in the form of the head injury criterion (HIC):

$$HIC = \text{Max} \left[\frac{1}{(t_2 - t_1)} \int_{t_1}^{t_2} a\, dt \right]^{2.5} (t_2 - t_1) \qquad (2)$$

where the times t_2 and t_1 are two arbitrary instances of the pulse history chosen so as to obtain the supremum (maximum) of the function. Equation (2) was supposedly an improvement over the SI by concentration on the most dangerous portion of the history, which was bound to take into account the rate of load application. The legal limit of 1000 was placed on the permissible HIC value, although diverse opinions have been voiced that this magnitude was either too high or too low relative to sustaining intolerable levels of injury. The HIC representation has also generated considerable controversy between its detractors and advocates. These arguments pro and con can be found in the recent consensus workshop proceedings and in the paper by Goldsmith and Ommaya.[7,32] One of the major drawbacks of the HIC is that it is a go–no go type of criterion; what is needed is a continuous scale.

Angular Motion Limits

There is a substantial difference between the mechanical processes involved in direct head impact and sudden torso motion inducing whiplash type of head motions, although both cause linear and angular acceleration of the head. Head impact requires consideration of both potential skull fracture and trauma due to load transmission to the brain, whereas whiplash need not address skull fracture but will be additionally concerned with neck response.

Extensive experiments testing the rotational or angular motion hypothesis involving subhuman primates were done not only to assess the effects of restraining collars on concussion but also to determine dosage limits in these animals that could be extrapolated to man.[1,12] Although both direct head impact and torso acceleration (whiplash) tests were performed, quantification and, particularly, scaling of the contact phenomenon between species was not considered to be sufficiently reliable, so that only whiplash experiments were scaled and extrapolated to the human case.[23] Damaging rotational velocities and accelerations for the brain were related by the assumption of a linear, elastic undamped single-degree-of-freedom system. Concussion was induced in the squirrel monkey at about 1200 to 1600 rad/second, in the Rhesus monkey at 300 to 600 rad/second, and in the chimpanzee at about 70 to 90 rad/second, as determined from physiologic responses and high-speed camera measurements of the motion. With the established brain frequency value for the Rhesus monkey (from calvarium measurements) of 5 to 10 Hz and an observed value of 4 to 5 Hz in a human determined from fluorocinematography of radiopaque clips on the brain surface, it is reasonable to delineate the natural frequency of both squirrel monkey and chimpanzee in the 5 to 10 Hz range. The value for the former was chosen as 10 and the latter as 5 from size consideration. The threshold for cerebral concussion in terms of head angular accelerations for the squirrel monkey, Rhesus monkey, and chimpanzee were then calculated as 20 krad/second,[2] 10 krad/second,[2] and 2.2 krad/second,[2] respectively. Employment of the scaling relation (inversely proportional to the two-thirds power of the brain mass) and the above data results in a rotational acceleration and velocity tolerance for man of 1.8 krad/second[2] and 20 to 30 rad/second, respectively; these limits are plotted in Figure 4. It should be reemphasized that this information shown as a band is considered to be reliable for the Rhesus, sketchy for the chimpanzee, and completely speculative for man.[12] It is expected, therefore, that when data from other sources and in particu-

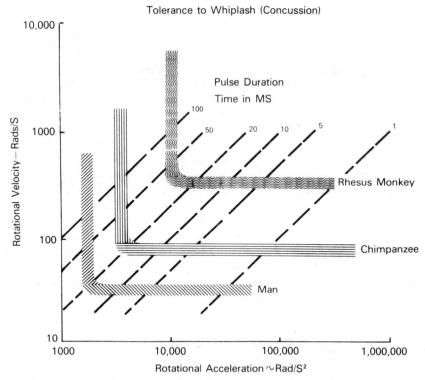

Figure 4. Theoretical scaling of rotational velocity and acceleration for 50 percent probability of onset of cerebral concussion from subhuman primates to humans.

lar from accident reconstructions in humans become available, these approximate bands of concussion thresholds for rotational motions will have to be revised.

Angular head acceleration has been proposed by Lowenheim as the cause of gliding contusions resulting from excessive strains in cerebral blood vessels.[34] The limiting dosage that will initiate such contusions can be quantified as a maximum angular acceleration of 4.5 krad/second^2 and an angular velocity change of 70 rad/second.[34]

A detailed consideration of the kinematics of head injury including the relevant biomechanics of translational and rotational acceleration responses has recently been published in collaboration with Advani.[35] The translational and rotational models described in this article provide an engineering link for quantifying critical head injury parameters. For the translational model these parameters are (1) the skull tangential stresses, (2) brain displacements and contrecoup pressures, and (3) brain shear strains at the surface and midbrain region. This model primarily describes inertial blood volume, cerebrospinal fluid, brain volume, and associated pressures. Contrecoup brain injury is interpreted by this model in terms of acceleration time pulses. The rotational model singles out the shear strain as a principal mechanism of injury. Peak surface brain shear strains are in the damage threshold range for angular accelerations around 2000 to 3000 rads per second.[2] Distortion of the intracranial contents is also characterized by the model. The models provide estimates of the deformable head motions, such as skull depressions, and selected cerebral blood vessel and ventricular system movements that can be experimentally verified by detailed high-speed cinefluorographic investigations on animals and human cadavers subjected to controlled impacts. Subsequent neuropathologic studies on the cerebral hemispheres, the cerebellum, the brain stem, and the cervical spinal cord can relate the sites and intensity of pathologic lesion patterns (e.g., hemorrhages with or without hemostasis, contusions, lacerations, with the predicted displacements and stresses). These models indicate that the skull–brain interface is vulnerable to the shear mode.

Finally, with regard to impact direction and resultant effects of interactions between the contact phenomena and resultant kinematic effects on injury levels, it has been suggested by Grecivik and Jacob that lateral blows cause less severe injuries as compared to fronto-occipital or vertex blows.[17] This observation agrees with the data of Ono et al.[33] but is not in accordance with the experimental data of Gennarelli et al. as cited earlier.[15] This emphasizes the need to discriminate between the loading on various regions of the head as well as in measurement of the resultant principal strains or pressure levels at various positions within the brain. An obvious necessity to address this requirement is the inclusion of a contact criterion in case of impacts that can take into account local variations in skull architecture and curvature.

A SUGGESTION FOR THE USE OF TWO HEAD INJURY CRITERIA

Starting with the Wayne State Tolerance Curve, a variety of empirical equations and lumped parameter models, such as the Gadd Severity Index, the Vienna Institute Index, the Revised Brain Model, the Effective Displacement Index, and the Head Injury Criteria, have tried to fit the Wayne State Curve. Experimental and/or analytic models of the head, such as the Michigan Strain Model, and several continuum and finite element models of the head have been suggested. Of these, the Mean Strain Criterion (MSC) would appear to lend itself well to use as a continuous criterion to address the contact and translation components of the impact. The reasons for this are as follows: of the simple analytic models available, only the MSC obtained its parameters directly from experimental efforts. All the others chose parameters that would make them best fit the Wayne State Curve. Moreover, additional experimental studies have allowed the MSC originators to propose outputs of the model associated with other levels of injury, i.e., to enable a more continuous type of criterion.[36]

It is generally accepted that the MSC and other similar criteria are most valid for the direct head impact situation and include only the translation component of inertial loading. Because rotation has been shown to be crucial for the concussive brain injuries (Tables 1 and 2) and for the rupture of the bridging veins between the skull and brain, the MSC *alone* is not adequate. The data identifying and quantifying rotational injury have been largely derived from animal experimentation and transformed into human thresholds via scaling techniques.[12,23] While there have been many studies of rotation-induced head injuries, these studies tend to either not report injury criteria or support the curve shown in Figure 4. Rotational acceleration may be measured in a dummy by using a nine linear accelerometer package. Angular velocity may be measured by integrating the angular acceleration.

It should be emphasized again that *two* criteria for brain injury are recommended to be used in tandem. Thus, if MSC predicts an AIS 1 and the rotational criterion predicts an AIS 4, the injury is an AIS 4. Most real-world accidents as well as our tests involve both direct contact impact (measured by the MSC) and rotational injury. Until an integrated criterion addressing both components is developed, we would recommend the combined use of the MSC and the angular acceleration tolerance levels as indicated below.[37]

For Contact Impacts and Translational Accelerations—Use MSC

AIS 1	MSC .00256	
AIS 2	MSC .00433	
AIS 3	MSC .00610	Values of mean strain
AIS 4	MSC .00787	
AIS 5	MSC .00964	
AIS 6	MSC .00964	

For Rotational Accelerations—Measure rotational acceleration ($\ddot{\theta}$) *about the center of gravity (cg) of the head, and the rotational velocity ($\dot{\theta}$) about the* head cg.

If $\dot{\theta} \geq 30$ rad/second and:

$\ddot{\theta} < 1700$ rad/second^2	AIS 2
$\ddot{\theta} < 3000$ rad/second^2	AIS 3
$\ddot{\theta} < 3900$ rad/second^2	AIS 4
$\ddot{\theta} < 4500$ rad/second^2	AIS 5

If $\dot{\theta}$ If < 30 rad/second and:

$\ddot{\theta} < 4500$ rad/second^2	AIS 0 or 1
$\ddot{\theta} \geq 4500$ rad/second^2	AIS 5

SUGGESTED AREAS OF RESEARCH

The following is an eclectic listing of what I consider to be some significant problems that need to be tackled within four methodologic categories for the study of closed head injuries.

I. Animal Models (including cellular and tissue studies)
   A. Inertial loading. This technique has provided very good data on combined rotational and translational acceleration-induced trauma but has not yet been used to validate scaling predictions for head injury tolerances based on other techniques. Only one report of pure translational loading in the squirrel monkey is available,[1] and these data require validation and extension in larger brains.
   B. Direct impact. Use of impact techniques to examine more closely the interaction between the contact phenomena and inertial loading effects is required.[12]
   C. Neuropathologic and neurophysiologic studies. The excellent studies of Adams and Graham require extension to the electron microscopy level now that the DAI lesion locations can be reproduced reliably in the coronal angulation mode.[15,16] Dynamic and functional studies of the morphology and neurophysiology of axonal transport in suitable invertebrate preparations, such as the squid axon, under controled loading conditions may well provide valuable insights into the field strain parameters that correlate with reversible and irreversible disruption of axonal functions by mechanical loads. Analog studies on neuronal cell body synaptic physiopathology and cerebral vascular physiology are also needed.
   D. Measurement of material and structural properties. Although some data on the mechanical properties of cranial tissues are available, much more are needed.[38]
II. Physical Models
   There are currently three groups working with physical assemblies modeling the head and neck, at Goteborg in Sweden (Dr. B. Aldman et al.), at Berkeley (Dr. W. Goldsmith), and at Philadelphia (Dr. L. Thibault). Improvements in these models in conjunction with work on the animal and tissue studies will be extremely valuable.
III. Human Studies
   A. Accident data banks. Information from such sources as well as from in-depth longitudinal studies of selected cases is invaluable in forming useful hypotheses for further investigation as well as to validate predictions made on the basis of experimental or theoretical work. In a recent consensus workshop organized by the National Highway Traffic Safety Administration (NHTSA) it was noted, however, that there is a significant mismatch of such data in terms of the biomechanical and medical data sets.[7] Thus, NIH-supported data banks

are rich in clinical and physiopathologic information but poor in biomechanical input data for these accidents. The opposite seems to hold for the NHTSA-supported data banks, such as the National Crash Severity Study and National Accident Sampling System files.[7] In collaboration with Dr. Howard Champion at the Washington Hospital Center, we are currently developing a new methodology seeking to correct this mismatch in a pilot study of 100 motor vehicle accident cases where the accident investigation and biomechanical data gathering are triggered by the arrival of the suitably injured patient at the trauma center. The data obtained will be used in improved accident reconstruction using physical and mathematical techniques.

IV. Theoretical Models

A large number of mathematical models have been developed, all of which have used linear theories involving only small strains. Nonlinearities associated with large strains, large rotations, or nonlinear material responses have not been modeled. Although it is true that the formalization of such complex properties can be computationally formidable it is necessary if we are not to remain saddled with trivial solutions. Thus, brain tissue is most probably anisotropic, inhomogeneous, and of a multiphase nature. Neuroanatomic data would indicate that the brain is certainly a composite material with variable orientation of the local principal directions of material response throughout its substance.[32]

The finite element method for mathematical modeling is a powerful tool that will undoubtedly prove to be even more useful than hitherto.[39] However, it would appear that the availability of better brain material property data and development of a model able to handle nonlinearities and large strains are essential.[40]

REFERENCES

1. Ommaya AK, Gennarelli TA: Cerebral concussion and traumatic unconsciousness. *Brain* 1974; 97:633–654.
2. Hitti PK: *History of the Arabs,* ed 10. New York, Macmillan, 1970.
3. Holbourn AHS: Mechanics of head injury. *Lancet* 1943; 2:438–441.
4. Gurdjian ES: Impact head injury. Springfield, Ill, Thomas, 1975.
5. Gadd CW: Use of a weighted impulse criterion for estimating injury hazard, in *Proceedings of the Tenth Stapp Car Crash Conference.* New York, Society of Automotive Engineers, 1966.
6. Versace J: A review of the severity index, in *Proceedings of the Fifteenth Stapp Car Crash Conference.* New York, Society of Automotive Engineers, 1971.
7. Ommaya AK (ed): *Proceedings of the Consensus Workshop on Head and Neck Injury Criteria.* Washington, DC, US Government Printing Office, 1983.
8. Strich SJ: Shearing of nerve fibers as a cause for brain damage due to head injury. *Lancet* 1961; 2:443–448.

9. Oppenheimer DR: Microscopic lesions in the brain following head injury. *J Neurol Neurosurg Psychiatry* 1968; 31:299–306.
10. Symonds CP: Concussion and its sequelae. *Lancet* 1962; 1:1–5.
11. Russell WR, Schiller R: Crushing injuries of the skull: clinical and experimental observations. *J Neurol Neurosurg Psychiatry* 1949; 12:52–60.
12. Ommaya AK, Hirsch AE: Tolerances for cerebral concussion from head impact and whiplash in primates. *J Biomech* 1971; 4:13–20.
13. Friede RL: Specific cord damage at the atlas level as a pathogenic mechanism in cerebral concussion. *J Neuropathol Exp Neurol* 1960; 19:266–279.
14. Mitchell DE, Adams JH: Primary focal impact damage to the brain stem in blunt head injuries: Does it exist? *Lancet* 1973; 2:215–218.
15. Gennarelli TA, Thibault LE, Adams JH, et al.: Diffuse axonal injury and traumatic coma in the primate. *Ann Neurol* 1982; 12:564–574.
16. Adams JH, Graham DI, Murray LS, et al.: Diffuse axonal injury due to nonmissile head injury in humans. *Ann Neurol* 1982; 12:557–562.
17. Grecivik N, Jacob H: Some observations on the pathology and correlative neuroanatomy of cerebral trauma, in *Proceedings of the Eighth International Congress of Neurology*. Vienna, Pergamon, 1965, Vol 1, pp 369–373.
18. Monakow C Von: Quoted by Sherrington in reference 19.
19. Sherrington CS: *The Integrative Action of the Nervous System*. London, Constable, 1910.
20. Kempinsky WH: Experimental studies of distant effects of acute focal brain injury. *Arch Neurol* 1958; 79:376.
21. Ommaya AK: Reintegrative action of the nervous system after trauma, in Popp AJ, Bonrice R, Nelson R (eds): *Neural Trauma*. New York, Raven Press, 1979.
22. Illis LS: The motor neuron surface and spinal shock, in Williams D (ed): *Modern Trends in Neurology*. Appleton-Century Crofts, 1967, Vol 4, pp 53–65.
23. Ommaya AK, Hirsch AE, Harris E, et al.: Scaling of experimental data on cerebral concussion in subhuman primates to concussive thresholds for man, in *Proceedings of the Eleventh Stapp Car Crash Conference*. New York, Society of Automotive Engineers, 1967, pp 47–52.
24. Nahum AM, Raasch F, Ward C: Impact responses of the protected and unprotected head, in Ommaya AK (ed): *Proceedings of the Consensus Workshop on Head and Neck Injury Criteria*. Washington, DC, US Government Printing Office, 1983.
25. Tarriere C, Walfisch G, Fayon A, et al.: Cerebral tolerance integration of experimental cerebral injuries obtained with cadavers. Brussels, Seminar on EEC Biomechanics Research Programme, 1983.
26. Schmidt G, Kallieris D, Barz J, et al.: Results of 49 cadaver tests simulating frontal collision of front seat passengers, in *Proceedings of the Eighteenth Stapp Car Crash Conference*. Society of Automotive Engineers, 1974.
27. Zimmerman RA, Bilaniuk LT, Gennarelli T, et al.: Cranial computed tomography in diagnosis and management of acute head trauma. *Am J Roentgenol* 1978; 131:27–34.
28. Gronwall D, Wrightson P: Delayed recovery of intellectual function after minor head injury. *Lancet* 1979; 2:605–609.
29. Levin HS, et al.: *Neurobehavioral Consequences of Closed Head Injury*. Oxford University Press, 1982.
30. Jane J, Rimel RW, Pobereskyn LH, et al.: Outcome and pathology of minor

head injuries, in Grossman R (ed): *Seminars in Neurologic Surgery. Proceedings of the Fourth Conference on Neural Trauma.* New York, Raven Press, 1982.

31. Rimel RW, Giordani R, Barth JT, et al.: Disability caused by minor head injuries. *Neurosurgery* 1981; 9:221–228.
32. Goldsmith W, Ommaya AK: Head and neck injury criteria and tolerance levels, in *Proceedings of the International Transportation School.* Amsterdam, North Holland (in press).
33. Ono K, et al.: Human head tolerance to sagittal impact. Reliable estimation deduced from experimental head injury using subhuman primates and human cadaver skulls, in *Proceedings of the Twenty-fourth Stapp Car Crash Conference.* SAE 801303. New York, Society of Automotive Engineers, 1981.
34. Lowenheim P: Mathematical simulation of gliding contusions. *J Biomech* 1975; 8:351–356.
35. Advani S, Ommaya AK, Yank WJ: Head injury mechanisms, in Ghista DN (ed): *Human Body Dynamics.* Oxford, Oxford University Press, 1982.
36. Stalnaker R: Personal communication.
37. Marcus J: Personal communication.
38. Ommaya AK: *Head Injury Mechanisms.* Final Report to NHTSA Contract DOT-HS-0811-106. 1973.
39. Khalil T, Viano DC: Critical issues in finite element modelling of head impact, in *Proceedings of the Twenty-sixth Stapp Car Crash Conference.* New York, Society of Automotive Engineers, 1982.
40. Chi M, Vaishnav RN, Shams T: A Preliminary Survey of Existing Head Models. Appendix 1. Progress Report No. 3. NHTSA Contract DTNH 22-83-C-07258, 1980.

CHAPTER 14

Face and Facial Bones:
Clinical Aspects

Robert Craig Bone

Maxillofacial trauma is occurring more frequently than ever before: Vehicular accidents, sports injuries, intentional wounding, work-related and recreational trauma have all increased during the last 20 years, and a high proportion of these incidents involve the soft tissue and bone of the head and neck.[1] Many of the problems encountered clinically as a result of these events are simple, straightforward exercises in wound repair employing conservative, well-tried means of treatment that yield excellent results. Less common, however, are a number of problems that are markedly more difficult to manage and in which the final outcome is not so gratifying. These injuries often occur in multiply-injured, severely traumatized victims who, in the past, might not have survived their initial period of injury but now, because of highly sophisticated, portable means of resuscitation, are potentially more likely to reach the emergency room in an ultimately salvageable state. In dealing with the biomechanical aspects of these facial wounds, one is actually discussing the forces involved in the creation and the repair of the facial bones that underly the soft tissues of facial expression and function. Therefore, without neglecting the importance of soft tissue injury and its management, we will confine ourselves largely to a broad outline of current state-of-the-art techniques and, where appropriate, areas in which a need is perceived for biomechanical research in the treatment of facial bone fracture and injury.

The primary goal in management of maxillofacial trauma is to initially recognize and treat patients in whom potentially life-threatening associated trauma is present. The neophyte physician in the emergency room is instructed early as to the likelihood of associated trauma below the clavicle occurring with a severe facial injury and that it is rarely the facial or neck wound itself that leads to a fatal outcome but the associated, acutely compromised thoracic, cardiac, or major vascular injury that may be first unrecognized if immediate attention is directed toward the maxillofacial area. Another potentially devastating occult injury that may coexist with maxillofacial injury is a cervical vertebral disruption. This specific injury must always be considered prior to positioning the patient for radiographic evaluation of facial fractures. Only after these potentially lethal injuries are considered and the cardiorespiratory as well as neurologic stability of the patient assured should attention be directed toward evaluation and management of the maxillofacial injury per se.

Once underway, however, management of the maxillofacial injury can be thought to have two specific goals: (1) the restoration of pretraumatic function and (2) restoration of appearance. Special functions to consider include speech, mastication, deglutition, unimpeded respiration, and salivary flow. Additional consideration needs to be given to the special senses in terms of support to the globe, transmission of sound energy by the ossicles, and unimpeded air flow to the olfactory region. It is indeed fortunate that these two important goals of restoration of function and appearance are rarely, if ever, mutually exclusive.

Several means of classifying facial injuries are available, but for purposes of this discussion simple regional subgrouping seems most appropriate. One needs to realize, however, that to some extent such grouping is artificial, and very often the injuries discussed below will occur multiply, involving several areas of the face and the facial skeleton synchronously.

INJURIES TO THE LOWER THIRD OF THE FACE (Mandibular)

The mandible can be biomechanically likened to a bony arch to which are attached two main sets of muscles that serve to approximate and distract the intact arch from the toothbearing bone of its maxillary counterpart. The archlike form of the mandible has practical relevance in regard to injury; it is generally true that a force applied to one part of the arch is transmitted equally throughout its circumference. Thus, weaker portions of the arch are at risk every time such energy is exerted, no matter where it is applied. It is therefore not surprising that (1) the highest incidence of fracture of the mandible occurs at its weakest structural point (the condyle or surgical neck) and (2) multiple fractures of the mandible are not uncommon.[2] Secondly, the traction of the two main groups of muscles attaching to the mandible can have one of two effects, either to reduce the fracture or to distract the fragments. In the first case, the fracture is termed "favorable" and in the second "unfavorable."

The mandible can be divided into several subregions, each with its own characteristic response to injury. Fractures of the symphyseal and perisymphyseal regions, in or near the midline, tend to produce highly unstable fractures and, because this is the site of attachment for the protrusive muscles of the tongue, may potentially result in airway compromise when the base of the tongue cannot be pulled forward out of the oropharynx. This is especially a problem in the semiconscious patient. Fractures occurring in the body–alveolus (the main horizontal and toothbearing portion of the mandible) may be favorable or unfavorable and considered closed or open (compound, depending upon whether or not the fracture passes through the tooth socket and lacerates the mucous membrane of the oral cavity). Ramus (the main vertical portion of the mandible) fractures can be handled nonoperatively simply with dietary restriction to soft food. Muscles medial and lateral to the ramus create a natural sling for this portion of the mandible that tends to act as a natural immobilizer and, therefore, to prevent fragment distraction. Fractures in the region of the surgical neck or condyle of the mandible are often difficult to see radiologically but should always be searched for because of their likelihood.

The basic decision to be made in the treatment of mandibular fractures is whether or not fixation is to be required or whether dietary restriction with or without splinting or external bandaging will be sufficient. Secondarily, if open fixation is to be used, what kind will be employed. Depending upon the status of the dentition (as points of fixation), the direction of the fracture line in relation to favorable vs unfavorable orientation, whether tooth sockets are involved in the fracture site, and the status of the overlying and surrounding soft tissue, as well as the region of the mandible involved, a decision is made as to the need for open reduction.[3] If such reduction is determined unnecessary but more than simple splinting and diet are required, the mandible may be reduced by wiring the teeth together (intramaxillary fixation) or, if the patient is edentulous, by fixing the patient's dental plates by means of wiring or screws to the mandible and maxilla. Finally, if no plates are available, they may be fashioned preoperatively by the prosthedontist. In all of these maneuvers, the goal is to bring the teeth into a stable prereduction occlusion. The proprioceptive nerves of the alveolus tell the patient whether or not "the teeth are lined up right" and provide the clinician with the most sensitive gauge available for judging the accuracy of his reduction (Fig. 1).

If open reduction is deemed necessary, a number of means of fixation of the fractured segments of the bone are available. These would exclude simple direct wiring, compression plates, biphasic external fixation, and Kirchner wiring.[4]

MIDTHIRD FACIAL FRACTURES (Midfacial)

Fractures of the midthird of the face are a diverse group of injuries, including maxillary, ethmoid, nasal, malar, and zygomatic types, either singly or in

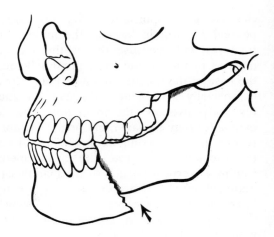

Figure 1. Schematic representation of the displacement of the edentulous proximal segment in a fracture of the mandible. The fragment is displaced upward, inward, and forward against the maxillary teeth by the muscles attached to the ramus. Effective treatment for this fracture consists of open reduction and interosseous wiring at the inferior border of the mandible. (*From Dingman and Natvig.*[2])

combination. For all of these injuries, long-term concerns are for restoration of pretraumatic appearance, support for the eye and dental structures, and an unimpeded nasal airway.

Maxillary fractures have been well studied and are known to generally follow predictable pathways. These pathways were originally investigated in cadaveric experiments by Rene Le Fort shortly after the turn of the century. Thus, the three main predictive patterns are today commonly given designations of Le Fort I, II, and III. Le Fort I fractures (also occasionally called Guerin fractures) occur through the palate but do not rise onto the face of the maxilla or the nasal pyramid (Fig. 2). These fractures are often difficult

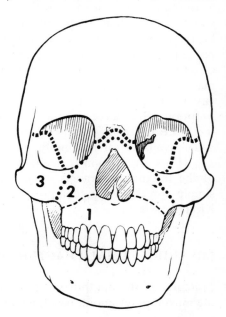

Figure 2. The Le Fort lines of fracture; types I, II, and III are indicated. (*From Killey HC:* Fractures of the Middle Third of the Facial Skeleton. *Bristol, England, John Wright & Sons, 1971.*)

to detect radiologically, and the best clue to their presence is a rocking, unstable palate noted on palpation. This fracture usually responds quite well to simple intramaxillary fixation and placement of a wire from the dental arch bar to a supramaxillary point of stability above the fracture line for a period of 4 to 6 weeks.

Le Fort II (or pyramidal) fractures occasionally begin at the maxillary buttress, extend across the face of the maxilla often through the infraorbital foramen, and cross the midline at the dorsum of the nose. A similar pathway on the opposite side then completes the course of the fracture line. Treatment for these injuries will often depend on whether or not associated fractures, for example, of the zygoma or the orbital floor, are present. Generally, however, one attempts to stabilize the fracture segment inferiorly by intramaxillary ligation. At the same time, superior fixation is provided above the unstable segment by wiring to the next highest point of stability, often the zygomatic or supraorbital arches. Dental appliances are usually removed at 6 to 8 weeks.[5]

Le Fort III fractures (craniofacial dissociation) are usually the result of a massive force exerted against the midface, as in vehicular accidents. The face virtually acts as an energy dissipator, protecting the cranium from exposure to a strongly disruptive force. As a result, this fracture usually occurs with the plethora of associated facial disruptions, lacerations, fractures, and other injuries. Craniofacial dissociation per se, however, involves a fracture dislocation at the frontozygomatic suture and an associated fracture running through the orbital floors, across the nasal pyramid superiorly, and a similar pathway on the opposite side. The type, degree, and severity of the associated nasal injury may vary considerably. In spite of the energy absorption role of facial structures in this type of injury, the great amount of energy usually expended will often incur an associated central nervous system injury that will take precedence in the management for the first 48 to 72 hours. A characteristic flattening and elongating of the face is noted on physical examination and is known as "dishpan face." Treatment of this fracture usually involves open reduction of the frontozygomatic separation bilaterally and appropriate treatment of the nasal fractures and intermaxillary fixation with associated ligation of the dental appliance to a stable area immediately above the frontozygomatic junction. This latter point of fixation is often a hole drilled directly into the supraorbital ridge.[6]

UPPER THIRD (Frontal)

Fractures of the upper portion of the facial skeleton are really those that involve the frontal sinuses and ethmoid labyrinth. Often both of these paranasal cavities are injured simultaneously. Function in this area is not as important as it is in the lower two thirds of the face. Any or all of the uppermost sinus cavities can be obliterated without loss of significant function if it is felt that restoration of appearance would be improved. The main risk in this type of trauma is that an associated neurologic injury to the contents of the anterior cranial fossa or base of the skull may be missed during the initial assessment.

Frontal sinus injuries are generally subdivided into three types: Anterior table fracture, posterior table fracture with or without associated anterior table injury, and injury to the nasofrontal duct. The basic decision faced in the management of all these injuries is whether or not to explore the fracture to assure (1) the best possible restoration of the bone fragments to their pretraumatic state and (2) the absence of dural disruption with associated brain damage. Radiographic assessment is crucial in these decisions, and computer-assisted tomography (CAT scan) has been a major advancement in the management of these problems.

Simple, nondisplaced fractures of the frontal bone may be observed without direct intervention. However, if the fracture line involves regions near the base of the sinus and the nasofrontal duct (by which normally secreted mucus drains from the sinus into the nose), prolonged, more careful observation will be required. Fractures in which fragmentation of the anterior table has occurred can often be explored through an accompanying laceration. If necessary, a so-called seagull incision or a coronal incision (across the top of the cranium) can be placed to allow for a complete exposure of the anterior frontal table. By either means, bone fragments are carefully repositioned and wired with fine ligatures back into their original positions, with great care being taken to leave the supporting and nourishing periosteum intact.[7] Additional support can be provided by entering the frontal sinus through a nonfractured site in its floor and packing underneath the site of injury with either a long thin strip of antibiotic-impregnated gauze or a balloon catheter. These supporting devices can then be left in place for 5 to 7 days and easily removed. Restoration of pretraumatic forehead contour should thereby result in the great majority of instances.

Fractures involving the posterior table with or without concomitant significant anterior table injury are more ominous because of the already mentioned possibility of serious, initially occult cerebral injury. Although nondisplaced fractures well away from the inferiorly located nasofrontal duct may be observed, those fractures of the posterior walls showing some degree of displacement or encroachment of the nasofrontal region should in most cases be surgically explored to rule out the possibility of duct transection or stenosis. The treatment of fractures of the posterior table involves reconstruction of bone fragments, or if damage is truly extensive, the posterior wall can be totally removed, allowing the brain to expand forward to obliterate the old frontal sinus region and contact the back of the anterior table of the frontal bone (Fig. 3).[8]

Nasofrontal duct injury can be best handled in one of two ways. The entire frontal sinus system can be carefully obliterated by removing all mucous membrane lining from the surrounding bone if it is felt that both

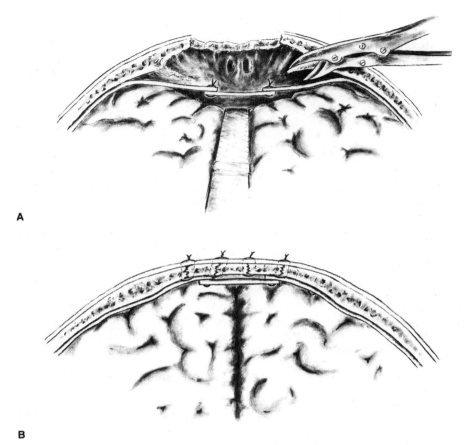

A

B

Figure 3. Dural repair and removal of the posterior frontal bone plate following severe frontal bone trauma. **A.** A small plate of autogenous fascia has been used to reconstruct the dural defect. **B.** A completed repair at the time of brain expansion is shown. Note the wiring of retained anterior table fragment immediately below the skin. *(From Donald and Bernstein.[8])*

nasofrontal ducts have been significantly injured. Thereafter, adipose tissue is harvested from the left lower abdominal quadrant and is placed as a free graft into the frontal defect. The anterior table of the frontal bone is then repositioned to restore frontal contour. Alternatively, one may undertake reconstruction of the nasofrontal duct by removal of obstructing bone with a large burr. This is followed by rotation of a large mucous membrane flap based on the superior portion of the nasal septum into the newly created passage to reconstitute a new duct. Operative technique depends more on personal preference and past experience than on any recognized difference in success or end results.

PROBLEM AREAS WITH RESEARCH POTENTIAL

One of the most obvious needs to a clinician who treats facial fractures is a means to provide consistent anterior traction to the fractured segment of the face. Many facial fractures result from either direct injury with a blunt object or rapid accelerative force. This subsequently results in a displacement of unstable fragments posteriorly, often producing a notable deformity. Present means of dealing with this problem involve halo devices that are solidly fixed to the skull by means of osseous screws and connected to the unstable, posteriorly displaced segment by means of wires or struts. Such a device is quite uncomfortable, especially when the patient is supine, and because the vectors of applied force are not universally predictable, end results are often not ideal. A similar but slightly different problem exists in the repair of the nasal ethmoid (midfacial) fractures. The nasal pyramid consists of a strong unyielding bone that is anatomically supported bilaterally on the ethmoid labyrinth. This latter structure is a loose, honeycombed complex of thin lamellar bone capable of resisting only a small amount of force prior to collapse. Thus, a severe blow to the nose can result in a generally intact nasal pyramid being forced back into the face as the ethmoid complex collapses. Thereafter, when distracted, there is little structural support for the nasal pyramid during healing. Some means of providing such support for this key portion of the profile that could be easily applied and comfortable and would yield dependable results would be a significant advance and of great assistance in the clinical management of these difficult injuries.

Brief mention was made above of the sometimes deleterious effects of muscular traction on fracture segments of the mandible. Such force exerted in the immediate postoperative period often produces distraction of an otherwise undisplaced fracture, thus requiring surgical (open) reduction of what is initially a well-aligned fracture site. Biomechanical research to create a means of counteracting these vectors would save considerable expense, operative time, and morbidity.

Finally, another area in which assistance would be helpful would be the creation of a dependable means of creating permanent support for the upper trachea after a crush injury to the cricoid cartilage of the larynx. The cricoid is the only circumferentially intact cartilage of the laryngotracheal complex and as such is of prime importance in maintaining luminal patency. Occasional injuries to this structure that produce rupture of this ring can result in a stenosed larynx and trachea that require the patient to be committed to a permanent tracheotomy in order to assure a safe airway.

REFERENCES

1. Schultz RC: *Facial Injuries.* Chicago, Year Book, 1977, pp 12–400.
2. Dingman RO, Natvig P: *Surgery of Facial Fractures.* Philadelphia, Saunders, 1969.

3. Bernstein L: Mandibular fractures. *JCEORL Allergy* November 1978; 5:15–19.
4. Edgerton MT, Hill E: Fractures of the mandible. A series of 434 cases. *Surgery* 1952; 31:933–950.
5. Bernstein L: Delayed management of facial fractures. *Laryngoscope* 1970; 80:1323–1341.
6. Dingman RO, Natvig P: *Surgery of Facial Fractures.* Philadelphia, Saunders, 1969, p 266.
7. Newman MH, Travis LW: Frontal sinus fractures. *Laryngoscope* 1973; 83:1281–1290.
8. Donald PJ, Bernstein L: Compound frontal sinus injuries with intracranial penetration. *Laryngoscope* 1978; 88:225–232.

CHAPTER 15

Biomechanics of Facial Bone Injury: Experimental Aspects

Dennis C. Schneider

INTRODUCTION

The origins of biomechanical investigations of facial bone fractures are contained in the descriptive writings of the early clinicians. The identification and classification of injury are necessary in any study of trauma and the mitigation of injury. Blunt trauma to the facial structures was a relatively common event arising from the high incidence of hand-to-hand combat in early military engagements. The advantages of completely enclosing the face in various types of protective armor were quickly recognized as injury-mitigating techniques. This solution was not universally practicable, however, and the recognition and treatment of facial skeletal trauma persisted as a clinical issue. The interested reader is referred to the work of Dingman and Natvig[1] for an overview of the historical development regarding surgery of facial fractures. At approximately the turn of the century, the demise of hand-to-hand combat in modern warfare was accompanied by the advent of high-speed transportation technology. Although the accidental crash rate is relatively small, trauma to the facial skeleton comprises a substantial segment of the total injuries sustained by vehicular occupants. Recent studies by Huelke and Compton[2] and by Karlson[3] are representative of the current frequency of facial injuries in the automobile crash environment. The National Crash Severity Study established by the National Highway Traffic Safety Administration was utilized by Huelke to project that over 29,500 moderate to severe facial injuries involving fractures or avulsions are sustained annually nationwide. The use of restraint systems was found to reduce the occurrence of serious facial injuries by approximately two thirds. Unfortunately, at the present time lap and shoulder belt restraints are not employed by a majority of vehicular occupants. Karlson's examination of hospital-treated facial trauma provides a similar annual estimate of 29,000 serious to severe facial fractures. Although these estimates are not directly comparable due to differences in injury classification and data bases, such data indicate the nature of the problem confronting the physician and biomechanical engineer. If these data are combined with the facial trauma produced annually in job-related and sporting activities, the magnitude of this health care issue becomes apparent.

The injury environments discussed above are man–machine interactions and thus provide the potential to mitigate this trauma through the rational design of structures that an individual might contact in an impact event. The first requisite step in this process is to define the biomechanical characteristics of the facial skeleton, including its mechanical response and tolerance to impact.

An examination of the face reveals a geometrically complex structure that exhibits symmetry only about the midsagittal plane. Inferior to the supraorbital ridge, a total of 14 major bony segments can be identified encompassing 13 suture lines. These bones vary in shape, thickness, and radii of curvature and include arched and membraneous configurations. For these reasons, in part, the biomechanical impact response of the face has received little attention in the scientific literature. The geometric complexity of the facial anatomy inhibits simplification in engineering analysis, and, therefore, analytic approaches to studying the biomechanics of the facial bones have largely been ignored. Rather, the research investigations to date have concentrated on developing experimental impact data that characterize the response of the facial structures to localized and blunt impact forces. Of particular interest has been the establishment of impact tolerance levels for forces applied to various regions of the face through contact surfaces of simple geometry. Human cadaver specimens both embalmed and unembalmed have been utilized almost exclusively in these studies. The following material represents an overview of the major findings in these studies of the biomechanics of trauma to the facial bones.

PRIOR INVESTIGATIONS

A pioneering investigation of trauma to the facial bones was performed by René Le Fort in 1900.[4] Le Fort reported his observations describing the types of maxillary fractures produced in human cadavers as a result of blunt dynamic loading by a foreign body. These experiments, numbering approximately 40, dealt with the effects of varying the location and direction of load application to the facial skeletal structures as well as the degree of head support during the loading event and their effect on fracture patterns produced. An understanding of Le Fort's observations regarding the consequences of varying bone quality and changes in impactor mass, velocity, and impulse duration can be obtained from the following excerpt[4]:

> Variations in the structure of the bone and the degree of ossification of the sutures will cause variations in the extent or intensity of the lesion, as also will variations in the weight of the injuring tool, the swiftness of the blow and the duration of the action of the injuring agent; but these differences do not modify the general type of fracture.

These experiments and subsequent work by Le Fort provided the basis for mapping the lines of fracture (Fig. 1) that most commonly result from blunt maxillary trauma. Now known as Le Fort's classification of maxillary fractures, this landmark nomenclature continues to be utilized, as evidenced in the current American Association for Automotive Medicine's Abbreviated Injury Scaling (AIS) guide.[5] Shown in Tables 1 and 2 are the injury classifications of the AIS-80 scale and their application to facial fractures.

One of the earliest attempts to quantify the impact response of the individual facial bones was reported by Swearingen.[6] This study was a retrospective analysis of automobile collision cases in which facial injuries were produced from contact with the instrument panel of the vehicle. Human cadaver skulls with soft tissues intact were then impacted into similar structures to reproduce the panel structural damage and correlate the cadaver injuries sustained with various dynamic responses of the skull. Measurement of the head acceleration was obtained, in addition to the acceleration of small blocks molded to fit the individual facial bones. However, the mass and geometry of the load-distributing blocks were not reported. Knowledge of the acceleration–time history of these isolated blocks does not allow conclusions to be drawn regarding the force or average pressure tolerance of the impacted bones. The tolerance data reported consequently are unique to the experimental conditions employed and have somewhat limited value in contributing to an understanding of the impact response of the bones of the face.

The mechanics of mandibular fractures was examined by Huelke in a series of publications[7-9] and summarized in 1961.[10] In vitro preparations of 27 human mandibles were used to study the stress distribution in that structure resulting from static and dynamic loads applied to the mid symphysis of the bone. Brittle lacquer (Stresscoat) techniques were employed to obtain a qualitative understanding of the dependence of mandibular deformation on the restraint provided by condylar fixation during impact. Conditions studied included both condyles free to move, both fixed, or one fixed with the contralateral condyle free to move. A 0.07 kg steel ball impinged on the bone at a velocity of 2.0 m/second to provide the impact load. The response of the mandible was likened by Huelke to an arch loaded at midspan with the condyles fixed and free, approximated by fixed and pin-jointed support

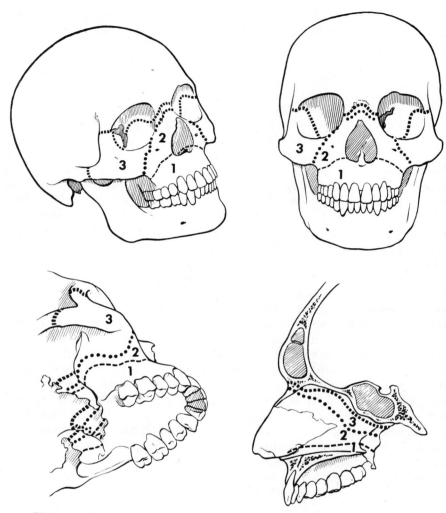

Figure 1. Fractures of the maxilla **1.** Le Fort I (transverse fracture). **2.** Le Fort II (pyramidal fracture). **3.** Le Fort III (craniofacial disjunction). (*From Dingman and Natvig.*[1])

conditions, respectively. Tensile stresses are produced along the outer surface of the bone with concentrations in the subcondylar and mental foramen areas. The inner aspect of the curve of the lingual chin area immediately opposite the impact site responded in a tensile stress mode. The location of high stresses in the subcondylar region corresponded with the results of accident studies performed previously,[11] in which the subcondylar region was found to be the most frequently fractured region of the mandible (36.3 percent).

The first comprehensive study of the biomechanical impact response of

TABLE 1. ABBREVIATED INJURY SCALE (AIS)— 1980 REVISION

AIS	Severity Code
1	Minor
2	Moderate
3	Serious
4	Severe
5	Critical
6	Maximum injury, virtually unsurvivable in AIS-80

From the American Association for Automotive Medicine.

the facial bones was undertaken by Hodgson et al. and reported in a series of papers from 1964 to 1968.[12-15] Because this continuing study examined the phenomenon of facial bone fracture in considerable depth, it merits discussion in some detail. The initial aspect of this study[12] examined the nature of the dynamic response of the supraorbital ridge, zygoma, and zygomatic arch to impact with another body. In this investigation, stationary human heads were struck by a rigid mass traveling at a known velocity. Seven unembalmed human cadaver subjects provided fundamental information relating impact response to impactor mass, velocity, energy, acceleration, padding conditions, and bone characteristics. These parameters were related to the mechanical response as well as fracture tolerance of the structures studied. Initially, static loadings were carried out on the zygoma and zygomatic arch, while simultaneous recordings of the applied force and displacement of the bone were made. Bone displacement was characterized by both gross deflection adjacent to the loaded site and local strain measurements derived from strain gauges that had been mounted near the point of load application.

TABLE 2. FACIAL BONE FRACTURE SEVERITY

	AIS
Mandible	
Ramus	1*
Body	2*
Subcondylar	2*
Maxilla	
Le Fort I	2
Le Fort II	3
Le Fort III	4
Orbit	2*
Zygoma	2*
Frontal	
Linear, closed	2
Comminuted, depressed	3
Open or brain exposed	4

*If open, displaced, or comminuted, add AIS-1.

Experimental data indicated that the zygoma and zygomatic arch respond as linear springs in the range of forces studied. The static exposures were followed by dynamic loadings employing a rigid mass weighing 1.0, 2.3, or 5.6 kg traveling at speeds ranging from 1.3 to 31.2 m/second. These impactor mass and velocity combinations when coupled with various padding interfaces produced pulse durations of 1.7 to 5.9 msec and head accelerations ranging from 91 to 372 g. Blows were delivered to the facial bones with and without the overlying soft tissues in place. It was determined that the maximum force whether applied statically or dynamically produced the same strain in the bone (Fig. 2). After determining that the motion of the head during impact was essentially translational, a two-mass linear spring-lumped

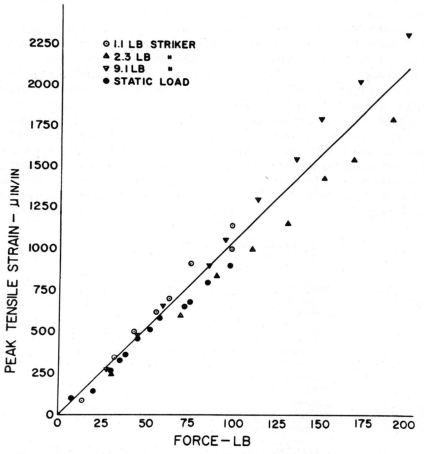

Figure 2. Relationship of peak dynamic force and static force to strain in the zygomatic arch. (*From Hodgson et al.*[12])

parameter model of the facial bones response was proposed. Results of this analog showed good correlation between measured and predicted peak impactor acceleration. Impactor energy was found not to be a good correlate of facial bone fracture when the head is allowed some degree of movement during the impact event. If the peak applied force is determined on a mass times acceleration basis for the impactor, an average fracture load for the zygoma of 3.05 kN can be obtained from these data. In a subsequent publication,[13] the impact parameters of applied force, pulse duration, and impact velocity, and their relationships to each other and the resulting bony fracture were studied more comprehensively. A total of 26 fracture data points were obtained from 12 embalmed human cadaver specimens. The majority of the tests involved the zygoma, while impacts were also delivered to the zygomatic arch, maxilla, mandible, and frontal bone in this tolerance study. A comparison of the static and dynamic loading response for various facial bones indicated that the more deformable facial bones react similarly, while the frontal bone, essentially a structural component of the cranium, moves with the skull and undergoes less local deformation. Therefore, its apparent mass is much greater, yielding a higher net stiffness when loaded dynamically. When padding was introduced over the blow site, the pulse duration was lengthened at low impact velocities, but at higher speeds the duration was independant of velocity due to the pad bottoming out, allowing the bone characteristics to dominate the response. In paired fracture tests with and without the overlying soft tissue in place, the energy required to fracture was significantly different. For example, an impact to the uncovered zygoma required 10.6 N-m of impactor kinetic energy to fracture, while the contralateral bone with soft tissue intact failed at an input energy of 16.3 N-m. This energy difference of approximately 50 percent was not reflected in recorded peak impact force—1.07 vs 1.16 kN, respectively. Based on these and other results, the criterion of damage was determined to be maximum impact force. Due to the nature of the test procedure a broad range of impact pulse durations were obtained. With the skin removed over the impact location, it was possible to achieve pulse durations as low as 1.0 msec in duration. Fracture loads for the zygoma ranged from 0.56 kN at 18 msec, to 4.35 kN at 3 msec. (Fig. 3). The higher fracture forces were concentrated in impacts whose duration fell between 1.5 and 4.0 msec and were predominantly tests with soft tissues removed. It should be noted that multiple impacts were administered to the individual bones, and the potential for lowering the fracture tolerance as a result of cumulative damage could not be assessed.

Further experiments using similar methodologies were reported,[14] which examined the fracture tolerance of the mandible and frontal bone. Fractures of the frontal bone generally propagated to one or both orbits, while mandibular fractures were produced in either or both necks. Failure forces for the mandibles ranged from 1.60 to 2.67 kN and 4.19 to 9.12 kN for the frontal bone. The effect of increasing the area of load application

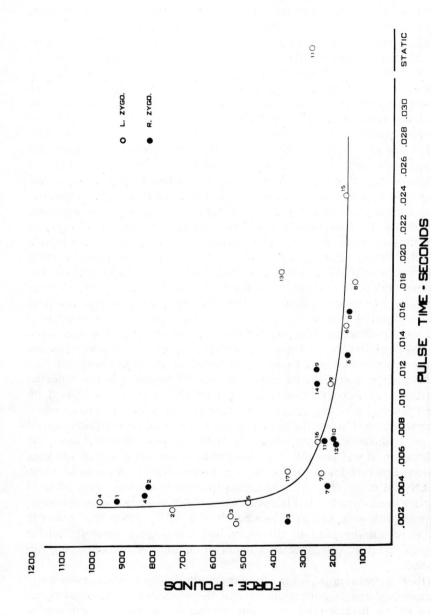

Figure 3. Force and pulse baseline time to cause linear fracture of the zygomatic bone. (*From Hodgson.*[14])

was studied in impact tests of the zygomatic bone. It was found that if the contact area was increased from 6.5 to 33.5 cm², the force to fracture increased 150 to 250 percent. The energy to fracture in those experiments also reflected a similar increase. Subsequently, Hodgson and Nakamura[15] employed mechanical impedance techniques to investigate the phenomenon of elevated fracture tolerance at short pulse durations. An impedance head was attached to the zygoma of human cadaver subjects and generated a sinesoidal input force, while the acceleration of the driven point of the bone was simultaneously recorded. These quantities together with the phase angle between them allow the computation of the mechanical impedance of the driven structure. Shown in Figure 4 is the variation in impedance determined when the forcing frequency is varied from 20 to 5000 Hz. The significan drop in impedance at approximately 300 Hz indicates that the zygoma begins to vibrate independently of the skull, and consequently the apparent mass of the structure is only a small portion of the total mass of the human head. Below 300 Hz the apparent mass is relatively constant and results in load transfer from the zygoma to the skull proper and produces rigid body motion during impact. These results, to a large extent, explain the increase in fracture tolerance observed previously[13] at short pulse durations, since a 1.5 msec half sine pulse contains large amplitude vibrational energy components in the frequency domain of 300 Hz.

Nahum et al.[16] reported the first quantitative information regarding the force to fracture tolerance of the facial bones employing embalmed and unembalmed human cadaver specimens. In this investigation, an unconstrained falling mass was allowed to impinge on the facial structures of stationary cadaver subjects. The contact surface area was a 6.5 cm² circular disk that was coupled to a load cell permitting a record of the impact force–time history. By varying the drop height, a range of impact velocities could be obtained. Of specific interest was the effect of pulse duration on the force to fracture tolerance of the facial skeleton. Aluminum tubes fitted over the end of the impactor provided a method of lengthening the duration of load application. These tubes rolled over the end of the impacting mass at a constant force during the dynamic event and increased the duration of the impulse delivered to the bone at a constant load. The rate of onset was also studied by comparing loads at fracture developed by large masses dropped from low heights to smaller masses released from greater elevations. For a majority of the experiments the head was supported by foam rubber, allowing a relatively free inertial reaction during the loading event. The use of both male and female subjects permitted initial observation of sex-related tolerance differences. Fractures were divided into the following categories:

0	None
1+	Minimal detectable change, usually a hairline crack not clinically significant
2+	readily detectable fracture that is clinically significant
3+	comminuted and/or depressed fracture

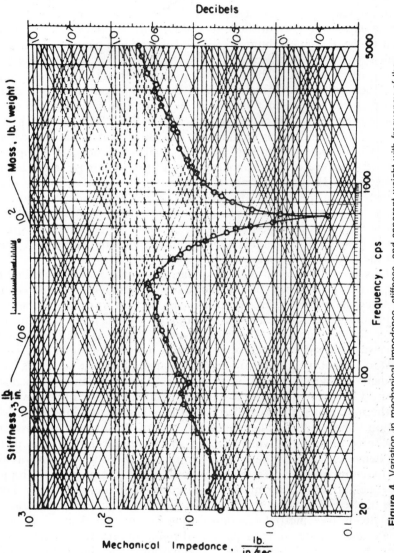

Figure 4. Variation in mechanical impedance, stiffness, and apparent weight with frequency of the zygoma. (*From Hodgson and Nakamura.*[15])

Shown in Figure 5 are the range of forces and types of fractures produced in the zygoma, temporoparietal, and frontal bones. These data were obtained in single-impact exposures to the various bones with overlying soft tissues in place. Based on these results, the authors proposed clinically significant fracture tolerances for the frontal, temporoparietal, and zygomatic areas of 4.89, 2.45, and 1.00 kN respectively. Pulse duration and rate of onset were found not to be critical factors in the determination of tolerance criteria. The effects of anatomic embalming did not appear to alter significantly the fracture tolerance. To distribute the applied load over the surface of the

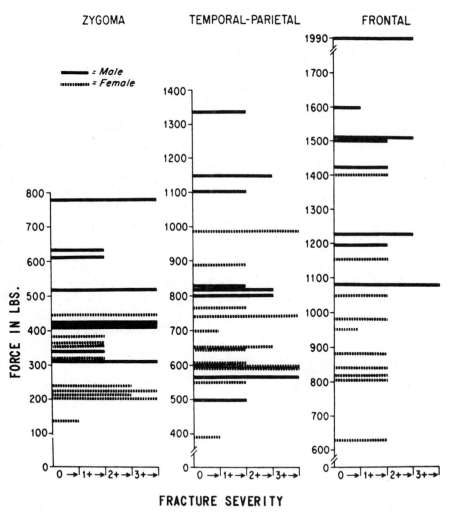

Figure 5. Force tolerances for the frontal, temporoparietal, and zygomatic bones. (*From Nahum et al.*[16])

impactor a time-independent, crushable metal foam material was attached that permitted the mapping of the local stress distribution at the impact site. These pressure distributions confirmed the wide variations in local stresses due to the irregular nature of the bony surfaces. Although the skin serves as an efficient load-distributing material, localized pressures were recorded that were 200 percent in excess of the average pressure that was calculated using the applied load and contact surface area.

Using similar methodologies, Schneider and Nahum[17] further defined the range of variation in impact tolerance of the facial bones. Employing both male and female, embalmed and unembalmed cadaver specimens, additional data were acquired describing the tolerance of the frontal and temporoparietal bones and zygomatic arch. New information was also presented regarding the maxilla and reaction of the mandible to midsymphysis and lateral body impact. Impactor masses ranged from 1.1 to 3.8 kg. Impulses were delivered over a 6.5 cm^2 contact area for all tests except the lateral mandible impacts, where a 25.8 cm^2 interface was employed. A total of 74 impact experiments were performed on 17 human cadavers. Force to fracture data obtained for the frontal, temporoparietal, and zygomatic areas corroborated the results reported by Nahum et al.[16] The dynamic loading response of the mandible was found to be, in part, a function of the path of the reaction forces developed by the skull proper. The mandible approximates a rigid semicircular link with pinned joints at its free ends. When an A-P load in the saggital plane is applied to the midsymphysis, instability is encountered unless the line of action of the force passes through the condylar processes. The mandible is a structure composed of regionally differing geometries, each with its own failure mode and force tolerance level. Therefore, considerable variation was observed in the force required to produce mandibular fracture when loads were applied in the saggital plane. Figure 6 demonstrates the range of forces producing fracture in anterior–posterior loading. Encompassed in these data was a single condylar fracture (1.89 kN), bilateral condylar fractures (2.38 kN), and a symphysis fracture (4.11

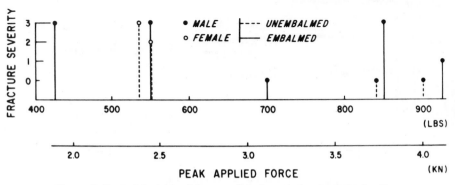

Figure 6. Force tolerance of the mandible for anterior–posterior loading.

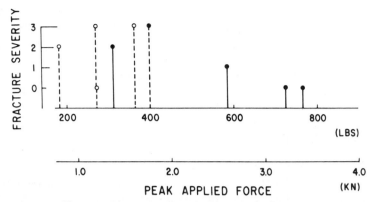

Figure 7. Force tolerance of laterally loaded mandible.

kN). It was suggested that the path of the reaction forces determines the ultimate mode of mandibular failure.

If the dentures are in contact and the direction of the impact has a sufficient inferior–superior component, the greater component of the force will be developed across the symphysis and denture contact line, causing eventual failure at high levels of force of the body and/or symphysis. If the direction of the applied force is approximately parallel with the denture contact line, the load is transmitted through the condylar processes that fail at significantly lower magnitudes of force. Lateral impacts to the mandible generally produced transverse fractures through the body over a wide spectrum of forces (Fig. 7). The maxilla was found to be one of the more frangible of the facial bones. Fractures tended to be depressed and comminuted, although this area is covered by thick subcutaneous and skin tissues. Minimal force tolerances were proposed for the anterior–posterior and laterally loaded mandible of 1.78 and 0.89 kN, respectively. Similar values for the maxilla and zygomatic arch were established at 0.67 and 0.89 kN, respectively.

The facial bones are capable of withstanding substantial forces if the load is distributed over a significant area of the face. Swearingen[6] reported that no facial fractures were sustained in a human cadaver test in which the applied contact force was distributed over the entire face via a molded form-fitting block, even though the skull experienced an accaleration of 300 g. Assuming a head weight of 4.5 kg this corresponds to an approximate 13.34 kN maximum dynamic load. Daniel and Patrick,[18] in a study of automotive instrument panel impacts, also confirmed the advantage of force distribution in prevention of facial fractures in human cadavers. In this investigation, forces up to 7.34 kN were generated on the facial structures of human cadavers by allowing them to impact a deformable automotive instrument panel at speeds of 17.9 m/second. No fractures were observed in any of the test subjects. Quantification of the impact response of the face to distributed loads was accomplished by Leung et al.[19,20] for simple geometries. Masks were constructed that spread the contact force over the facial structures inferior to the frontal bone in human cadavers. A fracture tolerance of 7.76 kN was observed for male cadavers when a force-distributing mask was utilized. This load to fracture is considerably greater than the tolerance of the individual bones supporting such a mask, such as the zygoma and maxilla (Table 3).

An important aspect of any biomechanical analysis of impact response is the geometric characterization of the anatomic system being studied. Although the relative size and shape of the facial bones vary greatly among individuals, progress has been made in defining the breadth of the variation as well as characterizing the dimensions of the hypothetical average or 50th percentile skull. The latter effort has particular significance in the construction of anthropometric test devices that can be used to evaluate the relative safety afforded the individual by various vehicular interior designs. Two major techniques for acquiring data of this type are direct physical measurement of the facial structures and dimensions obtained from lateral and anterior–posterior radiographs of the skull. The former approach has been employed extensively in the field of anthropology, while the latter has evolved into the science of cephalometrics and has considerable application in dental research studies of bone growth and development. Hubbard and McCleod[21] developed an anthropometric skull that reflected the geometric features of a 50th percentile male. These features included anatomically correct headform coordinate axes, head–neck articulation, cranial contours, and major facial landmarks. Shown in Figure 8 is the skull geometry model proposed. Indicated are the major anatomic landmarks that were located three-dimensionally to represent the 50th percentile male. The anthropologic data of Byars et al.[22] and Hertzberg et al.[23] were relied upon to develop this geometry. We are not aware of the implementations of radiographic cephalometric data in the construction of similar headforms. However, this technique is well established and provides a valuable resource in defining anatomic geometry. The works of Krogman and Sassouni[24] and of Salzman[25] provide thoroughly referenced introductions to this area.

TABLE 3. FACIAL RESPONSE TO DISTRIBUTED FORCES

Impact Condition	Force		Fracture	Source
	N	(Pounds)		
Automotive				
Instrument panel	7340	(1650)	No	Daniel and Patrick[18]
Molded mask	13,350	(3000)	No	Swearingen[6]
Molded mask				
(Not including frontal bone)	7340	(1650)	Yes	Leung et al.[19]

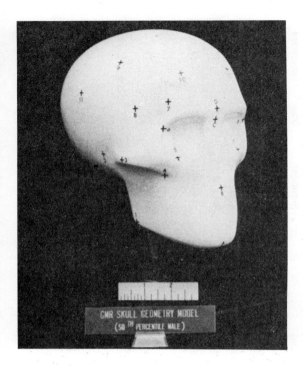

Figure 8. Anthropometric skull model of geometric features of 50th percentile male. (*From Hubbard and McCleod.*[21])

DISCUSSION

Considerable effort has been expended in quantifying the mechanical response and tolerance of various organ systems of the body. As evidenced by discussions in other chapters, the mechanics of closed skull brain injury, thoracic impact characteristics, and the response of the vertebral column to acceleration have all received well-justified attention in the scientific literature. Apparent from the foregoing review of facial bone biomechanics, however, is the lack of a similar effort, and consequently, a thorough quantitative understanding of the impact response of the facial skeleton. In spite of this, some preliminary conclusions can be drawn based on the limited experimental data at hand.

Extensive material property studies of isolated bone preparations have shown that their failure is associated with critical levels of tensile or shear stress generated in a loading event. The impact studies discussed herein have examined the structural properties of the facial bones. These investigations have identified maximum applied force as the failure descriptor for the facial bones.[14,16,17] Table 4 summarizes the prior work in the form of data means and ranges of the force to fracture of the individual facial bones studied.

TABLE 4. FACIAL BONE IMPACT TOLERANCE

Bone	Impactor Area cm² (in²)	Fracture Forces				Sample Size	Reference
		Mean		Range			
		N	(Pounds)	N	(Pounds)		
Zygoma	6.5 (1.0)	2594	(583)	614–3470	(138–780)	29	16, 17
Zygoma	6.5 (1.0)	1259	(283)	845–1665	(190–374)	5	14
Zygoma	33.2 (5.2)	2297	(516)	1600–3360	(360–756)	7	14
Zygomatic arch	6.5 (1.0)	1535	(345)	925–2110	(208–475)	17	17
Maxilla	6.5 (1.0)	1148	(258)	623–1980	(140–445)	13	17
Mandible							
Midsymphysis	6.5 (1.0)	3100	(697)	1890–4110	(425–925)	9	17
Lateral	25.8 (4.0)	1918	(431)	818–3405	(184–765)	9	17
Frontal	6.5 (1.0)	5287	(1188)	2670–9880	(600–2220)	31	16, 17

Impactor kinetic energy was found not to be a good correlate of fracture tolerance.[12,16,17] In the condition of a stationary subject impacted by a moving mass representative of most experimental protocols, whole head movement occurs, and, therefore, not all impactor energy is converted to strain energy of bone deformation. For a majority of the bones studied, a relatively small amount of energy is stored in the bone as strain energy prior to failure. Hodgson's force-deflection characteristics of the zygoma indicate that approximately 0.16 N-m of energy is transferred to the bone at failure for a 6.5 cm^2 impactor area. This represents two orders of magnitude less than the available energy. Additionally, depending on the contact surface area, the overlying soft tissues absorb varying amounts of energy. The functional role of soft tissues in impulsive loading are twofold: (1) to distribute the load over a broad surface area and thereby reduce stress concentrations in the irregularly shaped skeleton and (2) to absorb energy that would otherwise be converted to strain energy within the bone. The concept of load distribution is of primary importance in reducing the probability of skeletal fracture in blunt facial trauma. Hodgson[14] found nearly a twofold increase in fracture tolerance of the zygoma when the contact area was increased from 6.5 to 34.7 cm^2 (Table 4). If the load is distributed over the entire facial skeleton, significant dynamic forces can be developed before localized fracture occurs, as was shown in Table 3.

Neither rate of loading nor impulse duration influences the fracture tolerance for the spectrum of loading velocities and durations commonly encountered in blunt impact. Although Hodgson et al.[13] observed an elevated load to failure for the zygoma for pulses in the neighborhood of 3 msec duration, that response was peculiar to that bone and at a duration not commonly encountered in impacts with structures of significant mass. Nahum et al.[16] and Schneider and Nahum[17] specifically examined rate of loading onset and pulse duration in the 5 to 20 msec range and could observe no effect on the force to fracture tolerance. To summarize, of the impact variables studied, only the maximum force sustained correlated with fracture. If the force is distributed over a larger area, the tendancy for fracture to occur is reduced. Impactor kinetic energy, rate of load onset, and impulse duration do not affect the failure tolerance of the facial bones (Table 5).

TABLE 5. FACIAL BONE IMPACT TOLERANCE

Impact Variable	Tolerance Affected?	Consequences	
Kinetic energy	No	—	
Rate of loading	No	—	
Pulse duration	No	—	
Maximum force	Yes	Force ↑	Fx ↑
Contact area	Yes	Area ↑	Fx ↓

Fx, probability of fracture.

FUTURE INVESTIGATIONS

Much work remains to be accomplished in this area to gain a better understanding of the mechanics of facial bone fracture. Efforts to date have neglected to quantify the material characteristics of the skeletal bones tested. These characteristics would include descriptions of the geometry of the bones tested, such as thickness and radii of curvature. In addition, the bone quality as represented by mineral content would assist in evaluating the relative bone strength between test subjects and provide a potential basis for normalization of data. The sheer lack of data is immediately apparent. Our knowledge of the biomechanics of facial bone fracture is founded on a series of experimental studies encompassing some 30 human cadaver subjects of advanced age. When the variance in experimental protocol is taken into consideration, the basis for statistical comparison dwindles further. Therefore, the need for additional data is clear, particularly with respect to more youthful subjects. The injury mechanisms examined thus far have been produced by simple impactor geometries and have dealt primarily with fractures to isolated bones of the face. The classic Le Fort fracture patterns involving one or more bones have not been addressed even though these injuries are routinely treated by the clinician. A second common injury, the orbital wall fracture or so-called blowout fracture, has not been investigated from a biomechanics standpoint. Finally, the use of analytic techniques has largely been ignored. The application of finite element-modeling techniques (FEM) appears to be a logical approach that would permit study of the irregular bony structure of the face. FEM-generated stress and strain characteristics would aid in our understanding of both the fracture mechanism and the design of reconstructive surgical repair techniques.

This outline of future areas of research is certainly not all inclusive, and it is offered in the context of initial suggestions. Solutions to the problem of prevention and treatment of facial bone trauma will demand considerable attention by the scientific community. Innovative research concepts will undoubtedly be required and thus offer substantial challenges to the interested investigator.

REFERENCES

1. Dingman RO, Natvig P: *Surgery of Facial Fractures*. Philadelphia, Saunders, 1964.
2. Huelke DF, Compton CP: Facial injuries in automobile crashes. *J Oral Maxillofac Surg* 1983; 41:241.
3. Karlson TA: The incidence of hospital-treated facial injuries from vehicles. *J Trauma* 1982; 22:303.
4. Le Fort R: Etude experimentale sur les fractures de la machoire superieure. *Rev Chir* 1901; 23:208, 360, 479.
5. *The Abbreviated Injury Scale,* Park Ridge, Ill., American Association for Automotive Medicine, 1980 revision.

6. Swearingen JJ: *Tolerance of the Human Face to Crash Impact.* Federal Aviation Agency, Office of Aviation Medicine, Civil Aeromedical Research Institute, Oklahoma City, Okla., 1965.
7. Huelke DF: Preliminary studies on mandibular deformations and fractures. *J Dent Res* 1959; 38:663.
8. Huelke DF: Experimental studies on the mechanism of mandibular fracture. *J Dent Res* 1960; 39:694.
9. Huelke DF: The production of mandibular fractures: A study in high speed cinematography. *J Dent Res* 1961; 40:743.
10. Huelke DF: Mechanics in the production of mandibular fractures: A study with the "stresscoat" technique. I. Symphyseal impacts. *J Dent Res* 1961; 40:5,1042.
11. Hagen EH, Huelke DF: An analysis of 319 case reports of mandibular fractures. *J Oral Surg Anesth Hosp Dent Serv* 1961; 19:93.
12. Hodgson VR, Talwalker RK, Nakamura GS: Response of the facial structure to impact, in *Proceedings of the Eighth Stapp Car Crash Conference.* Detroit, Mich., Wayne State University Press, 1964.
13. Hodgson VR, Lange WA, Talwalker RK: Injury to the facial bones, in *Proceedings of the Ninth Stapp Car Crash Conference.* Minneapolis, Minn., University of Minnesota Press, 1965.
14. Hodgson VR: Tolerance of the facial bones to impact. *Am J Anat* 1967; 120:1,113.
15. Hodgson VR, Nakamura GS: Mechanical impedance and impact response of the human cadaver zygoma. *J Biomech* 1968; 1:73.
16. Nahum AM, Gatts JD, Gadd CW, Danforth J: Impact tolerance of the skull and face, in *Proceedings of the Twelfth Stapp Car Crash Conference.* New York, Society of Automotive Engineers, 1968.
17. Schneider DC, Nahum AM: Impact studies of the facial bones and skull, in *Proceedings of the Sixteenth Stapp Car Crash Conference.* New York, Society of Automotive Engineers, 1973.
18. Daniel RP, Patrick LM: Instrument panel impact study, in *Proceedings of the Ninth Stapp Car Crash Conference.* Minneapolis, Minn., University of Minnesota Press, 1966.
19. Leung YC, Walfisch G, Got C, et al.: Etude d'une face pour tete de mannequin. *Ingenieurs l'Automobile* 1979; 3:210.
20. Leung YC, Tarriere C, Fayou A, et al.: Simulation de la face humaine sur un modele de mannequin. *Ann Chir Plast* 1980; 25(4):311–318.
21. Hubbard RP, McLeod DG: A basis for crash dummy skull and head geometry, in King WF, Meitz HJ (eds): *Human Impact Response—Measurement and Simulation.* New York, Plenum Press, 1973, pp 129–152.
22. Byars EF, Haynes D, Durham T, et al.: Craniometric measurements of human skulls. American Society of Mechanical Engineers paper No. 70-WA/BHF-8, 1970.
23. Hertzberg HTE, Daniels G, Churchill E: *Anthropometry of Flying Personnel,* 1950. Wright Air Development Center Technical Report 56-621, Wright Patterson Air Force Base, Ohio, 1957.
24. Krogman WM, Sassouni V: *A Syllabus in Roentgenographic Cephalometry.* Philadelphia, Philadelphia Center for Research in Child Growth, 1957.
25. Salzman JA: The research workshop on cephalometrics. *Am J Orthodont* 1960; 46:834.

CHAPTER 16

The Vertebral Column: Clinical Aspects

Steven R. Garfin, Michael M. Katz

INTRODUCTION

The initial major goals in caring for a patient sustaining spinal column trauma must be to preserve life, protect the neurologic function of the patient against further injury, and attempt to restore and maintain the stability of the spine. The latter objectives can best be accomplished by understanding spinal column (bone and soft tissue) anatomy and biomechanics. Anatomy is critical in identifying structures damaged in the plane of injury (tissues often poorly delineated in routine radiographs). Knowledge of the basic biomechanics of vertebral column trauma aids in the assessment of specific mechanisms of injury to the spine. These evaluations, supplemented by available data on recovery and healing rates, allow the physician to determine the stability of the vertebral column acutely and later, after bone and soft tissue healing have been completed.

Within the confines of this chapter it is impossible to present in detail all the clinical aspects of biomechanics, injury patterns, and treatment considerations related to spine trauma. Instead, an overview of the concepts of clinical stability/instability, as applied to spine trauma, is provided. Additionally, the topic of spinal cord repair/regeneration is not included. This is not meant to imply that this area is insignificant. Although currently, little can be done to reverse the pathophysiologic processes of spinal cord injury, many modalities are being investigated, including biochemical manipulations (e.g., high-dose steroids, thyroid-releasing hormone, naloxone), grafting techniques, and electrical stimulation.

INCIDENCE OF SPINAL TRAUMA AND SPINAL CORD INJURY (CLINICAL SIGNIFICANCE)

The rate of occurrence of vertebral column injury resulting in hospitalization in the United States has been estimated to be 233 per million population, or approximately 54,000 individuals per year.[1] This includes patients with and without spinal cord injury. The incidence of acute traumatic spinal cord injury is approximately 50 per million population. Neurologic damage associated with spinal injuries occurs in 14 percent of hospitalized, traumatized individuals,[1] and 10 percent to 20 percent of spinal fractures have associated neurologic deficits.[1-5] The exact percentage of spinal fractures resulting in neurologic deficit is difficult to obtain, since minor injuries, such as compression fractures, may not require hospitalization and are not easily recalled in retrospective studies. Conversely, many patients sustaining injuries associated with neurologic deficit have vertebral column fractures or dislocations. When anterior and posterior element fractures occur at the same level, the incidence of spinal cord injury reportedly is as high as 61 percent.[1] The lowest incidence of neurologic injury (3 percent) is observed with isolated vertebral body (compression) fractures.

Motor vehicle accidents are the most common cause of vertebral column injuries with or without neurologic deficits, followed by diving or falling accidents.[1] Fortunately, only 13 percent of vertebral column fractures sustained in motor vehicle accidents and 5 percent of those from falls produce spinal cord injuries. However, 43 percent of spinal fractures sustained in recreational activity (e.g., diving and football) produce spinal cord injury, often involving the cervical spine.

Age and sex are also factors in spinal trauma statistics. The highest incidence of males sustaining spinal fractures, with or without neurologic deficit, is in the 15-to 24-year-old age group. The percentage of females with a vertebral column fracture and neurologic deficit is also highest in this same range. However, 35 percent of females with a vertebral fracture without neurologic deficit are generally older (sixth decade or greater), while only 12.5 percent of all females with a vertebral fracture and a neurologic deficit fall into this elderly population.[1] This probably is related to the higher incidence of osteoporosis and associated low-energy fractures that occur in females over 55.

HISTORICAL PERSPECTIVE

The earliest written record on spine injury is found in the Edwin Smith Papyrus (3000 BC). This archeologic document discusses one apparent case of spinal cord injury.[6-8] Later, Egyptian physicians noted that some patients with vertebral trauma often had paralysis of the arms and legs and urinary incontinence, suggesting an association between spinal cord damage, vertebral injuries, and loss of function. Such injuries were treated with poultices of fresh meat, grease, and honey applied to the neck of the victim.[9–11] (The success rates of these remedies are not reported.) In 1000 BC Homer described a high mortality rate in Greek warriors with traumatic lesions to the head and neck, and Hippocrates, some 600 years later, pessimistically described spinal trauma in detail and felt little could be done to effectively treat cervical injuries.[11–13]

Celseus made the next important contribution in describing spinal cord trauma and distinguished cervical from thoracolumbar spinal cord injuries.[14] He reported that fractures of the cervical spine produced respiratory embarrassment and vomiting, while trauma to the lower portion of the spinal column produced paralysis of the lower extremity and urinary incontinence. He also expanded on Hippocrates' concept of manual, extension reduction of spinal deformities.[6] Later, Galen distinguished between injuries of the cervical vertebrae.[12,14] He observed that injuries of C-1 and C-2 caused sudden death, injuries of C-3 and C-4 stopped respiration, and injuries below C-7 seemed to spare the upper extremities.

In the 16th Century, Ambrose Paré readdressed the problem of spinal injury.[12–14] He accurately described the symptoms of cord compression:

> Amongst the symptoms are the stupidity, or numbness and palsy of the arms, legs, fundament and bladder, which take away their sense and motion, so that their urine and excrements come from them against their wills and knowledge, or else are wholly suppressed. Which when they happen, saith Hippocrates, you may foretell that death is at hand, by reason that the spinal marrow is hurt. . . . Having made such a prognostication, you may make an incision so to take forth the splinters of the broken vertebrae, which driven in press the spinal marrow in the nerves thereof.[14]

The recent period of managing vertebral column trauma arrived with the development of anesthesia and radiography. Hadra attempted to stabilize a C-6 on C-7 fracture dislocation with wire in 1891.[15] In 1911 Hibbs performed the first successful spinal fusion.[16] In the 1920s, based heavily on the principles advocated by Sir Ludwig Guttman, the emphasis in the treatment of vertebral trauma was placed on closed reduction of fractures. Davis proposed a method of reduction in which the patient was anesthetized and placed in the prone position.[17] An overhead pulley suspension raised the lower limbs and produced marked hyperextension (Fig. 1). A manual thrust by the physician was then made over the fractured vertebra in an attempt to realign the fracture. When reduction was achieved, the patient was immobilized in a plaster jacket.[17] In 1931, Watson-Jones modified this technique by using tables of different heights to hyperextend the spine and obtain a reduction (Fig. 2).[18]

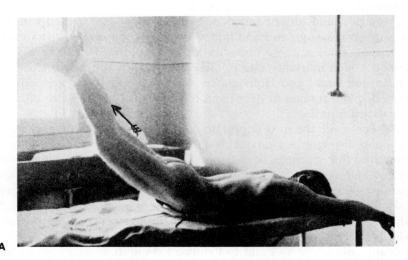

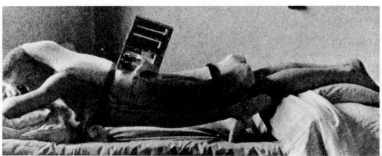

Figure 1. Photographs from Davis depicting **(A)** the overhead pulley suspension mechanism hyperextending the spine and **(B)** the postreduction plaster bed immobilization. (*From Davis.*[17])

Rigid internal fixation of thoracic and lumbar spinal fractures began after World War II with the development of spinous process plating for unstable fractures.[19-21] Later, Harrington revolutionized spinal care and rehabilitation with the introduction of his posterior spinal instrumentation devices.[22,23] Since then surgical techniques and instruments have proliferated and, hopefully, have continued to improve our ability to anatomically reduce and internally stabilize the injured spinal column.

Throughout the more modern era of spinal trauma management, rigid fixation has taken precedence. Despite the increasing use of internal immobilizing devices and fusions, neurologic recovery following traumatic spinal cord injury has not been impressive.[23-27] The major benefits of internal fixation of spinal fractures appear to be decreased hospital stay and early rehabilitation.[24]

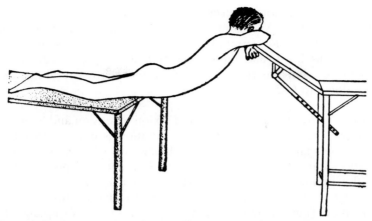

Figure 2. Drawing from Watson-Jones illustrating his technique of reducing flexion injuries using tables of different heights. (*From Watson-Jones*[18])

REGIONAL VARIATIONS AND ANATOMIC AND BIOMECHANICAL CHARACTERISTICS OF THE SPINE

Each region of the vertebral column has unique anatomic and biomechanical characteristics. Such variations must be understood in order to deal with spinal trauma effectively. (The reader is referred to the excellent text *Clinical Biomechanics of the Spine* by White and Panjabi for a more thorough and detailed discussion.[28])

The Cervical Spine
The cervical spine is adapted for mobility and support of the head. It is best discussed in two sections—the occipito–atlanto–axial complex and the lower cervical spine.

The Occipito–Atlanto–Axial Complex. This region of the spine serves as the transition zone between the vertebral column and the skull; the kinematics of this region are intricate. The occiput, atlas (C-1), and axis (C-2) allow significant motion to occur, particularly flexion-extension, lateral bending, and axial rotation. Despite approximately 50 percent of total cervical motion occurring here, the relatively wide diameter of the spinal canal at this level helps protect the underlying vital medullary structures from damage during physiologic activities.

At the atlanto–axial joint, 47 degrees axial rotation (50 percent of cervical rotation) can normally occur.[28] The transverse ligament passes posterior to the odontoid process and maintains the odontoid's position against the anterior ring of C-1. This, coupled with slightly laterally angulated facets, limits flexion-extension while allowing rotation. Though a significant amount of axial rotation is permitted at C-1-2, virtually none occurs at the occiput/C-1 junction. At this more cephalad location the primary motion is flexion-extension.[28] The vertebral arteries, along with the spinal cord, enter the skull at this level, and their integrity must be preserved. Loads borne through the occiput are transferred to the biconcave lateral masses of C-1 and then down to C-2.

The Lower Cervical Spine (C-3 to C-7). Flexion-extension and lateral bending occur across these segments, but axial rotation is limited. Rotation is inhibited by the ligamentum flavum, which begins at C-2 and is stiffer than the atlanto–occipital membrane, which joins C-1 and C-2 posteriorly. In addition, the oblique orientation of the facets (Fig. 3) combined with the contouring of the intervertebral disk space articulations found in the lower cervical spine (uncus) anatomically restrict rotation. Coupling of lateral bending with axial rotation does, however, occur in this region and is also related to these same anatomic configurations.[28]

The Thoracic Spine

The thoracic spine bridges the more mobile cervical and lumbar region. Less flexion-extension and lateral bending occur in the thoracic spine than in any other region. The rib cage provides much of this rigidity. Axial rotation in the thoracic spine is slightly greater than in the lumbar spine but less than in the cervical region. Though the ribs restrict the amount of rotation, the orientation of the facets (relatively vertical) allows for some motion in this plane (Fig. 3). The spinal canal in the thoracic region has the least amount of free space for the spinal cord. Teleologically, this limitation of space must be coupled with limited motion to protect the spinal cord. The upper thoracic spine is similar to the lower cervical spine in both the size of the vertebral bodies and the orientation of the facet joints.[29] In the lower thoracic spine the orientation of the facets and the size of the vertebral bodies are similar to those of the lumbar spine. The vertebral bodies gradually increase in size distally through all the anatomic regions. The spatial orientation of the facet joints, however, may change abruptly to the pattern found in the lumbar region anywhere from T-9 to T-12.[30,31] Because of this alteration in facet orientation, the upper thoracic spine exhibits axial rotation and lateral bend coupling similar to that found in the cervical spine, while the lower thoracic spine's motion characteristics more closely resemble those of the lumbar spine.[28] The thoracic spine vertebral bodies are somewhat wedge shaped, with the greater sagittal height posteriorly conferring a kyphotic curve to this region.

The rib cage and costovertebral junctions provide an important stabilizing unit for the thoracic spine.[32-34] Panjabi et al. found that when flexion was performed and posterior elements (including the posterior longitudinal ligament and the posterior half of the disk) were transected, the spine was on the verge of instability. Subsequent transection of the costovertebral

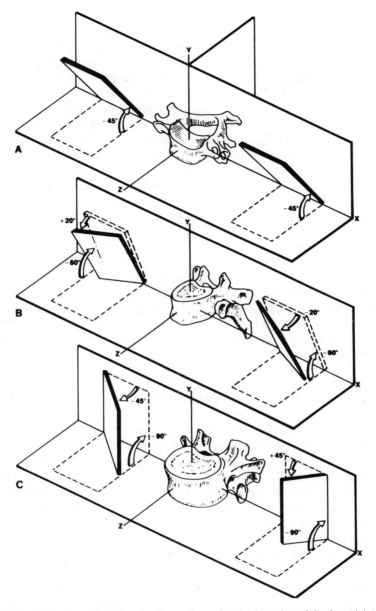

Figure 3. Drawings depicting the three-dimensional orientation of the facet joints of the spine in the three vertebral anatomic regions. **A.** A generalized example of the cervical spine. **B.** The thoracic region. **C.** A lumbar vertebra. These are generalized examples, and variations from individual to individual and within each segmental area frequently occur. (*From White et al.*[56])

joint produced failure. Similar results were produced when the spine was placed in extension and the anterior structures were sectioned. They concluded that in the clinical situation, when costovertebral joint disruption has occurred, the ability of the thoracic spine to carry normal physiologic loads should be questioned and appropriate treatment precautions instituted.[33]

The functional biomechanics of the rib cage have been well demonstrated.[28,34–36] Andriacchi et al., utilizing a computer model, found that the stiffness properties of the spine were greatly enhanced by the presence of the rib cage in all physiologic motions, especially extension.[36] Removal of the sternum decreased the stiffness values to those of the spine without the thoracic wall. Removal of one or two ribs, however, did not significantly affect the stiffness properties. Additionally, it has been observed that the rib cage increases the axial mechanical stability of the spine in compression by four times when compared to models without the rib cage.[36,37]

The Lumbar Spine
The lumbar spine, being the most distal, must bear the greatest loads in the vertebral column. The vertebral bodies are the largest, perhaps in response to this load-carrying capacity. The facets gradually attain a sagittal orientation in the progression from L-1 to L-5 (Fig. 3). This provides significant stability against rotation and bending in physiologic ranges. A normal lordotic curve is found in this region, partially created by wedge-shaped disks, larger anteriorly than posteriorly.

DEFINITION OF CLINICAL STABILITY

The clinical determination of spine stability is often difficult and controversial. Watson-Jones published a classification of fractures of the spine, together with a method of treatment of what he considered to be pure flexion fractures that were reduced and immobilized in plaster jackets.[18] He stated that accurate reduction was usually possible and a goal necessary to pursue. He felt strongly that consolidation would generally occur without deformity and that if all instructions were followed by the physician and patient, the functional results would be excellent.

Nicoll, in 1949, however, found this not to be the case.[38] In a large series of thoracolumbar fractures immobilized in hyperextension, anatomic union was not frequent and did not correlate with function. Since he felt that the anatomic results were unsatisfactory (and perhaps not essential to obtain), he suggested a new classification of thoracolumbar fractures, dividing them into stable and unstable patterns. In stable fractures the interspinous ligaments remained intact, while in the unstable configuration the ligaments were ruptured. Amount and type of wedging, fracture pattern, and widening of the interspinous process distance were critical factors in Nicoll's classification scheme (Table 1). His stable patterns were anterior and lateral wedge fractures and all lamina fractures above L-4. Fracture-subluxations or

TABLE 1. CHANGING DEFINITIONS OF STABILITY/INSTABILITY

Stable	Unstable
Nicoll[38]	

Anterior wedge fractures
Lateral wedge fractures
Lamina fractures above L-4
	Fracture-subluxations with
	  rupture of interspinous
	  ligaments
	Fracture-dislocations
	L-4 or L-5 lamina fractures

Holdsworth[40]	

Wedge compression fractures
Compression (comminuted)
  burst fractures
	Dislocations
	Extension fracture/dislocations
	Slice fractures (flexion-
	  rotations)

Whitesides et al.[41,83]	
Compression fractures Stable burst fractures	Unstable burst fractures Flexion-distraction fractures Dislocations Extension injuries Slice fractures
Jacobs et al.[26]	
Clinically unstable	1. "Any injury of the thoracolumbar spine sufficiently severe to be associated with a neurologic deficit" 2. Injuries that may result in neurologic damage 3. Injuries that may lead to chronic instability
White and Panjabi[28]	
Clinical instability	" . . . the loss of the ability of the spine under physiologic loads to maintain relationships between vertebrae in such a way that there is neither damage nor subsequent irritation to the spinal cord or nerve roots, and, in addition, there is no development of incapacitating deformity or pain due to structural changes."

With time, the definitions have become less precise, yet more encompassing. An understanding of clinical biomechanics is necessary to effectively employ the categorization developed by Jacobs and White and Panjabi.

dislocations and lamina fractures of L-4 or L-5 were considered "unstable," and compression fractures with "severe wedging" were listed as unstable. Nicoll advocated plaster immobilization for unstable fractures, with no attempt made to anatomically reduce the spinal deformity. His findings were confined to the thoracolumbar region.

Roaf, in what many consider a classic article, studied the mechanisms of spinal injuries and the behavior of the spine and its components under a variety of loads and stresses.[39] Using cadaveric spinal units as models, he found that the forces of pure compression were primarily absorbed by the vertebral body. The normal nucleus pulposus in its liquid (or gel) state was incompressible, and the anulus fibrosus bulged very little. The vertebral body always failed before the normal disk gave way. If the nucleus pulposus had lost its turgor, abnormal mobility developed between the vertebral bodies, and disk prolapse occurred prior to fracture. Similarly, pure flexion produced a fracture of the vertebral body with collapse. Roaf stated he "never succeeded in producing pure hyperflexion injuries of a normal intact spinal unit. Before the posterior ligaments rupture, the vertebral body always becomes crushed."[39] Similarly, with a pure extension mechanism, the posterior elements failed and ligamentous disruption was not produced.

In contradistinction, Roaf found that rotational forces in conjunction with either flexion or extension could produce disruption of ligaments either posteriorly (with flexion) or anteriorly (with extension). He proposed this mechanism—rotation coupled with flexion or extension—as the primary cause for a dislocation injury. He concluded that the intervertebral joints, disks, and ligaments were resistant to compression, distraction, flexion, and extension but vulnerable to rotation and horizontal (shear) forces.[39]

Holdsworth published in 1963 his classification of spinal fractures, one which is still used widely by many clinicians.[40] Employing Roaf's mechanism of injury, Holdsworth elaborated on Nicoll's original distinction between stable and unstable injuries and formulated a scheme applicable to cervical as well as thoracolumbar fractures. Similar to Roaf, Holdsworth felt that the spine could be subjected to four types of force: (1) flexion, (2) flexion and rotation, (3) extension, and (4) compression.

Stable injuries in the Holdsworth classification included wedge compression fractures and compression burst fractures.[40] As these are primarily bone injuries, both fracture patterns were thought to heal readily. He felt that while these fractures often united with some deformity, complete stability of the spine was the usual result. In an earlier article, co-authored with Hardy, Holdsworth felt that an anatomic reduction was important and advocated plate fixation of spinous processes to achieve this result.[19] Later, he noted the difficulties inherent in that type of surgical stabilization, as well as a lack of correlation between anatomic reduction, neurologic function, and/or pain. Holdsworth concluded that anatomic reduction may not be critical in overall results, but may be worth striving for, if possible.[40]

Holdsworth classified unstable injuries into dislocations, extension fractures, and rotational fracture-dislocations.[40] The basic component of these patterns was ligamentous disruption. With a flexion-rotation mechanism, the posterior ligaments ruptured, and if the amount of flexion was sufficient to disengage the articular facets, a dislocation occurred. The vertebral body would not be compressed as there was no fulcrum within the body for the posterior lever. He believed pure dislocation (without fracture) to be common in the cervical spine, for the amount of flexion necessary to disengage the horizontally positioned articular facets was relatively slight. In the lumbar spine, however, the facets supply more stability as they are sagittally and obliquely oriented, and fracture-dislocations are more commonly produced

(Fig. 3). The thoracolumbar junction is especially prone to this type of injury, as this is the first mobile segment beneath the rigid thoracic column. Holdsworth expressed the view that the rotational fracture-dislocation occurred frequently at the thoracolumbar junction and that the spinal cord and roots were endangered with these forces. He stated that 95 percent of all thoracolumbar level paraplegia occurs as a result of this mechanism of injury. Despite the postoperative difficulties encountered, Holdsworth felt surgical fusion of these unstable patterns was important. Though drawing our attention to acutely stable and unstable injuries, he ignored the potential for late instability.

Whitesides and Kelly recognized that unstable burst fractures were "the most common cause of neural injury in the thoracolumbar region."[41,42] They proposed a classification system based on a two-column construct—an anterior weightbearing column of vertebral bodies and a posterior column of neural arches resisting tension (Fig. 4). An advantage of this system was that injuries with late instability could be incorporated into their scheme.[42]

Recently, Dennis[43] and McAfee et al.[44] proposed a new classification scheme for thoracolumbar fractures utilizing a three-column system. A middle osteoligamentous complex consisting of the posterior longitudinal ligament, posterior portion of the vertebral body, and posterior anulus fibrosus, along with the spinal cord, is now considered in their classification of injuries (Fig. 4). The status of the middle osteoligamentous complex is the key anatomic determinant of stability and also of the surgical method chosen for fixation.

Jacobs et al. in 1980 amended Holdsworth's classification with regard to thoracolumbar injuries.[26] These authors consider any injury of the thoracolumbar spine associated with a neurologic deficit to be unstable and advocate internal fixation to adequately immobilize the injury, promote healing,

Figure 4. Vertebra overlying a two-column pillar system. One column consists of the vertebral body, and the second column is composed of the posterior elements. In the three-column concept, the middle column (the area between the pillars) is composed of the posterior longitudinal ligament, posterior anulus fibrosus, and posterior lip of the vertebral body, along with the spinal cord/canal and ligamentum flavum.

prevent late pain and deformity, and allow early ambulation/rehabilitation. Additionally, fractures that may result in neurologic damage secondary to extrusion of fragments into the neural canal, demonstrate evidence of posterior ligamentous disruption, or have the possibility of instability or neurologic deficit occurring late due to progressive deformity should have the benefit of early surgical stabilization. Jacobs cites the work of Nash and colleagues, observing the so-called stable wedge compression fracture progressing to produce neurologic deficits over long periods and emphasizes caution in treating these injuries as stable.[45] Jacobs considers a compression fracture with loss of at least 50 percent of its anterior vertebral body height to have a high likelihood of late mechanical instability.[26]

White and Panjabi perhaps best define clinical instability as:

> . . . the loss of the ability of the spine under physiologic conditions to maintain relationships between vertebrae in such a way that there is neither damage nor subsequent irritation to the spinal cord or nerve root and, in addition, there is no development of incapacitating deformity or pain from structural changes.[28]

This definition addresses both the acute and late stages of vertebral column trauma. White and Panjabi define physiologic loads as those loads incurred during normal activity, incapaciting deformity as gross deformity unacceptable to the patient, and incapacitating pain as discomfort uncontrolled by nonnarcotic analgesics.[28] This definition is broad and perhaps all encompassing, but it imposes on the treating physician the responsibility to distinguish between lesions that should heal from those with a high incidence of nonunion or potentially progressive problems.

A notable exception to Roaf's observation on the inability of a pure flexion injury to produce posterior disruption is found in the lumbar seat belt injury or Chance fracture.[46,47] The mechanism, well described by Smith and Kaufer and later Rennie and Mitchell, provides a good clinical example of spinal biomechanics.[48,49] In the thoracolumbar spine the flexion axis normally passes through the nucleus of the disk (Fig. 5). The distance from the flexion axis (fulcrum) to the tip of the spinous process is three to four times greater than the distance from the flexion axis to the anterior margin of the

Figure 5. Diagrammatic representations of the flexion axis of the **(A)** normal spine and **(B)** a spine involved with a flexion-distraction injury related to wearing a seat belt. **A.** The fulcrum is through the nucleus pulposus and the distance from the flexion axis to the posterior lip of the spinous process (line b) is approximately three to four times greater than the distance from the flexion axis to the anterior lip of the vertebral body (line a). **B.** When the flexion axis is moved to the anterior abdominal wall, the entire vertebral body and posterior elements are subjected to tensile forces, as opposed to the large compression force applied to the anterior aspect of the vertebral bodies shown in **A.** (*From Smith and Kauffer.*[48])

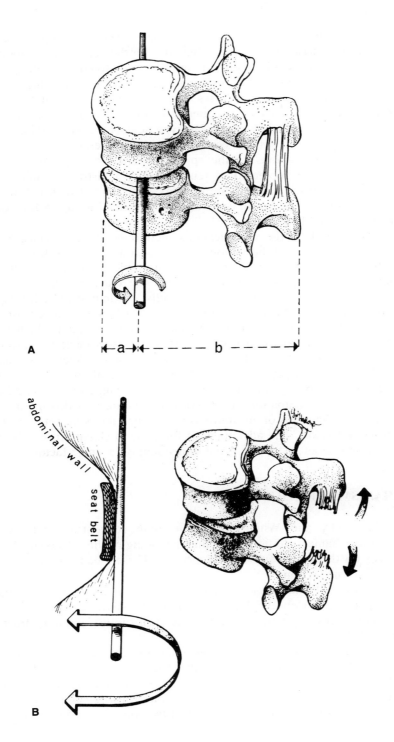

A

B

vertebral body. With flexion the anterior body receives a compressive load that is three to four times larger than the tensile load placed upon the interspinous ligaments. Therefore, an anterior body fracture occurs prior to posterior ligament failure. Larger forces produce more wedging. As the anterior body deformation progresses, energy is dissipated so that posterior disruption is minimized. However, abnormal lumbar mechanics occur during deceleration in individuals wearing lap type seat belts. The flexion axis is moved anteriorly to the point of contact between the belt and the abdominal wall (Fig. 5). All portions of the spine are, therefore, posterior to the flexion axis and are subjected to tensile stress; a distraction injury with disruption of the posterior structures occurs, typically between L-1 and L-3 (Fig. 6).[48] A similar pattern in cervical injuries related to lap-sash restraints has been described by Epstein et al.[50]

While several of the classifications mentioned have emphasized the mechanism of injury and the radiographic appearance as the sole criteria on which to base the label of "stable" and "unstable" fractures, White and Panjabi have developed a system in which the diagnosis of an unstable fracture is arrived at by completing a checklist encompassing radiographic, neurologic, and social factors.[28] Different checklists are designed for each area of the spinal cord, reflecting regional anatomic and biomechanical differences. These checklists provide a systematic assessment using various criteria, including status of the anterior and posterior elements, sagittal plane translation or angulation, the presence of spinal cord or nerve root damage, disk narrowing, and the anticipation of dangerous loading in the future (hazardous occupation or lifestyle). Points are awarded in each category, and the diagnosis of instability is made when the point total reaches a predetermined value. The advantage of this type of classification scheme is that the diagnosis of instability is not dependent on a single factor.

MECHANISMS OF INJURY

Employing Kelly and Whitesides' two-column concept of stability and Roaf's biomechanical studies, some understanding of the effects of the various forces on the spine can be developed.[39,41] These mechanisms are briefly described in this section.

Compression (axial) loading generally occurs as forces are transmitted to a straightened vertebral column. As shown by Roaf, vertebral body/endplate fractures develop before an intact disk will rupture and burst fractures are created.[39]

Flexion forces cause compression anteriorly at the bodies and disks, with tensile forces developed posteriorly. The ligaments usually do not fail immediately (particularly with rapid loading rates), but posterior avulsion fractures may develop. Anteriorly, as the bone fractures and more angulation develops, force is dissipated. With intact posterior ligaments (or an avulsion fracture), a stable fracture pattern most often results. However, facet capsule disruption may occur, leading to instability.

Flexion-rotation forces are similar to those described above for pure flexion, but the ligaments and facet capsules tend to fail as rotation increases, disrupting both anterior and posterior columns. An unstable fracture pattern frequently develops.

Extension forces are created when the head or upper trunk is thrust posteriorly and produces the reverse pattern from pure flexion injuries. Tension is applied anteriorly to the strong anterior longitudinal ligaments and anterior portion of the anulus fibrosus, while compression forces are transmitted to the posterior elements. This may result in lamina, spinous process, and perhaps pedicle fractures.[51] Avulsion fractures off the anterior portions of the vertebral bodies (previously called "teardrop") may occur but are not pathognomonic of extension injuries as previously thought.

Distraction forces have been described in a previous section. In this injury pattern the fulcrum is moved anteriorly (usually to the abdominal wall), and the entire vertebral column is subjected to large tensile forces. The vertebrae and disks essentially are torn or avulsed, not crushed as typically occurs in most spinal fractures. These are potentially very unstable injuries, especially if the anterior longitudinal ligament also fails.

Unfortunately, the mechanisms just described do not often occur as singular events. Some combination of forces is frequently applied to the spine as the human frame is whipped around the motor vehicle, twists, and tumbles down heights, and so on. Because of this combination of forces, careful assessment must be made to determine which columns are involved, whether bone and/or ligaments have failed, and what remaining structures are intact to employ in maintaining or obtaining a reduction. Only after making such an assessment can the physician intelligently plan surgical stabilization. For example, an anterior approach to the spine for a flexion-rotation injury should be avoided, if possible, as it would violate the only remaining intact structures (e.g., anterior longitudinal ligament), thereby making postoperative immobilization extremely tenuous.

As is evident in this discussion, the overall concept of spinal stability following trauma is slowly evolving. When first introduced by Nicoll and amplified by Roaf and Holdsworth, the mechanisms of injury seemed clear and easy to apply.[38–40] Additionally, these authors felt comfortable listing the fracture patterns associated with the stable and unstable categories. More recently, the work of Jacobs and of White and Panjabi have made it difficult to categorize general types of fractures as either stable or unstable.[26,28] As radiographic techniques improve, biomechanics laboratories test more theories, and long-term follow-up of patients increases, the definitions become less precise. Despite this, the concept of stability is important in preparing a treatment plan and managing the fracture.

Stability can be examined or predicted for both acute and late clinical settings (Table 2). If significant deformity exists, a neurologic deficit is

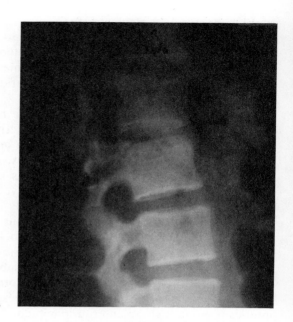

A

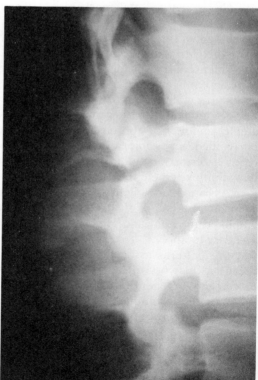

Figure 6. An example of a flexion-distraction injury. This 27-year-old male was involved in a motor vehicle accident while wearing a seat belt. **A.** A fracture through the spinous process, pedicle, and the posterior portion of the vertebral body can be seen. Presumably this injury then extended out through the L-1-2 disk space. Narrowing anteriorly between L-1 and L-2 can be seen. **B.** A cone-down view is shown. The fracture can be seen extending through the spinous process. (*cont.*)

B

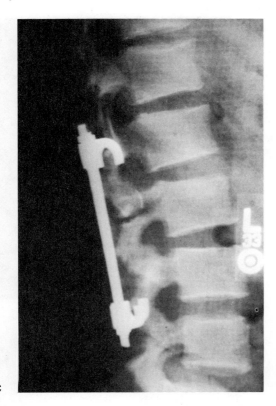

C

Figure 6. (*cont.*) **C.** Harrington compression rods were used to stabilize this injury.

present or was present by history, and/or significant translation or angulation is noted, one can assume that early (acute) instability exists. In this case the spine must be protected from further damage by some form of relatively rigid immobilization (surgery, halo-immobilizer, skeletal fixation, traction, Stryker frames, etc.). If, however, the injury can be classified as "stable" and involves primarily bone with mild vertebral body compression, isolated lamina, or facet fractures, less rigid immobilization may be employed. Weightbearing, gravity, or other normal physiologic forces should not alter the configuration of these fractures.

TABLE 2. STABILITY AND TIME

Stable acute	→	Stable late
Unstable acute	→	Stable late
Stable acute	→	Unstable late
Unstable acute	→	Unstable late

The patterns of stability/instability relative to time from injury that may occur.
Type of immobilization and surgical recommendations are dependent on understanding these relationships and possible early/late changes.
Late is greater than 8–12 weeks from traumatic event.

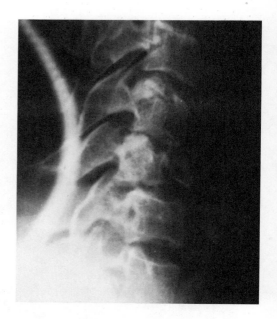

A

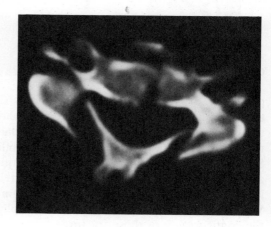

Figure 7. A 19-year-old female with quadriparesis from a C-5 fracture with retropulsion of bone into the canal. **A.** This lateral radiograph demonstrates the comminution of C-5 with retropulsion and retrolisthesis of the body. **B.** The bony involvement is better demonstrated on the CAT scan. Fractures of the lamina and vertebral body, as well as spinal canal encroachment, can be well seen in this cut. (*cont.*)   **B**

As stated previously, the physician must not only assess the early stability of a given spinal injury, but also must be able to predict the late stability of that injury. Early instability does not preclude late stability. Similarly, an injury that is stable early may demonstrate late instability. For example, a cervical body burst fracture with additional fractures of the lamina and perhaps a pedicle is acutely unstable. However, it is almost entirely a bony injury, which would be expected to heal and can be considered stable-late (Fig. 7). Ligaments, on the other hand, do not have such high healing rates. A flexion-rotation injury causing wedging (fracture) of the body anteriorly with disruption of posterior ligaments may at first appear innocuous and

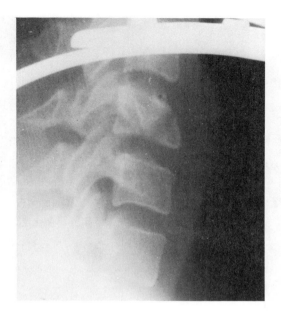

C

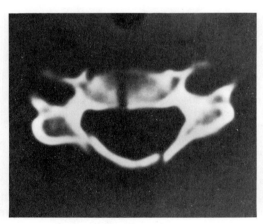

D

Figure 7. (*cont.*) **C.** Postreduction radiograph, after applying skeletal traction, shows realignment of the vertebral column. **D.** This nearly anatomic reduction is confirmed on the accompanying CAT scan. Although this injury is acutely unstable, it mainly involves bone and will heal primarily. Because of this, the patient was treated in a halo-cervical immobilizer without surgical fusion.

stable. However, the expected rate of ligamentous healing in spinal injuries varies but may be as low as 50 to 60 percent.[28] For many individuals this healing rate is unacceptable and implies an almost 50 percent chance of pain and progressive slipping or deformity. This injury pattern, therefore, could be classified as unstable-late. Stauffer and others have shown examples of apparently stable compression fractures developing progressive late deformities, ultimately requiring surgical stabilization and fusion (Fig. 8).[45,52] If the physician is able to predict which injury patterns will go on to late instability, early fusion/stabilization may be recommended.

It is interesting to note that early authors did not explicitly comment on

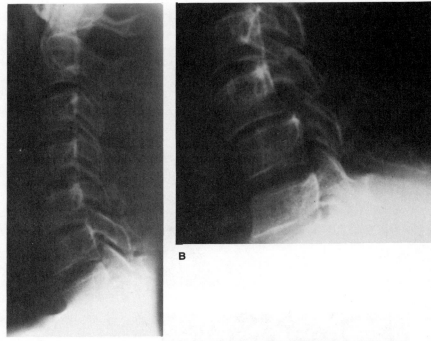

Figure 8. Example of a 34-year-old male with a cervical spinal injury. He initially presented with neck pain and paresthesia in his left arm in approximately the C-6 nerve root distribution. **A.** Original radiographs were relatively unremarkable, except for narrowing at the C-5-6 disk space. Flexion-extension films were performed and were felt to be normal. Because of the neck pain and history of paresthesia, the patient was treated in a cervical orthosis. He returned at 2 weeks for follow-up with, again, normal radiographs. **B.** At 1 month, he returned with a lateral radiograph demonstrating subluxation of C-5 on 6. This was reduced and fused posteriorly along with interspinous process wiring.

the development of late instability. Even though current definitions appear vague compared to the specifics of Nicoll, Holdsworth, and Whitesides, it is this appreciation of late changes that has altered our understanding of spinal trauma. Biomechanical and anatomic factors and considerations are essential for the understanding and treatment of spinal trauma (Table 1).

DIAGNOSTIC TECHNIQUES IN SPINAL TRAUMA

The evaluation of spinal injuries begins with a thorough history and physical examination. Clues to the mechanism of injury may be obtained by questioning either the patient or witnesses to the accident. The location of contusions and abrasions along the back, shoulder, and trunk may provide additional information as to the mechanism of injury in thoracic and lumbar injuries, while associated facial and scalp lesions may prove useful in evaluating cervical spinal injuries. A complete neurologic examination must be performed as soon as possible to assess the degree of spinal cord injury, help determine the level of injury, and provide a baseline for future examinations.

Radiographic techniques provide the most useful information for assessing the extent of bony injury and degree of stability present at the site of injury. Standard anteroposterior and lateral radiographs should be obtained whenever significant trauma is suspected. In addition, for cervical injuries, open-mouth views of the odontoid process should be obtained when possible.[53]

Conventional polytomography and computerized tomography provide additional definition in evaluating vertebral fractures.[44,54,55] Keene et al. have shown that computerized tomography combined with standard radiographs is at least equal to conventional polytomography in providing information on spinal trauma, while computerized tomography is superior to all methods in demonstrating impingement on the neural canal.[54] These authors conclude that computerized tomography should replace conventional polytomography as the initial study to augment standard radiographs in the assessment of thoracic and lumbar fractures, while conventional polytomography should be reserved for precise evaluation of the pars interarticularis and/or facets. With newer, lateral reconstruction formats, even this recommendation may soon be outdated. Additionally, computerized tomography with metrizamide is probably superior to myelography in localizing the site of neural canal compromise in acute thoracolumbar injuries. Hematoma, edema, and so on can obstruct dye flow and obfuscate the standard myelographic study in the assessment of acute spinal injuries.

Radiographs also help evaluate the continuity of the ligamentous restraints of the spine. Lateral views taken of the spine during flexion and extension can be compared, directing attention to sagittal translation or angulation of one vertebra upon another. This test should be performed with a conscious, alert, and cooperative patient, since neurologic compromise may result from unsupervised or overzealous motion of the injured spine. Attention to the locus of motion noted on the radiograph is also important in judging the adequacy of this technique. In the cervical spine, cranio-occipital flexion may be all the motion obtained in the acute setting because of cervical muscle spasm. Similarly, in assessing lumbar instability, the physician should evaluate the film and not accept flexion attempts that occur only at the hips. Using a standardized lateral radiographic technique to maintain consistent magnification, translation between segments of 3.5 mm or more or angulation of 11 degrees greater than adjacent segments is suggestive of cervical instability (Fig. 9).[28,56,57] These values were determined in the laboratory by sectioning various ligaments across a motion segment until instability was noted. Absolute values vary depending on the area studied.

Based on Breig's findings that the spinal cord demonstrates elasticity and

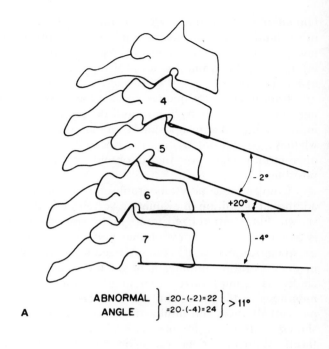

ABNORMAL ANGLE } =20-(-2)=22 } > 11°
               =20-(-4)=24

A

Figure 9. Measurements of instability are made on lateral radiographs as shown in **(A)** and **(B)**. These examples of instability were determined in the laboratory as related to the cervical spine. **A.** Angulation between adjacent segments of greater than 11 degrees or translation **(B)** greater than 3.5 mm is significant. These drawings demonstrate how measurements are obtained. Values may vary for different regions and levels of the spine. (*From White et al.*[56])

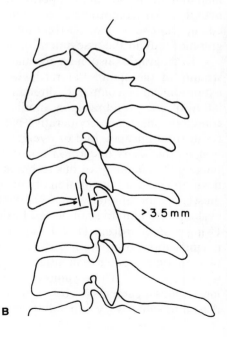

> 3.5 mm

B

compliance in the axial direction but is unable to accommodate deformation in the horizontal direction, White and Panjabi have proposed an axial stretch test for evaluating acute injuries[28,56-58] This is felt to represent less risk of neurologic impairment than present with conventional flexion-extension views. Progressive increments of traction are applied to the cervical spine in an effort to define ligament disruption. This test should only be performed according to the originators' protocol, with careful monitoring of the patient and radiographs at each 10 pounds addition of weight. It is most useful in determining the anatomic level of instability when a neurologic deficit is present and the preceding radiographic evaluation has provided no clues as to the level of instability. If the axial stretch test is also normal, the diagnosis of cord contusion without evidence of spinal column instability may be made to explain a neurologic abnormality (Fig. 10).

COMMON SPINE FRACTURES AND DISLOCATIONS

Atlanto–Occipital Joint
Injuries to this articulation are frequently fatal and often missed clinically. In addition to the stability inherent in the cup-shaped articulation of the atlanto–occipital joints, additional anatomic stability is obtained through the tectorial membrane, the alar ligaments, and apical ligament. Wiesel and Rothman have shown that the normal range of sagittal plane translation in flexion and extension is less than 1 mm.[59] White et al. suggest that any subluxation or dislocation be considered unstable.[28,60] More than 5 mm between the tip of the dens and the base of the occiput or more than 1 mm translation in flexion-extension is evidence of clinical instability.

Atlanto–Axial Joint
The most important structure affecting clinical stability of this articulation is the transverse ligament. The apical and alar ligaments act as secondary stabilizers of the C-1, C-2 joint.[28] The dens itself is also important in the overall stability of this joint, since if the dens is hypoplastic, congenitally not intact (os odontoideum), or fragmented, the transverse ligament may not provide stability. Fielding's studies revealed that an intact transverse ligament prevented significant anterior displacement of C-1 on C-2.[61] Sagittal displacement greater than 3 mm indicates rupture of the transverse ligament. Sagittal motion more than 5 mm implies transverse ligament disruption as well as failure of the alar and apical ligaments.

A comminuted fracture of the ring of C-1 (Jefferson fracture) with overlap of the lateral mass of C-1 on C-2 of 7 mm or more (demonstrated on open-mouth radiographic views, tomograms, or CT scans) indicates transverse ligament disruption along with the C-1 fracture and should be considered clinically unstable.[62,63] Additionally, any C1-2 injury with neurologic signs or symptoms should be considered unstable. Hangman's fractures

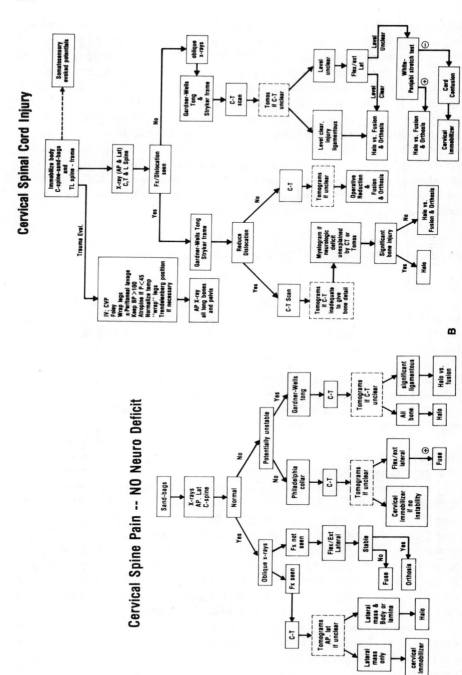

Cervical Spinal Cord Injury

Immobilize body C-spine-sand-bags and TL spine - frame — Somatosensory evoked potentials

Trauma Eval.

IV; CVP
Foley
Wrap legs
±Peritoneal lavage
Keep BP >100
Atropine if P <45
Normalize temp
"wrap" legs
Trendelenburg position
if necessary

AP X-ray all long bones and pelvis

X-ray (AP & Lat) C,T & L Spine

Fx/Dislocation seen

No → oblique x-rays → Gardner-Wells Tong & Stryker frame → C-T scan → Tomes if C-T unclear
- Level unclear → Flex/ext Lat → Level Unclear → White-Panjabi stretch test
  - ① → Cord Contusion → Cervical Immobilizer
  - ⊕ → Halo vs. Fusion & Orthosis
  - Level Clear → Halo vs. Fusion & Orthosis
- Level clear, injury ligamentous → Halo vs. Fusion & Orthosis

Yes → Gardner-Wells Tong Stryker frame → Reduce Dislocation
- No → C-T → Tomograms if unclear → Operative Reduction & Fusion & Orthosis
- Yes → C-T Scan → Tomograms if C-T inadequate to give bone detail → Myelogram if neurologic deficit unexplained by CT or Tomos → Significant bone injury
  - No → Halo vs. Fusion & Orthosis
  - Yes → Halo

B

Cervical Spine Pain -- NO Neuro Deficit

Sand-bags → X-rays AP. Lat C-spine → Normal

No → Potentially unstable
- Yes → Gardner-Wells tong → C-T → Tomograms if C-T unclear
  - significant ligamentous → Halo vs. fusion
  - All bone → Halo
- No → Philadelphia collar → C-T → Tomograms if C-T unclear
  - Flex/ext lateral → ⊕ → Fuse
  - Cervical immobilizer if no instability

Yes → Oblique x-rays
- Fx seen → C-T → Tomograms AP. Lat if unclear
  - Lateral mass & Body or lamina → Halo
  - Lateral mass only → cervical immobilizer
- Fx not seen → Flex/Ext Lateral → Stable
  - No → Fuse
  - Yes → Orthosis

A

Thoraco-Lumbar Spinal Cord Injury

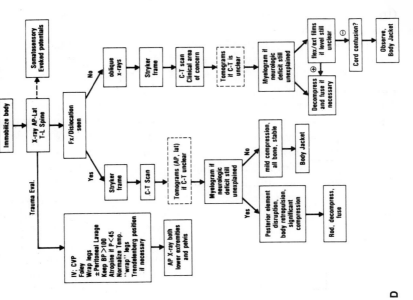

D

Thoraco-Lumbar Spine -- NO Neuro Deficit

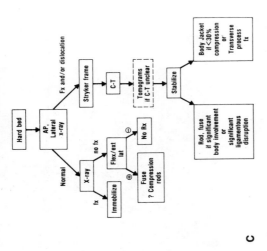

C

Figure 10. These flow charts are used in the Trauma Unit at the University of California, San Diego Medical Center. These are not rigid protocols, but they do provide guidelines to be followed by the various treating physicians involved in the care of patients with spine injuries. These flow charts have been developed for **(A)** patients with cervical spine pain but no neurologic deficit, **(B)** individuals with spinal cord injuries at the cervical level, **(C)** thoracic or lumbar spine pain without neurologic deficit, and **(D)** a spinal cord injury emanating from the thoracic or lumbar area. (*We appreciate the assistance of Dr. L.F. Marshall, Department of Neurosurgery, at the University of California, San Diego, in the preparation of these flow charts.*)

spondylolisthesis of the axis) occur through the pars interarticu-
. Though initially unstable, they tend to heal without surgery,
canal is widened from the injury, neurologic deficits seldom

Lower Cervical Spine

The most important anterior stabilizing structures in the lower cervical spine
are the anulus fibrosus, with its sturdy attachments to the vertebral bodies,
and the anterior and posterior longitudinal ligaments. Posteriorly, the facet
joint capsules and articulations are critical. If a motion segment has all its
anterior elements plus one additional structure intact or all of its posterior
elements plus one additional structure intact, it will most likely remain
stable under physiologic loads.[28] For a clinical margin of safety, however,
any motion segment in which all anterior elements or all posterior elements
are either destroyed or unable to function should be considered potentially
unstable.[28,65]

There is some correlation between neurologic deficit and the radiographic appearance of the spine after trauma. Burst fractures are frequently
associated with spinal cord injury.[28] Unilateral facet dislocations tend to be
stable, although bilateral facet dislocations are usually unstable.[28,66] Beatson,
in an often quoted work, established criteria to differentiate unilateral from
bilateral facet dislocations on lateral radiographs.[66] Twenty-five percent or
less anterior displacement of one cervical vertebral body on another implies a
unilateral facet dislocation (Fig. 11). Fifty percent or more translation is

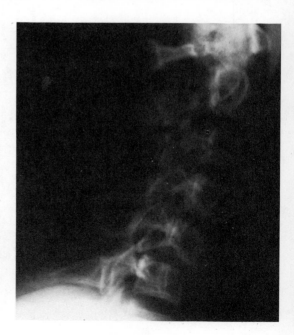

Figure 11. Nineteen year-old
male with unilateral locked facet
at C5-6. There was a fracture of
the inferior tip of the right facet
that is not demonstrated on this
radiograph. An open reduction
was required, followed by posterior wiring and fusion.

indicative of bilateral dislocations. In the former instance, anterior ligaments may be stretched but intact, and usually one capsule is intact. This may, therefore, be stable and not progress, though it frequently leads to pain and may cause nerve root compromise. The mechanism of injury is flexion-rotation. Bilateral facet dislocations, however, are unstable and frequently associated with neurologic injury. Anterior and posterior ligaments are disrupted, and the incidence of spontaneous healing, even if reduced and immobilized, is relatively low.[66]

Thoracic and Thoracolumbar Spine

As discussed earlier, the rib cage provides a unique contribution to the biomechanics of this region. This, coupled with the relatively sagittal orientation of the thoracic facets, creates a fairly rigid construct. Generally, significant force is required to fracture the thoracic spine, and rotational, ligamentous disruptions are rare. This is fortunate, as the canal size in the thoracic region is relatively small in relation to the size of the spinal cord. Thus, ligament deformation within the elastic range may permit enough displacement to deliver a detrimental impact to the cord.

Anterior wedge, lateral wedge, and compression fractures of less than 30 percent are usually stable unless the posterior interspinous ligaments are torn. Compression fractures and wedging of greater than 50 percent are usually associated with posterior element disruption and should be considered unstable.[67] Dislocations and fracture-dislocations should also generally be considered unstable.[68-70] White and Panjabi's checklist for clinical instability of this region reflects Riggins and Kraus' finding that there is a tendency for structural damage in the thoracic spine to be associated with neurologic deficits.[1,60]

Lumbar Spine

Clinical instability of lumbar fractures presents unique problems to the physician. Riggins and Kraus reported that only 3 percent of patients with lumbar spine dislocations and fracture-dislocations had neurologic deficits.[1] In addition, neurologic injuries at this level tend to be less debilitating than at higher levels, since roots rather than spinal cord are involved. The lumbar spine, however, must bear such high physiologic loads that subsequent pain and deformity are of significant consideration in treating lumbar fractures.

Anterior element stability in lumbar spine fractures depends largely on the well-developed anterior longitudinal ligament and anulus fibrosus. The posterior longitudinal ligament is less important in this region. As discussed earlier, the posterior facet joints play a critical role in maintaining stability. The interspinous ligaments are less important and occasionally absent.[71]

The presence of a neurologic deficit associated with a lumbar injury is a strong indicator of clinical instability. The presence of such a deficit within the relatively capacious and mobile lumbar spinal canal indicates that a large displacement must have occurred and significant bony or ligamentous damage taken place. Wedge fractures in the lumbar spine are common. As in other areas, compression of 50 percent or more usually implies instability. Futhermore, after the acute lumbar compression fracture heals, late deformity, pain, and perhaps neurologic compromise may develop due to the effects of gravity and weight on the spine above the kyphotic element. As the amount of compression increases, the bending moment at the body increases, and progressive deformity results as the tensile forces on the posterior interspinous ligaments elevate and exceed their capacities (Fig. 12).

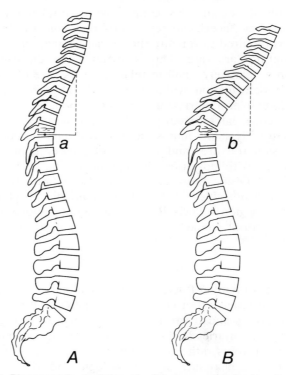

Figure 12. Diagrammatic depiction of mild compression fracture in **(A)** and / fracture with anterior wedging and significant posterior ligamentous disruption **(B)**. Line *a* is the distance from the flexion axis and a plumb line drawn from the vertebral body above which reassumes a more normal mechanical relationship to the lower vertebral segments. The plumb line in **(B)** is drawn from the same vertebral body. The body weights above this point are assumed to be equal. The force applied to the posterior elements at the fracture level in **(A)** is equal to the compression force anterior to the flexion axis. This force or bending moment can be calculated by multiplying the body weight above the fracture by the moment arm. As shown here, the moment arm *b* is significantly greater than *a*. Therefore, the tensile forces in **(B)** exceed those in **(A)** and may, in fact, exceed the limits of the posterior ligaments and cause further kyphosis.

TREATMENT METHODS FOR SPINAL TRAUMA

The treatment options for trauma to the spinal column are many and somewhat controversial. Initial experience with internal fixation (wiring of spinous processes in the cervical and lumbar spine, Meurig-William plates, and so on) proved less than ideal and led many physicians to advocate a nonsurgical approach to treat spinal trauma. Perhaps the foremost proponent of the nonoperative treatment was Sir Ludwig Guttman. His articles, teachings, and students strongly advocated postural reduction and bedrest as the safest and perhaps surest method of managing spinal trauma with or without neurologic deficits.[72-75] Prior to the advent of modern spinal instrumentation, which began with the Harrington system, there was limited argument against Guttman's teaching. Now, however, improved surgical fixation systems and techniques, combined with increased knowledge of the biomechanics of injury and the need for better reduction of fractures, have made operative intervention to stabilize, reduce, and fuse the traumatized spine widely accepted.

In this discussion the cervical spine will be considered separately from thoracic and lumbar spinal injuries. Postural reduction and bedrest will not be explored, as these treatment modalities are relatively clear and require little explanation (though carrying out the techniques is demanding and requires considerable attention to detail). Techniques of stabilization will not be described, as the details involved would be too extensive and beyond the scope of this chapter.

Cervical Spine

Rigid immobilization is the key to success in treating cervical spinal injuries, particularly with neurologic involvement. External immobilizing devices are available and have been studied by Johnson et al.[76] Various collars and orthoses are available, progressing from the limited stabilization afforded by soft collars (only 5 percent reduction in motion) to the more rigid and confining cervicothoracic orthoses. Though easy to apply and obtain, these orthoses do not supply complete rigidity, are better in controlling flexion-extension than lateral rotation, and without a thoracic extension do not adequately immobilize at the cervicothoracic junction nor the more proximal cervical spine.[28,76] The halo device to provide skeletal (skull) fixation, described by Perry and Nickel, is the gold standard to which all other devices are compared.[77] This device, depending on the superstructure chosen and the vest or plaster cast applied, can limit almost 95 percent of cervical motion and supplies rigid external fixation while allowing the patient out of bed. It may aid in returning an individual to an early, relatively normal lifestyle and help initiate rehabilitation. Problems do exist with this device related to pressure sores under the vest and lack of ideal skull–pin fixation (evidenced by relatively high infection and loosening rates). In addition, it is somewhat cumbersome to wear. Nonetheless, it does provide rigid immobilization and is relatively easy to apply and manipulate, if necessary.

Surgical techniques to stabilize the cervical spine can be divided into anterior or posterior approaches. If the injury is primarily a flexion or flexion-rotation problem (evidenced by anterolisthesis of one vertebral body on another, widening of the spinous process distance, and disruption of the facet joints), the posterior approach for the lower cervical spine is indicated. The techniques include wiring the spinous processes with fusion,[78] wiring of the facets individually or to bone graft or metal rods, or wiring of a bone graft to the spinous process or laminae.[79] This latter approach is usually performed for unstable odontoid fractures, type 2 odontoid fractures with a high incidence of nonunion, or transverse ligament ruptures.[80] When wiring is used and a relatively stable reduction of the spine has been obtained, external immobilization devices are usually chosen to supplement the immobilization. Wiring alone is not sufficient to maintain the reduction, and fractures either of the wire or the bony elements to which they are inserted may occur.[81] When the reduction is unstable, further extension or distraction is required, or the fracture is so comminuted that the alignment of the anterior elements cannot be maintained, the internal fixation should be supplemented with halo external immobilization.

An anterior approach to the cervical spine is indicated if there is retropulsion of bone fragments from the vertebral body into the spinal canal or if there is significant loss of vertebral body height anteriorly that cannot be restored mechanically. Additional indications for this approach are disk retropulsion centrally or inability to maintain stability. Frequently, an iliac crest graft or fibula is used to replace the removed vertebral body or disk. The shape of the graft is dependent upon the particular surgical requirements (a block of bone, a horseshoe-shape graft, a keystone graft, T-shaped configuration, etc.).[82] These bone grafts lock into place as the neck comes out of extension and usually remain in place until fusion by the forces of the cervical spine in flexion. If only limited resection has been necessary, limited external immobilization is required. If a number of vertebral bodies are resected, halo immobilization usually supplements the procedure postoperatively.

Thoracic and Lumbar Fractures
Thoracic and lumbar external immobilizers are not as well studied and do not appear to rigidly fix the spine in the degree shown for cervical orthoses. Because of this, except for compression fractures where symptomatic relief is frequently obtained by Jewett braces, Knight-Taylor orthoses, or Kydex/plaster body jackets, external immobilizers are not generally employed as the sole treatment of unstable thoracic or lumbar vertebral column fractures.

Surgically, the approaches to the thoracic and lumbar regions, as in the cervical spine, are anterior or posterior. Anterior approaches are usually reserved to strut significant deformities and/or when bone encroaches anteriorly upon the spinal canal.[83] In many cases, as in the cervical spine, no internal fixation is necessary, but keyed-in iliac crest or fibular grafts are employed and locked into place by the compression forces supplied by

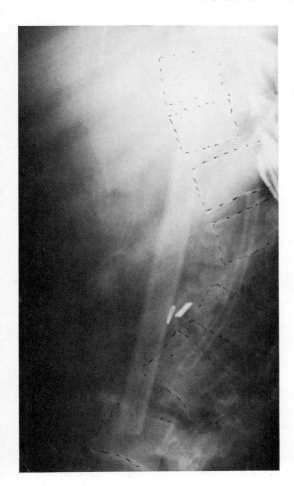

Figure 13. A 38-year-old male with a significant kyphotic deformity due to marked angulation at T-12. To prevent further deformity, anterior strut grafting and fusion were required. As shown here, a fibular graft was employed to strut anteriorly across the deformity. Rib and iliac crest were employed across the apex. The disk spaces were also fused with iliac crest bone. This construct gives mechanical stability anteriorly. This procedure was later followed by posterior compression rodding and fusion.

the spine (Fig. 13). However, internal fixation plates and screws, wire instrumentation and screws,[85,86] and more rigid devices[84,87] are available, if necessary, for anterior fixation. These techniques tend to be used more frequently when deformities require mechanical correction, as with scoliosis or late untreated spinal fractures. The Dwyer instrumentation has been well studied, and though it adequately corrects deformities, the cable does not provide rigid internal fixation and allows a certain degree of rotation, as well as flexion and extension, to the side of the cable. Additionally, the screw–bone interface can be the site of weakening and loss of fixation, particularly in osteoporotic spines.[88,89] The Zielke apparatus and Dunn devices also allow significant corrections but reportedly provide more rigid internal fixation, as the cable component has been eliminated.[84,87,90]

Posteriorly, current options include Harrington distraction/compression rodding and/or the Luque instrumentation technique, which involves mul-

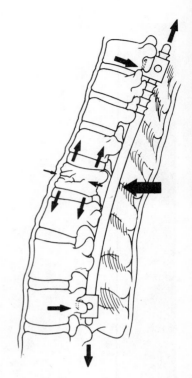

Figure 14. A Harrington distraction system is illustrated. Three-point fixation is shown, with one point at the proximal hook and a second point at the distal hook. The third point is the interface of the lamina with the rod at the fractured vertebra creating an anterior force. The resultant is a distraction force across the bodies, as shown by the arrows, which helps realign the vertebral body and allows it to regain height.

tiple level sublaminar wires fixed to rigid, contoured rods.[22,92–95] Harrington instrumentation has proven successful over a number of years and has been the first choice of most spine surgeons. Reduction of anterior vertebral body fractures is dependent not only on the distraction forces created by the device but by three-point fixation obtained by contact with the rod and the lamina adjacent to the fractured and displaced vertebra (Fig. 14).[26] The distraction rod can be used successfully to stabilize and reduce flexion-rotation injuries, burst fractures, and compression fractures that are considered unstable due to posterior element disruption. Again, the rods must obtain a three- or four-point fixation to not only distract but reduce the vertebral body anteriorly and close the disruption posteriorly. Supplemental bone graft, either autogenous or allograft, is added. The internal fixation device will eventually fail if the fusion does not become solid. To avoid hook failure and dislodgment, external immobilization is essential to obtain uniformally good results. This generally can be provided by a Kydex body jacket or a plaster shell. Studies by Wenger et al.,[96,97] Dunn,[90] Harrington,[22] and Waugh[98] have shown that the hooks should be placed at the strongest portion of the posterior elements to prevent fracture at the metal–bone fixation. The transverse processes and spinous processes do not provide as rigid or strong resistive forces as do the laminae (Table 3).

The Harrington compression system also plays a role (particularly the

TABLE 3. POSTERIOR ELEMENT FIXATION STRENGTHS

	Dunn et al.[90]	Harrington[22]
Anatomic Structure		
Lamina (thoracic)	70 kgf	100–300 pounds
Lamina (lumbar)	140 kgf	100–300 pounds
Base of transverse process (thoracic)	35 kgf	50–100 pounds
Tip of transverse process	7 kgf	10–25 pounds
Spinous process	15 kgf	

Values of vertebral bone strength at various locations along the posterior elements. The absolute numbers differ between the authors, but relative strength relationships are consistent.

larger of the two compression rods) in stabilizing flexion distraction injuries (Fig. 6). With a minimal amount of deformity anteriorly, but posterior distraction as evidenced by widened facets and spinous processes (Chance type fractures), compression rods will control the instability, close the fracture, and provide rigid stabilization while the fusion becomes solid.[99,100] Once again, in this case, external immobilization also is required. Compression rods fixed to hooks around the lamina have been shown by Stauffer and Neil to provide a more rigid stabilization and stronger construct than the distraction system when tested against flexion loading (Fig. 15).[101]

By understanding the mechanism of injury and the nature of the particular instability, the proper selection of Harrington distraction or compression rod can be made. For years many surgeons routinely chose the Harrington distraction system for all spinal fractures. This is not and should not be the case. Furthermore, beyond 3 weeks postinjury, spinal disruptions and deformities are difficult to adequately reduce solely by posterior instrumentation. To correct a burst fracture or compression injury anteriorly, when the injury is more than 3 weeks old, an anterior decompression and fusion are required, supplemented by posterior instrumentation (usually compression rodding).

The Luque system, first described in 1976,[92–97,102] provides even more rigid fixation than does Harrington instrumentation. Sublaminar wires are employed and fixed to a countoured Luque L rod. A fusion is then performed posteriorly. This device has been shown by Wenger et al. to provide significantly increased stability over the Harrington system and does not require external immobilization to prevent failure.[97] However, probably because of the lack of distraction, many instances of fracture stabilization-fixation by the Luque technique fail to demonstrate an anatomic reduction of the fractured vertebral body and spinal column. Though the fusion becomes solid posteriorly and the bone heals anteriorly, the deformity has not been corrected, and the spinal canal (at least as demonstrated by some reproduced examples) remains compromised and narrowed.[95]

This technique, however, may be the procedure of choice for thoracic or lumbar spinal fractures with neurologic injuries resulting in parapalegia.

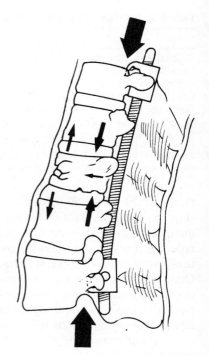

Figure 15. Harrington compression rods also help develop reduction forces anteriorly through a three-point system. The mechanical lever arm, however, is less in this system than with the distraction device. The forces producing distraction anteriorly are less than those compressing the posterior aspect of the vertebral body. Although the system may be more stable than the distraction instrumentation, it is less efficient in realigning a compressed vertebral body.

Since no operation has been shown to reverse a complete neurologic deficit, rigid stabilization for rehabilitation is important. The Luque technique provides stability. It does not require extensive external immobilization to maintain stability, and therefore rehabilitation of the acute paraplegic can begin almost immediately following surgery and medical stabilization, without external encumbrances.

Finally, in the discussion of internal fixation, the strength of the various devices often is considered. Without question, the Luque system provides the most rigid immobilization, especially when employing the larger ¼ inch rods and 16 gauge sublaminar wires. The strength of this sytem lies not only in the rigid, unstressed (no notches, scoring) rod but in the distribution of forces widely across all the instrumented laminae. These devices withstand more torque in forward bending and rotation than other posterior instrumentation devices.[96,97]

Prior to the Luque system, the Harrington instrumentation appeared to be the most stable. Though many studies demonstrated the compression system to be superior to the distraction in resisting flexion forces, some authors disputed this.[103,104] Harrington distraction systems tend to fail by dislodgment at the rod–hook site, hook–lamina separation, or fracture at the ratchet–rod junction. Recommendations to prevent this include hook placement solidly under the lamina (both cortices), some rod extension beyond the proximal hook, and minimizing the amount of the racheted rod present between the hooks. The compression rods, apparently superior in resisting flexion and perhaps rotation forces, tend to fail by fracturing the laminae to which the hooks are attached. We try to use the larger of the two threaded rod systems for fractures, but most papers studying the instruments do not clearly specify which of the compression rod sizes were employed.

As a final point on this topic, location of bone–metal fixation site is critical. Though the absolute numbers vary, data from Harrington, Waugh, and Dunn demonstrate the superior strength of the lamina as compared to transverse processes and spinous processes (Table 3).[22,90,98] In fact, the strength of the transverse process diminishes with progression toward its tip. Accurate hook–metal placement and choice of location are crucial factors in determining the success or failure of whatever system is employed.

Included in this section on treatment should be some mention of laminectomy. Except for repair of dural tears or to allow access to the spinal canal to remove impinging bone or disk, laminectomy has little role in the acute treatment of acute spinal cord trauma. Numerous studies have demonstrated the clinical deterioration of patients following laminectomy, and laboratory analysis has shown an alteration of tensile stresses following laminectomy—which may lead to further collapse anteriorly after the sole tension-resisting posterior structure has been removed.[68,105]

IS PREVENTION POSSIBLE?

Prevention of spinal and spinal cord injuries is a very difficult problem. In certain instances prevention is impossible. For example, falls from horses, motorcycle accidents, hang glider accidents, falls from heights, and similar accidents are essentially unavoidable, unpredictable, and probably unpreventable. However, injuries from motor vehicle accidents, particularly automobiles and trucks, may be controllable. Certainly, the addition of the lap seat belt decreased some injuries, but it created a new type of spinal injury (flexion-distraction), which fortunately seems to result in less spinal cord injury or significant damage than the more devastating burst fractures. The addition of a three-point restraint system, such as a lap and a sash belt, further minimizes the flexion-rotation and even flexion-distraction component of automobile injuries. Hopefully, we will see a continued diminution in spinal trauma from these events. Statistically, however, the total number of spinal injuries seems to be increasing, and the percentage of spinal cord injuries relative to spinal trauma has remained constant over the years. Severe cervical spinal injuries are appearing more frequently. A number of years ago the incidence of parapalegia to quadriplegia was approximately 2 or 3 to 1. Currently, the incidence ratios are almost equal.[106]

Finally, in terms of prevention, a better understanding of the mechanics of the spine is necessary not only to prevent such tragic injuries but also to reduce the impact of the injury on the individual as well as society. Further work on modeling of cervical and thoracolumbar injury mechanisms is necessary not only to test prevention techniques but to increase our understanding of mechanisms of injuries.

Techniques and instrumentation to stabilize the spine are expanding rapidly, but biomechanical testing of these devices is still needed to help determine efficacy and allow the surgeon to make the proper choice of instrumentation. As we are becoming more aggressive in the surgical decompression and stabilization of the spine, the long-term results must be assessed to determine if our results are better than those described years ago by Guttman, employing nonoperative, closed reduction, bedrest, or plaster immobilization techniques to treat spinal column injuries.

Although certainly complete prevention of spinal trauma is impossible, perhaps the better understanding of causes of instability can diminish late problems. To accomplish this, checklists, flow charts (Fig. 10), and so on have been developed to aid physicians treating acute spinal column injuries. Whether these techniques or others are employed, at least some scheme should be followed to cogently manage spinal problems as they present and prevent late complications and sequelae.

REFERENCES

1. Riggins RS, Kraus JF: The risk of neurological damage with fractures of the vertebrae. *J Trauma* 1977; 17:126.
2. Castellano V, Bocconi FL: Injuries of the cervical spine with spinal cord involvement (myelic fractures). Statistical considerations. *Bull Hosp Joint Dis* 1970; 31:188.
3. Cheshire DJ: The stability of the cervical spine following conservative treatment of fractures and fracture dislocations. *Paraplegia* 1969; 7:193.
4. Griffith HB, Gleave JR, Taylor RG: Changing patterns of fractures of the dorsal and lumbar spine. *Br Med J* 1966; 1:891.
5. Watson-Jones R: *Fractures and Joint Injuries,* ed 4. Baltimore, Williams & Wilkins, 1960.
6. Breasted JH (ed): *The Edwin Smith Papyrus.* Chicago, University of Chicago Press, 1930.
7. Osler W: *The Evolution of Modern Medicine.* New Haven, Yale University Press, 1921.
8. Bohlman HH, Ducker TB, Lucas JT: Spine and spinal cord injuries, in Rothman RH, Simeone FA (eds): *The Spine,* ed 2. Philadelphia, Saunders, 1982, pp 661–757.
9. Gurdjian ES, Webster JE: *Head Injuries: Mechanism, Diagnosis, and Management.* Boston, Little, Brown, 1958.
10. Sachs E: *The History and Development of Neurological Surgery.* New York, Paul B. Hoeber, 1952.
11. Yashon D: *Spinal Injury.* New York, Appleton-Century-Crofts, 1978.
12. Bishop WJ: *The Early History of Surgery.* London, Robert Hale, Ltd., 1960.
13. Green JR: *Medical History for Students.* Springfield, Ill., Thomas, 1968.
14. Bick EM: *Source Book of Orthopaedics.* Baltimore, Williams & Wilkins, 1937.

15. Hadra B: Wiring of the vertebra as a means of immobilization in fracture and Pott's disease. *Medical Times and Register,* May 23, 1891
16. Hibbs RA: An operation for progressive spinal deformities. *NY Med J,* May 27, 1911.
17. Davis AG: Fractures of the spine. *J Bone Joint Surg* 1929; 11:133.
18. Watson-Jones R: Manipulative reduction of crush fractures of the spine. *Br Med J* 1931; 1:300.
19. Holdsworth EW, Hardy AG: Early treatment of paraplegia from fractures of the thoraco-lumbar spine. *J Bone Joint Surg* 1953; 35B:540.
20. Wilson PD, Straub LR: Lumbosacral fusion with metallic plate fixation, in *American Academy of Orthopaedic Surgeons Instructional Course Lectures.* Ann Arbor, Mich., J. W. Edwards, 1952, Vol 9.
21. Williams EWM: Traumatic paraplegia, in Mathews DN (ed): *Recent Advances in the Surgery of Trauma.* London, Churchill, 1963, p 171.
22. Harrington PR: Treatment of scoliosis. *J Bone Joint Surg* 1962; 44A:591.
23. Dickson JH, Harrington PR, Erwin WD: Results of stabilization of the severely fractured thoracic and lumbar spine. *J Bone Joint Surg* 1977; 59A:185.
24. Convery FR, Minteer MA, Smith RN: Fracture dislocation of the dorsal lumbar spine. *Spine* 1978; 3:160–166.
25. Bedbrook GM, Hon OBE: Treatment of thoracolumbar dislocation and fractures with paraplegia. *Clin Orthop* 1975; 112:27.
26. Jacobs RR, Asher MA, Snider RK: Thoracolumbar spinal injuries. *Spine* 1980; 5:463.
27. Flesch JR, Leider LL, Erickson DL, et al.: Harrington instrumentation and spine fusion for unstable fractures and fracture-dislocations of the thoracic and lumbar spine. *J Bone Joint Surg* 1977; 59A:143.
28. White AA, Panjabi MM: *Clinical Biomechanics of the Spine.* Philadelphia, Lippincott, 1978.
29. Davis PR: The medial inclination of the human thoracic intervertebral articular facets. *J Anat* 1959; 93:68.
30. Davies DV (ed): *Gray's Descriptive and Applied Anatomy,* ed 34. London, Longmans, Green and Co., 1967.
31. Rockwell H, Evans FG, Pheasant HC: The comparative morphology of the vertebral spinal column; its form as related to function. *J Morphol* 1938; 63:87.
32. Shultz A, Benson D, Hirsch C: Force deformation properties of human costo-sternal and costovertebral articulations. *J Biomech* 1974; 7:311.
33. Panjabi MM, Brand RA Jr, White AA: Three-dimensional flexibility and stiffness properties of the human thoracic spine. *J Biomech* 1975; 9:185.
34. Agostini E, Magnoni G, Torri G, et al.: Forces deforming the rib cage. *Respir Physiol* 1966; 2:105.
35. Nahum AM, Gadd CW, Schneider DC, et al.: Deflections of human thorax under sternal impact, in *1970 International Automobile Safety Conference.* New York, Society of Automotive Engineers, 1970, pp 797–807.
36. Andriacchi TP, Schultz AB, Belytschko TB, et al.: A model for studies of mechanical interactions between the human spine and rib cage. *J Biomech* 1973; 7:497.
37. Lucas D, Bresler B: *Stability of Ligamentous Spine.* Biomechanics Laboratory Report 40. San Francisco, University of California, 1961.
38. Nicoll EA: Fractures of the dorso-lumbar spine. *J Bone Joint Surg* 1949; 31B:376.

39. Roaf R: A study of the mechanics of spinal injuries. *J Bone Joint Surg* 1960; 42B:810.
40. Holdsworth FW: Fractures, dislocations and fracture-dislocations of the spine. *J Bone Joint Surg* 1963; 45B:6.
41. Kelly RP, Whitesides TE: Treatment of lumbodorsal fracture-dislocations. *Ann Surg* 1968; 167:705.
42. Whitesides TE: Traumatic kyphosis of the thoracolumbar spine. *Clin Orthop* 1977; 128:78.
43. Dennis F: Updated classification of thoracolumbar fractures. *Orthop Trans* 1982; 6:8.
44. McAfee PC, Yuan AA, Frederickson BA, et al.: The value of computed tomography in thoracolumbar fractures. *J Bone Joint Surg* 1983; 65A:461.
45. Nash CL, Schatzinger LH, Brown RH, et al.: The unstable stable thoracic compression fracture. *Spine* 1977; 2:261.
46. Chance GQ: Note on a type of flexion fracture of the spine. *Br J Radiol* 1948; 21:452–453.
47. Rogers LF: The roentgenographic appearance of transverse or Chance fractures of the spine: The seat belt fracture. *Contemp Orthopaed* 1983; 7:7.
48. Smith WS, Kaufer A: Patterns and mechanisms of lumbar injuries associated with lap seat belts. *J Bone Joint Surg* 1969; 51A:239.
49. Rennie W, Mitchell N: Flexion dislocation fractures of the lumbar spine. *J Bone Joint Surg* 1973; 55B:662.
50. Epstein BS, Epstein JA, Jones MD: Lap-sash three point seat belt fractures of the cervical spine. *Spine* 1978; 3:138.
51. Forsyth HF: Extension injuries of the cervical spine. *J Bone Joint Surg* 1964; 46A:1792.
52. Stauffer ES, Mazur JM: Unrecognized spinal stability associated with seemingly simple cervical compression fractures. *Orthop Trans* 1982; 6:179.
53. Weir DC: Roentgenographic signs of cervical injury. *Clin Orthop* 1975; 109:9.
54. Keene JS, Goletz TH, Lilleas F, et al.: Diagnosis of vertebral fractures. *J Bone Joint Surg* 1982; 64A:586.
55. Post JL, Green BA, Quencer RM, et al.: The value of computed tomography in spinal trauma. *Spine* 1982; 7:431.
56. White AA, Johnson RM, Panjabi MM, et al.: Biomechanical analysis of clinical instability of the cervical spine. *Clin Orthop* 1975; 109:85.
57. White AA, Panjabi MM, Saha S, et al.: Biomechanics of the axially loaded cervical spine: Development of a safe clinical test for ruptured cervical ligaments. *J Bone Joint Surg* 1975; 57A:582.
58. Breig A: *Biomechanics of the Central Nervous System: Some Basic Normal and Pathological Phenomena.* Stockholm, Almquist and Wicksell, 1960.
59. Wiesel SW, Rothman RH: Occipitoatlantal hypermobility. *Spine* 1979; 4:187.
60. White AA, Panjabi MM, Posher L, et al.: Spinal stability: Evaluation and treatment, in *American Academy of Orthopaedic Surgeons Instructional Course Lectures.* St. Louis, Mosby, 1981.
61. Fielding JW, Cochran GVB, Lansing JF, et al.: Tears of the transverse ligament of the atlas: A clinical biomechanical study. *J Bone Joint Surg* 1974; 56A:1683.
62. Spence KF, Decker S, Sell KW: Bursting atlantal fracture associated with rupture of the transverse ligament. *J Bone Joint Surg* 1970; 52A:543.
63. Sherk HH, Nicholson JT: Fractures of the atlas. *J Bone and Joint Surg* 1970; 52A:1017.

64. Garfin SR, Rothman RH: Traumatic spondylolisthesis of the axis, in Cervical Spine Research Society (ed): *The Cervical Spine.* Philadelphia, Lippincott, 1983, pp 223–231.
65. White AA, Southwick WO, Panjabi MM: Clinical instability in the lower cervical spine: A review of past and current concepts. *Spine* 1976; 1:15.
66. Beatson TR: Fractures and dislocations of the cervical spine. *J Bone Joint Surg* 1963; 45B:21.
67. Soreff J, Axdorph G, Bylund P, et al.: Treatment of patients with unstable fractures of the thoracic and lumbar spine. *Acta Orthop Scand* 1982; 53:369–381.
68. Roberts JB, Curtiss PH: Stability of the thoracic and lumbar spine in traumatic paraplegia following fracture or fracture-dislocation. *J Bone Joint Surg* 1970; 52A:1115.
69. Stanger KJ: Fracture-dislocation of the thoracolumbar spine. *J Bone Joint Surg* 1947; 29:107.
70. Bradford DS, Akbarnia BA, Winter RB: Surgical stabilization of fracture and fracture dislocations of the thoracic spine. *Spine* 1977; 2:185.
71. Rissanen PM: The surgical anatomy and pathology of the supraspinous and interspinous ligaments of the lumbar spine with special reference to ligament ruptures. *Acta Orthop Scand* 1960; 46 (suppl):1.
72. Guttman L: Surgical aspects of the treatment of traumatic paraplegia. *J Bone Joint Surg* 1949; 31:399.
73. Guttman L: Spinal deformities in traumatic paraplegics and tetraplegics following surgical procedures. *Paraplegia* 1969; 7:38.
74. Cheshire DJ: The complete and centralised treatment of paraplegia. *Paraplegia* 1968; 6:59.
75. Frankel HL, Hancock DO, Hyslop G, et al.: The value of postural reduction in the initial management of closed injuries of the spine with paraplegia and tetraplegia. *Paraplegia* 1969; 31:179.
76. Johnson RM, Hart DL, Simmons EF, et al.: Cervical orthoses. *J Bone Joint Surg* 1977; 59A:332.
77. Perry J, Nickel VL: Total cervical spine fusion for paralysis. *J Bone Joint Surg* 1959; 41A:37.
78. Rogers WA: Treatment of fracture-dislocation of the cervical spine. *J Bone Joint Surg* 1942; 24:245.
79. Murphy MJ, Southwick WD: Posterior approaches and fusions, in Cervical Spine Research Society (ed): *The Cervical Spine.* Philadelphia, Lippincott, 1983, pp 496–512.
80. Brooks AL, Jenkins EB: Atlanto-axial arthrodesis by the wedge compression method. *J Bone Joint Surg* 1978; 60A:279.
81. Munro D: The role of fusion or wiring in the treatment of acute traumatic instability of the spine. *Paraplegia* 1965; 3:97.
82. Vanden Brink DK, Edmonson AS: The spine, in Edmonson AS, Crenshaw AH (ed): *Campbell's Operative Orthopaedics,* ed 6. St. Louis, Mosby, 1980, pp 1939–2155.
83. Whitesides TE, Ali Shan SG: On the management of unstable fractures of the thoracolumbar spine: Rationale for use of anterior decompression and fusion and posterior stabilization. *Spine* 1976; 1:99.
84. Zielke K, Stunkat R: Derotation and fusion: Anterior spinal instrumentation. *Orthop Trans* 1978; 2:209.

85. Dwyer AF: Experience of anterior correction of scoliosis. *Clin Orthop* 1973; 93:191.

86. Dwyer AF, Newton NC, Sherwood AA: An anterior approach to scoliosis: A preliminary report. *Clin Orthop* 1969; 62:192.

87. Dunn HK: Internal fixation of the spine: A new implant system. *Orthop Trans* 1979;3:47.

88. Dunn HK, Bolstad KE: Fixation of Dwyer screws for treatment of scoliosis. *J Bone Joint Surg* 1977; 59A:54.

89. Hall JE, Gray J, Allen N: Dwyer instrumentation and spinal fusion: A follow-up study. *J Bone Joint Surg* 1977; 59B:117.

90. Dunn HK, Allen BL, Chan DPK: Spinal instrumentation, in *American Academy of Orthopaedic Surgeions Instructional Course Lectures*. St. Louis, Mosby, 1983, p 192.

91. Allen BL, Ferguson RL: The Galveston technique for L-rod instrumentation of the scoliotic spine. *Spine* 1982; 7:276.

92. Luque ER: Segmental spinal instrumentation: A method of rigid internal fixation of the spine to induce arthrodesis. *Orthop Trans* 1980; 4:391.

93. Luque ER, Cardoso A: Sequential correction of scoliosis with rigid internal fixation. *Orthop Trans* 1977; 1:136.

94. Luque ER, Cardosa A: A treatment of scoliosis without arthrodesis or external support: Preliminary report. *Orthop Trans* 1977; 1:37.

95. Luque ER, Cassis N: Segmental spinal instrumentation in the treatment of fractures of the thoracolumbar spine. *Spine* 1982; 7:312.

96. Wenger DR, Carollo JJ, Wilkerson JA: Biomechanics of scoliosis correction by segmental spinal instrumentation. *Spine* 1982; 7:260.

97. Wenger DR, Carollo JJ, Wilkerson JA, et al.: Laboratory testing of segmental spinal instrumentation versus traditional Harrington instrumentation for scoliosis treatment. *Spine* 1982; 7:265.

98. Waugh TR: The biomechanical basis for the utilization of methylmethacrylate in the treatment of scoliosis. *J Bone Joint Surg* 1971; 53A:194.

99. Grantham SA, Malberg MI, Smith DM: Thoracolumbar spine flexion-distraction injury. *Spine* 1976; 1:172.

100. Yosipovitch Z, Robin GC, Makin M: Open reduction of unstable thoracolumbar spinal injuries and fixation with Harrington rods. *J Bone Joint Surg* 1977; 59A:1003.

101. Stauffer ES, Neil JL: Biomechanical analysis of structural stability of internal fixation in fractures of the thoracolumbar spine. *Clin Orthop* 1975; 112:159.

102. Herring JA, Wenger DR: Segmental spinal instrumentation: A preliminary report of 40 consecutive cases. *Spine* 1982; 7:285.

103. Pinzur MS, Meyer PR Jr., Lautenschlager EP, et al.: Measurement of internal fixation device support in experimentally produced fractures of the dorsolumbar spine. *Orthopaedics* 1979; 2:28.

104. Armstrong GW, Connock SHG: A transverse loading system applied to a modified Harrington instrumentation. *Clin Orthop* 1975; 108:70.

105. Balasubramanian K, Ranu HS, King AI: Vertebral response to laminectomy. *J Biomech* 1978; 21:813.

106. Nickel V: Personal communication.

Chapter 17

The Vertebral Column:
Experimental Aspects

Albert I. King

INTRODUCTION

The human vertebral column is the principal load-bearing structure of the head and torso. There are also secondary functions performed by each portion of the spinal column. The cervical spine provides the head with a limited degree of mobility and a protected pathway for the proximal segment of the spinal cord. The thoracic spine offers the same protection to the cord, while it offers mobility to the upper torso and rib cage. The lumbar segment provides the lower torso with mobility and encloses the distal end of the spinal cord. The protective role of the vertebral column is analogous to the function served by the skull to protect the brain. However, anatomic requirements dictate that the spine be flexible and yet strong so that it can serve a multitude of functions. Like the skull, it is strong but not strong enough to withstand mechanical insults of modern day transportation systems. Injuries that affect the function of the spinal cord can result in death, quadriplegia, or paraplegia. Those who survive suffer permanent disabilities that cannot be restored as yet by modern medicine. Other biomechanical motivations to study the mechanical response of the spine include neckache and backache, osteoporosis, and sciatica.

This chapter deals with the biomechanics of the spine, with particular emphasis on injury mechanisms and mechanical response to impact acceleration. Although spinal injuries are relatively uncommon in automotive accidents, they can often be rather severe and disabling. They are more common in aircraft accidents and constitute a special problem in aircraft ejection, which is the cause of anterior wedge fractures of the thoracolumbar spine.

Huelke et al.[1] reviewed cervical injury data collected by the US Department of Transportation (DOT) under the National Crash Severity Study (NCSS). Data representing occupants who sustained severe, serious, critical-to-life, and fatal cervical injuries were reviewed. The frequency of such injuries was 0.4 percent for front seat occupants. It rises to 7 percent for those who were ejected. They are most common in frontal and side impacts, and the age group most susceptible to these injuries is 16 to 25 years. It is estimated that fatal cervical injuries make up about 20 percent of all occupant fatalities (5940 cases) and that about 500 cases of quadriplegia per year result from automotive accidents. In a review of thoracolumbar spinal injuries by King,[2] the frequency of injury was 0.8 percent for all levels of injury, according to NCSS data files, but there were no fatalities that were attributable to injuries of the thoracolumbar spine. The most frequent cause of spinal injury was rollovers (26 percent), followed by ejection (22 percent) and frontal impacts (20 percent). It should be noted that there were very few motorcycle cases in the data file.

ANATOMY OF THE VERTEBRAL COLUMN

Familiarity with the anatomy of the vertebral column is a necessary condition for the understanding of the biomechanics of the spine and its response to load. The ability to model this response also calls for an appreciation of the function of the various components of the column. From a macroscopic point of view, the vertebral column is made up of 24 individual bones, called vertebrae, that are joined together by several different types of soft tissue. The primary types of soft tissue are the intervertebral disks, ligaments, and skeletal muscle. As shown in Figure 1, the 7 vertebrae supporting the head constitute the cervical spine, while the 12 vertebrae below it form the thoracic spine. The lumbar spine is the most inferior segment and is made up of 5 vertebrae. The entire column is supported by the sacrum, which is anatomically a part of the pelvic girdle. The thoracolumbar spine is located along the midline of the posterior aspect of the torso, and the cervical spine is along the posterior aspect of the neck. In general, each vertebra consists of a body, neural arch or pedicles, laminae, facet joints, spinous process, and transverse processes. The body is a cylindrically shaped bone consisting of a core of spongy bone surrounded by a thin layer of cortical or compact bone. The endplates above and below the centrum are cartilaginous. The sides of the body are usually slightly concave, with a narrow waist at midlevel. Figure 2 shows a typical lumbar vertebra, viewed laterally. The pedicles arise from the posterolateral aspects of the body and are directed rearward. They form the lateral aspects of the spinal canal that

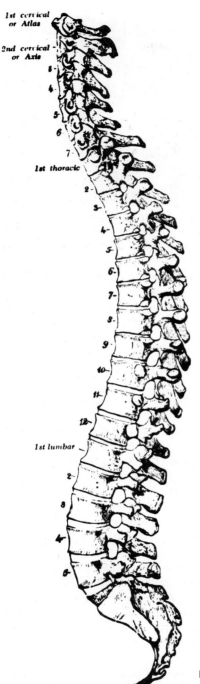

1st cervical
or Atlas

2nd cervical
or Axis

3

4

5

6

7

1st thoracic

2

3

4

5

6

7

8

9

10

11

12

1st lumbar

2

3

4

5

Figure 1. Lateral view of the spine. (*From Gray's Anatomy.*[3])

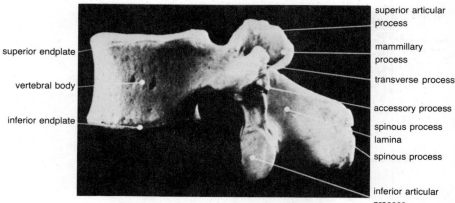

superior articular process

mammillary process

transverse process

accessory process

spinous process
lamina

spinous process

inferior articular process

superior endplate

vertebral body

inferior endplate

Figure 2. Lateral view of a lumbar vertebra.

surrounds the spinal cord and affords it mechanical protection. The laminae are quadrilaterally shaped pieces of compact bone that form the posterior aspect of the spinal canal. At the junction between the pedicles and the laminae are the articular facets. Each vertebra has four facets, two superior and two inferior. A posterior view of the facets of a lumbar vertebra is shown in Figure 3. These bony projections articulate with mating projections (facets) of the vertebrae above and below. The joints formed by the facets are true synovial joints, encapsulated by capsular ligaments. The orientation of the facet joint surfaces varies from vertebra to vertebra and is of biomechanical interest because the facets form a load-bearing function of the spine with the vertebral bodies. The geometry of the facets will be described below.

Continuing with the general description of a typical vertebra, the transverse and spinous processes complete the posterior structure. They act as attachment points for muscles and ligaments and can be considered as short cantilever beams with free ends. The vertebrae gradually increase in size caudally, roughly in proportion to the weight they are expected to support. The precise description of each vertebra can be found in a text on human anatomy. The lateral view of the entire column in Figure 1 shows three principal spinal curves, the lordotic cervical and lumbar curves and the kyphotic thoracic curve. The normal spine is straight when viewed frontally. Abnormal lateral curves found in scoliotic spines tend to develop in adolescence and are more common in females than males. Mechanical explanations for this form of instability are not completely satisfactory. It should also be noted that the thoracic spine supports the posterior section of the rib cage. A pair of ribs arise from each thoracic vertebra. These ribs articulate with the vertebrae near the junction of the pedicles with the vertebral bodies and at the tips of transverse processes.

A few of the special features of the spine will now be discussed. The

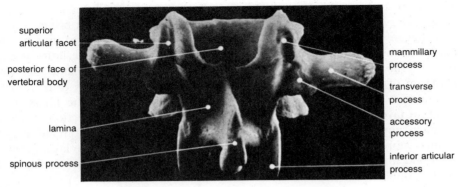

superior articular facet

posterior face of vertebral body

lamina

spinous process

mammillary process

transverse process

accessory process

inferior articular process

Figure 3. Posterior view of a lumbar vertebra.

first cervical vertebra (C-1) is called the "atlas." It does not have a true vertebral body but is made up of two arches. The facets articulate with the skull, allowing a nodding motion of the head. There is, thus, no intervertebal disk between the skull and C-1. The second cervical vertebra (C-2) is called the "axis." It articulates with C-1 to permit axial rotation of the head, and, again, there is no disk between C-1 and C-2. In fact, the body of C-2 has a superior projection called the "odontoid process." It is located at the level of C-1 and is held in place against the inner anterior aspect of the C-1 arch by several ligaments.

In order to describe the orientation of the facet joint surfaces, it is convenient to use a unit vector to establish the approximate orientation of these surfaces. Some of the surfaces are slightly curved, and the description of their orientation assumes the unit normal to be located at the center of the surface. For the cervical vertebrae, the unit vector is directed superiorly and posteriorly. Those for C-1 and C-2 have a slight rearward tilt that increases to a maximum of about 30 degrees at the level of C-4–C-5. The tilt is less at the C-5–C-6 and C-6–C-7 junction but is again about 30 degrees between C-7 and T-1, the first thoracic vertebra. The unit normal for thoracic superior facets is directed generally posteriorly with a variable lateral component of about 30 degrees and an upward tilt of about 20 degrees. The lumbar surfaces are slightly curved, but the unit normal at the center is directed medially. Its orientation tends to shift to a posteromedial direction for the lower lumbar vertebrae. However, the vector tends to lie in a horizontal plane. A pictorial description of the orientation of facet surfaces can be found in Gray.[3] Another special feature of note is the inclination of the fifth lumbar vertebra (L-5). The endplates are inclined at the L-4–L-5 and L5–S-1 level due to the lordotic curvature of the lumbar spine. The forward inclination of L-5 can be more than 30 degrees in some individuals, resulting in high shear loads at these lower lumbar joints.

The vertebrae are joined together by soft tissue, anteriorly by ligaments and intervertebral disks, and posteriorly by ligaments and facet joint capsules. Intervertebral disks are cartilaginous in origin and consist principally of collagen, proteoglycans, and water. The disk can be divided into two main regions, the nucleus pulposus and the anulus fibrosus. The latter is a ring of primarily type I collagen (the type found in skin, tendon, and bone), made up of dense layers of collagen fibers that have an intricacy of pattern that almost defies description. In general, the direction of the fibers in adjacent layers cross each other at an oblique angle, but the direction of the fibers in any given layer can also change or the fibers can bifurcate and assume more than one direction. In the lumbar region, 12 to 16 layers can be found anteriorly. Type II collagen (the type found in hyaline cartilage) can be found in the nucleus, which has a higher concentration of proteoglycans, giving it a gel-like character. Proteoglycans have an affinity for water and are responsible for the maintenance of tension in the annular collagen fibers. The anatomy and function of the disk are affected by age. Disk degeneration begins at a very young age, and normal healthy disks are the exception rather than the rule in spines over the age of 25. The number and size of collagen fibrils increase with age, and the distinguishing features of the nucleus disappear as age transforms the entire disk into fibrocartilage. A detailed description of the anatomy of the disk can be found in Peacock[4] and in a more recent version by Buckwalter.[5]

The articular facets are enclosed by a joint capsule and appear to allow the spine to flex freely while acting as motion limiters in spinal extension or rearward bending. The joint surfaces are lined with articular cartilage and are lubricated by synovial fluid.

There are three spinal ligaments that run along the entire length of the spine. They are the anterior and posterior spinal ligaments, which line the anterior and posterior aspects of the vertebral bodies, and the supraspinous ligament, which joins the tips of the spinous processes. The ligamentum flavum, or yellow ligament, is a strong band that connects adjacent laminae behind the spinal cord. The interspinous ligament is a thin membrane located between adjacent spinous processes.

The spine is maintained in an erect posture with the help of the skeletal musculature. The extensor muscles of the thoracolumbar spine can be divided into two main groups: the superficial transversocostal and splenius group and the deeper transversospinal group. The former group contains muscles that arise from the pelvic region and insert at various levels from the 6th to the 12th rib. Others arise from the lower ribs and insert at the upper ribs or along the cervical spine. The deeper group contains muscles that join one vertebra to another or span one or more vertebrae. The principal flexors of the thoracolumbar spine are the internal oblique muscles and the rectus abdominus. The cervical spine is extended by muscles that are a continuation of the extensors of the thoracolumbar spine. It also has a group of flexors located anteriorly.

INJURY MECHANISMS

Injuries to the vertebral column can be roughly classified into seven different categories:

1. Anterior wedge fractures of vertebral bodies
2. Burst fractures of vertebral bodies
3. Dislocations and fracture-dislocations
4. Rotational injuries
5. Injuries to the upper cervical spine
6. Chance fractures
7. Hyperextension injuries

Anterior Wedge Fractures

These injuries occur at all levels of the spine and are common in both aircraft and automotive accidents. The mechanism of injury is combined flexion and axial compression. It is a mild form of spinal injury commonly identified with the pilot ejection problem. The region most susceptible to anterior wedge fractures during ejection is between T-10 and L-2, although they can occur in the upper thoracic region as well (T-4–T-6). Kazarian[6] postulated that the mechanism of injury to the T-4–T-6 segment is forcible exaggeration of the normal upper spinal curvature. The fact that very little vertical ($+g_z$) acceleration is experienced in an automotive crash does not mean that wedge fractures cannot occur. Begeman et al.[7] have shown that subjects restrained by a lap belt and an upper torso belt, in a $-g_x$ environment, develop high spinal loads that can cause wedge fractures similar to ejection seat injuries.

Burst Fractures

These injuries are due to higher levels of input acceleration or applied load, causing the vertebral body to break up into two or more segments. The integrity of the cord is threatened by the motion of the segments posteriorly into the spinal canal. The cord can also be injured by the retropulsion of the disk into the canal, particularly in the cervical spine.

Dislocations and Fracture-Dislocations

These are generally flexion injuries accompanied by rotation and postero-anterior shear. Unilateral dislocations require an axial rotational component, while bilateral dislocations can be due solely to flexion. The essential difference between a simple wedge fracture and a fracture-dislocation is, according to Nicoll,[8] in the rupture of the interspinous ligament. This observation is biomechanically significant and will be discussed later. There are varying degrees of dislocation from a simple upward subluxation of the facets to perching of the facets, forward dislocation with fracture of the facets or the neural arch, and forward dislocation with locking of the facets. There is a high probability of neurologic damage in this type of injury because the cord is subject to high shearing and stretching forces. If there is dislocation without wedging, the mechanism of injury is a high shear load in the posteroanterior direction.[6]

Rotational Injuries

If the spine is twisted about its longitudinal axis and is subjected to axial and/or shearing loads, lateral wedge fractures can occur.[8] Other forms of injury include uniform compression of the vertebral body and fracture of the articular facets and lamina. Kazarian[6] indicated that lateral wedge fractures seem to gravitate to two spinal regions: T-2–T-6 and T-7–T-10. The damage to the posterior intervertebral joint is on the concave side, and this injury is often accompanied by fracture of the transverse process on the convex side. Unlike the anterior wedge fracture, this injury may result in neurologic deficit, including paraplegia.

Upper Cervical Spine Injuries

The upper cervical spine consists of the first three cervical vertebrae and the area of the skull forming the atlanto–occipital joint. Since the anatomy of the atlas (C-1) and the axis (C-2) is different from the other vertebrae and the atlanto–occipital joint is unique, the injuries to this segment of the spine fall into a separate category. Ring fractures of the base of the skull are generally caused by a violent impact to the head and neck. Axial tension or compression is needed to fracture the skull around the foramen magnum. If ring fractures do not occur, the atlanto–occipital joint can be dislocated either by a torsional load applied axially or a shear load applied in the anteroposterior direction or vice versa. A large axial compression can also cause the arches of C-1 to fracture, breaking it up into two to four secitons. The odontoid process of C-2 is also a vulnerable area. Extreme flexion of the neck is a common cause of odontoid fractures, and a large percentage of these injuries are related to automotive accidents.[9] Fractures through the pars interarticularis of C-2, commonly known as hangman's fractures, are the result of a combined axial compression and extension (bending) of the cervical spine. Impact of the forehead and face of unrestrained occupants with the windshield can result in this injury. Garfin and Rothman[10] provide an interesting discussion of this injury and traced the history of execution by hanging. It was estimated by a British judiciary committee that the energy required to cause a hangman's fracture was 1260 foot-pounds (1708 Nm).

Chance Fractures.

This injury was first described by Chance[11] as being a lap belt-related syndrome in which a lumbar vertebra is split in the transverse plane, beginning with the spinous process. Subsequent studies attribute the injury to the improper wearing of the lap belt while involved in a frontal ($-g_x$) collision.

The belt rides over the iliac wings and acts as a fulcrum for the lumbar spine to flex over it, causing a marked separation of the posterior elements without any evidence of wedging.[12] When the lap belt is used in conjunction with an upper torso restraint, this injury does not occur.

Hyperextension Injuries

Hyperextension injuries of the cervical spine result in avulsion of the anterior aspect of the vertebral bodies, sometimes termed "teardrop fractures." Kazarian et al.[13] reported the occurrence of hyperextension injuries of the thoracic spine resulting from ejection from F/FB-118 aircraft. The superior lip of one or more vertebrae is avulsed along with the rupture of the anterior longitudinal ligament. This injury is sometimes accompanied by loss of posterior vertebral body height. When this occurs, there may be injury to the articular facets, pedicles, and/or the laminae. The incidence was 23 percent over a 10-year period. The powered inertial reel and the seat back were considered responsible for this rare injury.

This brief discussion does not do justice to the many forms of spinal injury caused by impact forces and accelerations. For a more detailed treatment of cervical injury mechanisms, the reader is referred to the work of DePalma,[14] Moffat et al.,[15] and Huelke et al.[16]

BIOMECHANICAL RESPONSE OF THE VERTEBRAL COLUMN

Because of its flexibility, the vertebral column is frequently subjected to bending loads that are superimposed upon the axial load it bears to support the head and torso. There is no question that impact accelerations in the horizontal plane exert bending loads on the spine. However, vertical ($+g_z$) acceleration is also capable of subjecting the spine to a high level of bending due to the fact that the vertebral column is located along the posterior aspect of the torso.

It is perhaps interesting to trace the progress made in experimental research on spinal injury, beginning with this bending hypothesis made by King et al.[17] The development of countermeasures to prevent anterior wedge fractures from occurring in pilots who eject from disabled aircraft was somewhat hampered by simple spinal models of Latham[18] and Hess and Lombard.[19] While they are admirable modeling efforts for their time and are sound from an engineering viewpoint, they unfortunately led subsequent researchers away from looking at the anatomy of the spine. The models were capable of simulating axial loading only. Experimental studies on the spine during whole-body acceleration of cadavers in the +g impact acceleration mode revealed that the spine was subjected to high bending loads even though it was restrained by a shoulder harness, and the input acceleration was in the seat-to-head (vertical) direction. This led to a more detailed study of the load-carrying capacity of the spine during +g acceleration. Ewing et al.[20]

tested a series of embalmed cadavers on the Wayne State University vertical accelerator, using three different restraint configurations, the hyperextended, erect, and flexed modes. In the hyperextended mode the spine was pulled back at the shoulders by a pair of military-type harnesses, while the thoracolumbar spine was placed in extension by inserting a block of wood 50 mm thick behind the spine at the L-1 level. In the erect mode, the spine was in its natural configuration while seated in a rigid seat, with the shoulder belts tightened manually to a tension of approximately 300 N. The shoulder harness was loosened in the flexed mode, permitting the torso to flex forward freely. The objective of the study was to determine the fracture level of the spine as a function of its spinal configuration. The results are shown in Table 1. By hyperextending the spine, the fracture g-level increased some 80 percent, and the observed difference was significant at the 95 percent level. In a subsequent search for this dramatic increase in spinal strength, it was determined that the spine did not receive external support from the hyperextension block and that the reason was an internal redistribution of the load borne by the spine.

Prasad et al.[21] embarked on a study to prove the hypothesis that the spine had two load paths and that the articular facets were indeed capable of transmitting load from one vertebra to the next. This facet load was difficult to measure directly but could be computed indirectly if the load borne by the disk was determined. An intervertebral load cell was designed to replace the inferior portion of a lumbar vertebra that was cut out by means of a double-bladed rotary saw. The IVLC shown in Figure 4 is 10 mm thick and has a diameter of about 40 mm. It is capable of measuring axial compression and the eccentricity of that load in the midsagittal plane. Figure 5 shows the facet load computed by subtracting the intervertebral load from the total load borne by the spine. The latter was assumed to be proportional to the measured seat pan load, with the proportionality constant equal to the ratio of the weight of the torso above the IVLC to the total weight of the body. At the beginning of the acceleration pulse, the facets were in compression, sharing the inertial load with the vertebral body and disk. As the head and torso flexed forward, the facets went into tension. These results were confirmed by Hakim and King,[22] who reproduced the IVLC loads on excised spinal segments in an MTS materials testing machine. By hyperextending

TABLE 1.   AVERAGE RELATIVE ROTATION OF T-1 WITH RESPECT TO T-12 AND T-12 WITH RESPECT TO THE PELVIS

Subjects	Muscle Tone	Unit Rotation ±1 SD T-1/T-12 (degrees)	Unit Rotation ±1 SD T-12/Pelvis (degrees)
Male	Tense	2.47 ± 0.54	4.10 ± 1.15
Male	Relaxed	3.51 ± 0.89	4.40 ± 0.95
Female	Tense	2.25 ± 0.49	3.77 ± 1.13
Female	Relaxed	2.52 ± 0.37	4.60 ± 0.37

From Cheng et al.[42]

Figure 4. Photograph of an intervertebral body load cell (IVLC) (thickness = 10 mm).

the spine, the facets were prevented from going into tension, thus increasing the fracture level of the most vulnerable vertebral bodies in the thoracolumbar spine. Furthermore, the facet load hypothesis provided an explanation for the frequently observed anterior wedge fractures. The additional compression borne by the bodies was needed to balance the flexion moment caused by forward rotation of the head and torso. Since the moment arm is of the order of 25 mm and the flexion moment can be as high as 40 Nm, this additional compression is over 1000 N. Such excessive compressive loads are the cause for anterior wedging of the vertebral bodies. Injury data from subhuman primates, obtained by Kazarian et al.,[23] indicate that derangement of facets was due to locking of the facet joints to act as load paths during +g acceleration. Although injuries to the posterior elements are rare in pilot ejection, these observations corroborate the load-bearing hypothesis of the facets.

Recently, Patwardhan et al.[24] measured contact pressure between the articular surfaces of lumbar facets and computed a facet force, reporting it to be the vertical facet force that was measured indirectly by Prasad et al.[21] This was felt to be erroneous since the articular surfaces are quite incapable of transmitting large shear loads. Yang et al.[25] performed loading experiments on isolated facet joints and obtained results that can explain the mechanism of load transmission through the facet joint. The posterior ele-

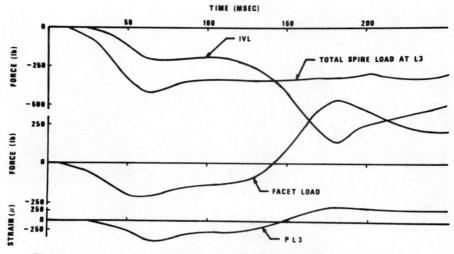

Figure 5. Facet load computed from measured IVLC data. (*From Prasad et al.*[21])

ments were separated from the body by cutting through the pedicles. The two facets were then loaded axially in their normal configuration in an Instron testing machine. In compression, they acted as a stiffening spring, as shown in Figure 6. In tension, however, they afforded very little resistance. Most of the tensile resistance was provided by the ligamentum flavum and the interspinous and supraspinous ligaments. Figure 7 shows the tensile load deflection curve of the isolated facets, with all ligaments severed. The mechanism of load transmission in compression is thus different from that in tension, and it is postulated that high compressive loads can be generated in the facet joint when the inferior tip of the inferior facet bottoms out on the pars interarticularis of the vertebra below it. In tension, the resistance is provided by soft tissues, such as the ligaments and the extensor muscles of the back. Figure 8 is a cutaway drawing of a facet joint and the postulated

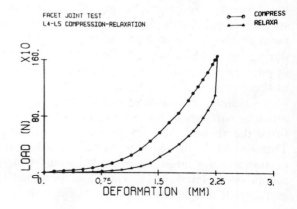

Figure 6. Response of isolated facet joints to a compressive load.

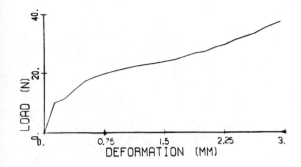

FACET JOINT TEST
L4-L5 TENSION W/O SP. PROCESS

Figure 7. Response of isolated facet joints to a tensile load.

bottoming-out effect, which needs to be demonstrated anatomically and biomechanically for a full understanding of the load-bearing role of the articular facets. It should be noted that the axial loading experiment on isolated facet joints was carried out because very little rotation of the facet joint could be detected when complete spinal segments were subjected to combined axial compression and forward flexion. Thus, for the isolated facets, axial compression is equivalent to spinal extension and axial tension to spinal flexion. These results have a significant impact on the understanding of injury mechanisms of the spine. The observation made by Nicoll[8] that dislocations occur if the interspinous ligaments are ruptured is equivalent to saying that the capsules cannot provide much resistance in flexion and that facets can be easily subluxed if the posterior ligaments are torn. In fact, the geometry of the facet surfaces can be an important consideration as far as dislocation is concerned. Those with surfaces that are almost horizontal would be easier to dislocate than those with vertical faces, particularly in the presence of torsional loads and horizontal shear forces. The high frequency of dislocations at the C-5–C-7 level can be attributable to the facet geometry of those vertebrae.

The biomechanics of a compressive force being generated in the vertebral column during a frontal impact ($+g_x$ acceleration) need to be discussed. This phenomenon was initially discovered with a two-dimensional model of the spine developed by Prasad and King.[26] Cadaver tests were carried out by Begeman et al.[7] to verify the existence of this force, since vertical forces are not expected to be generated in a horizontal crash. Large seat pan loads were measured if the subject was restrained by an upper torso restraint, such as a cross-chest belt. This was the net force after accounting for all lap belt forces. Dummies did not generate this seat pan load. Subsequent tests by Begeman et al.[27] involving volunteer subjects confirmed these results. It was postulated that the seat pan load was a manifestation of spinal compression due to the tendency of the spine to straighten out during $-g_x$ acceleration. The reverse situation exists during a rear-end collision. The spine goes

Figure 8. Cutaway drawing of inferior facet impinging the pars interarticularis.

into tension and the body tends to ramp up the seat back. It may be necessary to increase the height of some of the head rests to prevent hyperextension of the neck.

The biomechanics of the cervical spine have been studied extensively by Mertz and Patrick[28,29] and by Ewing et al.[30] Reference 30 contains a bibliography of a long series of papers published by these authors over a 10-year period. Mertz et al.[31] elected to quantify response in terms of rotation of the head relative to the torso as a function of moment at the occipital condyles. Loading corridors were obtained for flexion and extension (Figs. 9 and 10). These corridors do not take into account neck axial loads. Pure bending does not occur during $\pm g_x$ acceleration, and it appears that multiple corridors are needed for different levels of axial load in the neck. The suggested tolerance limit in terms of moment at the occipital condyles is 190 Nm (140 foot-pounds) for neck flexion, based on the data by Mertz and Patrick.[29] However, a recent report by Cheng et al.[32] indicated that the moment at the occipital condyles was about 340 Nm (250 foot-pounds) for catastrophic separation of the upper cervical spine. This abrupt change in injury level from an AIS of 2 to 6 cannot result from a 150 Nm (90 foot-pounds) increase in moment. The fact that the computed axial load at failure was in excess of 6.5 kN (1460 pounds) may have contributed to these fatal injuries. No axial force limit was provided by Mertz and Patrick[29] in conjunction with their moment data.

The voluminous data acquired at the Naval Biodynamics Laboratory in New Orleans constitute a valuable resource of neck response data of volunteers who were tested to relatively high g-levels. It will be necessary to analyze the data in detail before they can be used to define human biomechanical response. Such an analysis was performed recently by Wismans and Spenny[33] for lateral flexion.

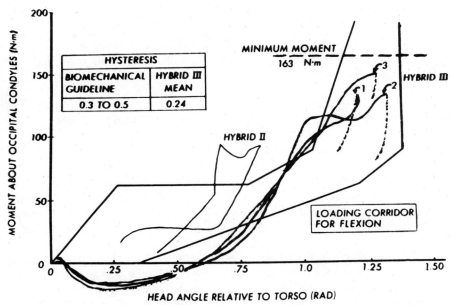

Figure 9. Neck flexion torque-angle responses are compared to the Mertz et al. biomechanical response corridor. (*From Foster et al.*[38])

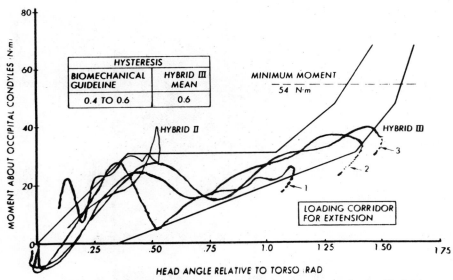

Figure 10. Neck extension torque-angle responses are compared to the Mertz et al. biomechanical response corridor. (*From Foster et al.*[38])

Neck response to crown impacts was studied by Culver et al.,[34] Hodgson and Thomas,[35] and Nusholtz et al.[36,37] Both embalmed and unembalmed cadavers were used in these studies. The failure mechanism was combined axial loading and bending, which can be either flexion or extension. The curvature of the cervical spine, its relationship to the head and torso, and the precise direction of impact were all difficult to control, resulting in a variety of observed injuries.

MECHANICAL SURROGATES OF THE VERTEBRAL COLUMN

Surrogates for the Cervical Spine

The principal aim in the design of an anthropomorphic test device (ATD) for crash tests is to have it respond in a humanlike manner. The neck of the Hybrid III dummy is an example of a surrogate that was designed to simulate the response obtained by Mertz and Patrick.[29] A description of the neck and its response to frontal $(-g_x)$ impact are given by Foster et al.[38] The one-piece neck is made of butyl elastomeric disks and aluminum plates (Fig. 11). The butyl elastomer has high damping characteristics that are needed to meet the hysteresis requirements of the neck. The axial strength of the neck is provided by a steel cable, which runs through the center of the device. Its response to flexion and extension are also shown in Figures 9 and 10 for tests on three different necks. The data show an improvement over those

Figure 11. The neck is one-piece, butyl elastomer and aluminum component. A special transducer measures forces and moments about the occipital condyles. (*From Foster et al.*[38])

obtained from an earlier version used in the Hybrid II ATD (Part 572 ATD). It is not known what the response of the neck would be if the biodynamic response parameter was changed to head acceleration or a combination of axial force and moment at the occipital condyles.

The Thoracolumbar Spine

The current ATDs (Part 572 and Hybrid III) have rigid thoracic spines and flexible lumbar segments. The latter is also made of an elastomer, and its reponse is sensitive to temperature variation. The spine was designed to support the head and rib cage and for simulation of the sitting posture. No biomechanical response data were available at the time it was designed. Such response data were first reported by Nyquist and Murton[39] and by Mallikarjunarao et al.[40] for static bending and by Mital et al.[41] and Cheng et al.[42] for frontal impact. Cadavers and volunteers were used in these studies to determine spinal configuration in a variety of bending modes. For the dynamic tests, two modes of restraint were used for the measurement of spinal response. The test subject was either restrained by a three-point belt or was clamped to a test seat at the pelvic level so that the motion of the spine could be monitored as it flexed during a series of $-g_x$ sled impacts. The data were then used to design a surrogate spine that had more humanlike responses than the current Part 572 ATD. Figure 12 shows a schematic of the static test device used by Mallikarjunarao et al.[40] to obtain static data of spinal flexion, extension, lateral bending, and oblique bending. Figure 13 shows a typical corridor for rotation of T-1 with respect to the pelvis for spinal flexion. The data were from eight different volunteers who were asked to resist the applied moment with active muscular contraction. The knees were fully extended (0 degrees knee angle). A total of 139 corridors were generated for the four bending modes. It was concluded that relaxed corridors tend to be bilinear, indicating a stiffening effect or muscular stretch response, whereas the tense corridors did not have a knee. The corridors compared favorably with previous data acquired by Nyquist and Murton,[39] who used a completely different method of drawing the corridors. The observed difference between the tense and relaxed states was found to be statistically significant ($P<0.05$). This was true for data acquired directly from film before they were subjected to any arithmetic operations. It was also found that the knee angle had a significant effect on the rotation of the pelvis with respect to the femur. Its effect on the rotation of T-1 with respect to the pelvis was minimal.

In order to design a surrogate spine, dynamic data are also needed. The tests to be described below are those carried out in the clamped mode. The pelvis, femur, and tibia of the test subject were tightly clamped to a test seat, and the spine was free to flex forward during a simulated frontal impact. The instrumentation consisted of accelerometer packages at T-1, T-12, and the pelvis as well as photo targets at the same locations. The peak sled acceleration for volunteers was 8 g, while that for cadavers was 30 g.

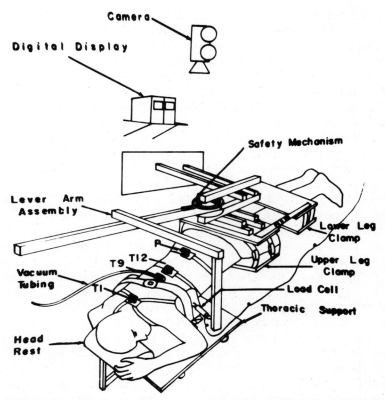

Figure 12. Schematic of static test device for spinal bending. (*From Mallikar-junarao et al.*[40])

The seat pan was equipped with a load cell that could measure the applied force and moment about all three orthogonal axes. After the sled had come to rest, it was possible to use the load cell data to calculate the moment at the hip joint, using equations of static equilibrium. The kinematic response of the spine is shown in Figure 14. The motion of T-1 with respect to T-12 was found to be quite substantial in comparison with that of T-12 with respect to the pelvis. In fact, the rotation of T-12 relative to the pelvis was much less than anticipated, since the lumbar spine was expected to be more flexible than the thoracic spine. Actual rotation data are provided in Table 2. The average rotation of thoracic vertebrae in the relaxed state was 3.51 degrees. That of the lumbar vertebrae was 4.4 degrees for relaxed male subjects.

The kinematic data were used to design a surrogate spine. An optimization technique for fitting the response of a single link to the acquired data was used. The result was a 486 mm long rigid link, pivoted at T-1 and the pelvis. This link was designed to replace the existing spine in the Part 572

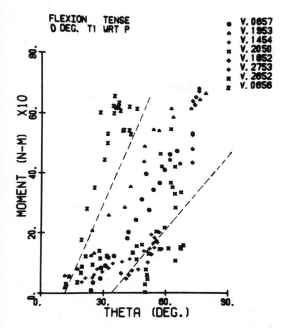

Figure 13. Typical corridor for the rotation of T-1 with respect to the pelvis during static spinal flexion.

ATD and chosen for the sake of repeatability and reproducibility of response. It is shown in Figure 15 and in Figure 16 after it was installed in the ATD. A series of sled runs was performed to validate the response of the new spine. Figure 17 shows the motion of the new spine and that of the Part 572 ATD relative to the human corridors. A large percentage of the Part 572 data did not fall within these corridors (targets with circles at the

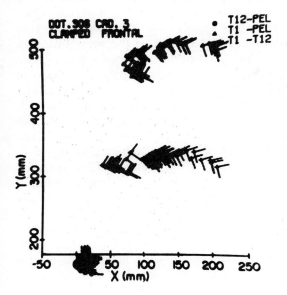

Figure 14. Relative motion of T-1 with respect to T-12 and pelvis, and T-12 with respect to the pelvis.

TABLE 2. SUMMARY OF PEAK ACCELERATION VALUES AT FRACTURE IN THE THREE SPINAL MODES

	Fracture Level (g)	No. of Cadavers	Average Age (yr)
Extended	17.75 ± 5.55	4	61.5
Erect	10.4 ± 3.79	5	61.0
Flexed	9.0 ± 2.00	3	54.3

From Ewing et al.[20]

corner). The acceleration response was also more humanlike. However, the moment at the hip needs to be adjusted with friction elements to match the data obtained from cadavers and volunteers.

MATHEMATICAL SURROGATES OF THE VERTEBRAL COLUMN

Mathematical models can often be used as an experimental tool to study the response of a system to a variety of input conditions. The premise is that the model must provide reliable predictions before this type of study can be of value. In other words, models that have been validated against experimental data are required. King and Chou[43] reviewed mathematical models of impact developed before 1975 and discussed in detail models of the spine that were available at that time.

Two-dimensional models developed after 1975 include discrete parameter models by Tennyson and King[44] and by Pontius and Liu.[45] Both were extensions of previous models without muscles, and both had a capability of simulating a delayed response of the musculature following a stretch stimulus. Validation against human volunteer impact data was provided by Tennyson and King.[44,46] Other two-dimensional models of the spine include a head and neck model by Reber and Goldsmith,[47] who also developed a discrete parameter model that was capable of transmitting loads via the articular facets. Models investigating the role of the added mass of a helmet were formulated by King et al.[48] and by Huston and Sears.[49] The study by King et al.[48] showed that if there was a simultaneous impact in the $+g_z$ and $-g_x$ directions and if the helmet resulted in an anterior shift of the combined center of gravity of the head and helmet, there was a significant increase in neck loads as well as a greater amount of cord stretch and an increased movement of the odontoid process into the vertebral canal. It was suggested that the stretching of the cervical cord and the impact of the cord by the dens could be two mechanisms of cord concussion that may be responsible for the loss of many pilots who ditch their aircraft during carrier landing attempts.

Since 1975, many three-dimensional models have been developed to simulate the response of the spine to impact acceleration. The model by

Figure 15. Rigid link surrogate spine.

Huston and Advani[50] simulated cervical spine response and had 54 degrees of freedom. Belytschko and Privitzer[51] demonstrated the capability of a 3-D model of the entire spine. It is basically a discrete parameter model in which the vertebrae are represented by rigid bodies interconnected by deformable elements. It was capable of simulating ejection seat dynamics as well as spinal response to a horizontal crash ($-g_x$ impact acceleration). There was, however, no validation against experimental data. Recently, Belytschko and Williams[52] formulated a 3-D head and neck model for the study of cervical spinal response to high acceleration environments. This is a finite element model of the seven cervical vertebrae and T-1 and utilizes a large deflection finite element code capable of handling both material and geometric non-linearities. However, the vertebrae were still assumed to be rigid bodies. The thoracolumbar spine was represented by three beam elements. The cervical musculature was simulated to 22 muscle groups of extensors and lateral flexors. The results of the model were compared with experimental data obtained by Ewing and Thomas.[53] This model has the potential of being a useful tool as an adjunct to experimental studies of 3-D cervical response. A finite element model of a motion segment was developed by Yang et al.[54]

Figure 16. New surrogate spine installed in a Part 572 ATD.

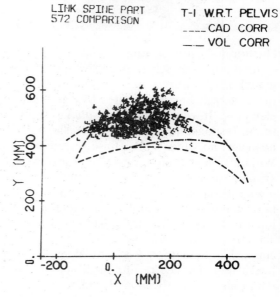

LINK SPINE PART 572 COMPARISON

T-1 W.R.T. PELVIS
____ CAD CORR
_._ VOL CORR

Figure 17. Comparison of motion of T-1 using two different surrogate spines.

to simulate the response of the articular facets, using the data detailed by Yang et al.[25] This model is based on that of Hakim and King[55] for a single vertebra and computes stress distribution in the body, disk, and posterior structures. The disk has a fluid nucleus surrounded by a low modulus annular material. It is still a static model, which can be made to respond to dynamic inputs at a later date. The computed intradiskal pressure compared favorably with experimental data. In a parametric study, it was found that the sensitive variables were the modulus of elasticity of the anulus fibrosus and spongy bone.

DISCUSSION

Because of the fact that spinal injuries are relatively infrequent in automotive accidents, research on the biomechanics of spinal injury and response is not as advanced as that for body regions that are frequently injured, such as the head or the thorax. There is a paucity of data on tolerance of the neck to impact acceleration. For the thoracolumbar spine, the tolerance information is over 20 years old for +g acceleration, and there has been no new information since the publication of the curves by Eiband.[56] The fact that tolerance is dependent upon the restraint system used and the age of occupant renders the problem of defining it as a single parameter virtually impossible. Furthermore, the configuration of the spine plays an important role in the injury pattern it sustains as a result of an impact in a given direction. In the absence of a restraint system, more spinal injuries are likely to occur, especially for motorcycle riders and occupants involved in rollovers. Thus, it is also extremely difficult to arrive at a limited set of injury criteria for the spine, particularly since the failure of the spinal components is not restricted to the bony portions of the spine. Much of the spinal resistance to bending and torsional loads is provided by soft tissues—ligaments, muscles, and cartilage (disk). For each set of loading conditions, injury criteria need to be formulated. Such a task is indeed formidable.

The mechanisms of injury to the spine are relatively well understood. Failure of the various spinal components can be attributed to a combination of axial and bending loads. The central role played by the articular facets cannot be overemphasized. Together with the vertebrae, they provide a dual load path for the transmission of axial load. In forward bending, the compressive load may be borne entirely by the body. However, new data being acquired from spinal segments appear to indicate that the facets are load-bearing until there is excessive flexion. In the living spine, the role of the musculature of the back acts to increase facet loading. Torsional loads and shear loads are also resisted to a great extent by the facets. In the lumbar region, Patwardhan et al.[24] may have measured this torsional resistance, using pressure-sensitive Fuji film.

CONCLUSIONS

1. Spinal anatomy is extremely complex. A thorough knowledge of its construction and function is essential to understanding its biomechanical role in load bearing and protection of the spinal cord.

2. The articular facets play a central role in the mechanism of spinal support and in the mechanisms of injury to the spine. The orientation of the facet surfaces may explain the tendency of subluxation and dislocation of certain vertebrae. It should be noted that subluxation is an unstable condition that endangers the integrity of the cord.

3. The manner in which facets resist compressive and tensile loads has been studied by means of tests on isolated facet joints. It can be hypothesized that compressive resistance is generated by the bottoming out of the tip of the inferior facets onto the pars interarticularis of the vertebra below. Proof of this hypothesis will require additional research. The facets offer virtually no resistance to tensile loads. The smooth surfaces of the facets and the relatively weak capsular ligaments are unable to provide the tensile resistance necessary for countering such loads. It is postulated that the ligaments and muscles along the posterior spine assume this role.

4. Although a limited amount of data are available for spinal response to impact acceleration, they have not been used in the design of a dummy spine that has humanlike responses. The problem of designing a dummy spine compatible with other regions of the ATD that can provide humanlike reponses in a variety of impact situations is a formidable task that needs to be tackled. For $+g_z$ acceleration, response data for the entire thoracolumbar spine are needed to build an ejection seat dummy. For $\pm g_y$ acceleration, there are virtually no response data for the thoracolumbar spine.

5. Tolerance data are woefully lacking. A carefully worked-out research plan is needed to address this complex problem of injury tolerance and injury criteria. In particular, there is a need to study the tolerance of the facets to dislocation, which is a high-risk injury. The types of dislocation and the injury mechanisms involved need to be identified at each spinal level and for each combination of applied loads.

6. The use of mathematical models to study the response of the spine is a viable approach because of the flexibility of the models and the low cost involved in comparison with that incurred in experimental research. Whenever possible, the combined approach of using models and obtaining experimental data based on model predictions will enhance the understanding of spinal response.

REFERENCES

1. Huelke DF, O'Day J, Mendelsohn RA: Cervical injuries suffered in automobile crashes. *J Neurosurg* 1981; 54:316–322.

2. King AI: Spine: Anatomy, types of injury, kinematics, mechanisms of injury, tolerance levels and injury criteria, in Aldman B (ed): *The Biomechanics of Impact Trauma.* Rome, International Center for Transportation Studies (in press).
3. Gray H: *Anatomy of the Human Body.* Philadelphia, Lea & Febiger, 1973, pp 109–111.
4. Peacock A: Observations on the postnatal structure of the intervertebral disc in man. *J Anat* 1952; 86:162–179.
5. Buckwalter JA: The fine structure of human intervertebral disc, in White AA, Gordan SL (eds): *Idiopathic Low Back Pain.* St. Louis, Mosby, 1982, pp 108–143.
6. Kazarian LE: Injuries to the human spinal column: Biomechanics and injury classification. *Exercise Sports Sci Rev* 1982; 9:297–352.
7. Begeman PC, King AI, Prasad P: Spinal loads resulting from $-g_x$ acceleration, in *Proceedings of the Seventeenth Stapp Car Crash Conference.* (SAE Paper No. 730977). Warrendale, Pa, Society of Automotive Engineers, 1973, pp 343–360.
8. Nicoll EA: Fractures of the dorso-lumbar spine. *J Bone Joint Surg* 1949; 31B:376–393.
9. Pierce DA, Barr Jr JS: Fractures and dislocations at the base of the skull and upper cervical spine, in Bailey RW, et al. (eds): *The Cervical Spine.* Philadelphia, Lippincott, 1983, pp 196–206.
10. Garfin SR, Rothman RH: Traumatic spondylolisthesis of the axis (Hangman's fracture), in Bailey RW et al. (eds): *The Cervical Spine.* Philadelphia, Lippincott, 1983, pp 223–232.
11. Chance GO: Note on a type of flexion fracture of the spine. *Br J Radiol* 1948; 21:452–453.
12. Steckler RM, Epstein JA, Epstein BS: Seat belt trauma to the lumbar spine: An unusual manifestation of the seat belt syndrome. *J Trauma* 1969; 9:508–513.
13. Kazarian LE, Beers K, Hernandez J: Spinal injuries in the F/FB-111 crew escape system. *Aviat Space Environ Med* 1979; 50:948–957.
14. DePalma AF: *The Management of Fractures and Dislocations,* ed 2. Philadelphia, Saunders, 1970, Vol 1, pp 272–295.
15. Moffat EA, Siegel AW, Huelke DF, et al.: The biomechanics of automotive cervical fractures, in *Proceedings of the Twenty-second AAAM Conference,* 1978, pp 151–168.
16. Huelke DF, Mendelsohn RA, States JD, et al.: Cervical fractures and fracture dislocations sustained without head impact. *J Trauma* 1978; 18:533–538.
17. King AI, Vulcan AP, Cheng LK: Effects of bending on the vertebral column of the seated human during caudocephalad acceleration, in *Proceedings of the Twenty-first ACEMB,* 1968, p 32.
18. Latham FA: A study in body ballistics: Seat ejection. *Proc Roy Soc* [B] 1957; 147:121–139.
19. Hess JL, Lombard CF: Theoretical investigations of dynamic response of man to high vertical accelerations. *Aviat Med* 1958, 29:66–75.
20. Ewing CL, King AI, Prasad P: Structural considerations of the human vertebral column under $+g_z$ impact acceleration. *J Aircraft* 1972; 9:84–90.
21. Prasad P, King AI, Ewing CL: The role of articular facets during $+g_z$ acceleration. *J Appl Mech* 1974; 41:321–326.
22. Hakim NS, King AI: Programmed replication of in situ (whole-body) loading

conditions during in vitro (substructure) testing of a vertebral column segment. *J Biomech* 1976; 9:629–632.

23. Kazarian LE, Boyd D, von Gierke H: The dynamic biomechanical nature of spinal fractures and articular facet derangement, AGARD Publication CP-88-71, Paper No. 19, 1971.

24. Patwardhan A, Vanderby R Jr, Lorenz M: Load bearing characteristics of lumbar facets in axial compression, in Thibault L (ed): 1982 *Advances in Bioengineering,* ASME, 1982, pp 155–160.

25. Yang KH, Tzeng CR, King AI: Response of the articular facet joint to axial loads, *Proceedings of the Ninth International Congress of Biomechanics,* 1983, in press.

26. Prasad P, King AI: An experimentally validated dynamic model of the spine. *J Appl Mech* 1974, 41:545–550.

27. Begeman PC, King AI, Levine RS, Viano DC: Biodynamic response of the musculoskeletal system to impact acceleration, in *Proceedings of the Twenty-fifth Stapp Car Crash Conference* (SAE Paper No. 801312), 1980, pp 477–509.

28. Mertz HJ, Patrick LM: Investigation of the kinematics and kinetics of whiplash, in *Proceedings of the Eleventh Stapp Car Crash Conference,* 1967, pp 267–317.

29. Mertz HJ, Patrick LM: Strength and response of the human neck, in *Proceedings of the Fifteenth Stapp Car Crash Conference,* 1971, pp 207–255.

30. Ewing CL, Thomas DJ, Lustick L, et al.: Effect of initial position on the human head and neck response to +Y impact acceleration, in *Proceedings of the Twenty-second Stapp Car Crash Conference.* Warrendale, Pa, Society of Automotive Engineers, 1978, pp 101–138.

31. Mertz HJ, Neathery RF, Culver CC: Performance requirements and characteristics of mechanical necks, in King WR, Mertz HJ (eds): *Human Impact Response—Measurement and Simulation.* New York, Plenum Press, 1973, pp 263–288.

32. Cheng R, Yang KH, Levine RS, et al.: Injuries to the cervical spine caused by distributed frontal load to the chest, in *Proceedings of the Twenty-sixth Stapp Car Crash Conference,* 1982, pp 1–40.

33. Wismans J, Spenny CH: Performance requirements for mechanical necks in lateral flexion, in *Proceedings of the Twenty-seventh Stapp Car Crash Conference.* (SAE Paper No. 831613), 1983.

34. Culver RH, Bender M, Melvin JW: *Mechanisms, tolerance and response obtained under dynamic superior-inferior head impact.* Final Report, HSRI, 1978.

35. Hodgson VR, Thomas LM: Mechanisms of cervical spine injury during impact to the protected head, in *Proceedings of the Twenty-fourth Stapp Car Crash Conference.* 1980, pp 15–42.

36. Nusholtz GS, Melvin JW, Huelke DF, et al.: Response of the cervical spine to superior-inferior head impact, in *Proceedings of the Twenty-fifth Stapp Car Crash Conference.* (SAE paper No. 811005). Warrendale, Pa, Society of Automotive Engineers, 1981, pp 195–237.

37. Nusholtz GS, Huelke DF, Lux P, et al.: Cervical spine injury mechanisms, in *Proceedings of the Twenty-seventh Stapp Car Crash Conference.* (SAE Paper No. 831616), 1983, pp 179–198.

38. Foster JK, Kortge JO, Wolanin MJ: Hybrid III—A biomechanically based crash test dummy, in *Proceedings of the Twenty-first Stapp Car Crash Conference,* 1977, pp 972–1014.

39. Nyquist GW, Murton CJ: Static bending response of the human lower torso, in *Proceedings of the Nineteenth Stapp Car Crash Conference.* Warrendale, Pa, Society of Automotive Engineers, 1975, pp 513–541.
40. Mallikarjunarao C, Padgaonkar AJ, Levine RS, et al.: Kinesiology of the human spine under static loading, *1977 Biomechanics Symposium,* ASME, 1977, pp 99–102.
41. Mital NK, Cheng R, Levine RS, et al.: Dynamic characteristics of the human spine during $-g_x$ acceleration, in *Proceedings of the Twenty-second Stapp Car Crash Conference.* (SAE Paper No. 780889), Warrendale, Pa, Society of Automotive Engineers, 1978, pp 139–165.
42. Cheng R, Mital NK, Levine RS, et al: Biodynamics of the living human spine during $-g_x$ impact acceleration, in *Proceedings of the Twenty-third Stapp Car Crash Conference.* (SAE Paper No. 791027). Warrendale, Pa, Society of Automotive Engineers, 1979, pp 721–763.
43. King AI, Chou CC: Mathematical modelling, simulation and experimental testing of biomechanical system crash response. *J Biomech* 1976; 9:301–317.
44. Tennyson SA, King AI: A biodynamic model of the human spinal column. Trans SAE, Paper No. 760771, 1976.
45. Pontius UR, Liu YK: Neuromuscular cervical spine model for whiplash, in SAE Publication No Sp-412, *Mathematical Modeling Biodynamic Response to Impact* (SAE Paper No. 760770). Warrendale, Pa, Society of Automotive Engineers, 1976, pp 21–30.
46. Tennyson SA, King AI: Mathematical models of the spine, in Avala XJR (ed): *Proceedings of the First International Conference on Mathematical Modeling,* 1977, Vol II, pp 977–985.
47. Reber JG, Goldsmith W: Analysis of large head-neck motions. *J Biomech* 1979; 12:211–222.
48. King AI, Nakhla SS, Mital NK: Simulation of head and neck response to $-G_x$ and $+G_z$ impact, in von Gierke HE (ed): *Models and Analogues for the Evaluation of Human Biodynamic Response, Performance and Protection* (AGARD Conf Proc No. 253), 1978, pp A7–1 to A7–13.
49. Huston RL, Sears J: Effect of protective helmets on head/neck dynamics, in Van Buskirk WC (ed): *1979 Biomechanics Symposium* (AMD-Vol. 32), 1979, pp 227–229.
50. Huston JC, Advani SH: Three-dimensional model of the human head and neck for automobile crashes. Trans SAE Paper No. 760769, 1976.
51. Belytschko T, Privitzer E: A three-dimensional discrete element dynamic model of the spine head and torso, in von Gierke HE (ed): *Models and Analogues for the Evaluation of Human Biodynamic Response, Performance and Protection* (AGARD Conf Proc No. 253), 1978, pp A9–1 to A9–15.
52. Belytschko T, Williams JL: A 3D model for head-neck dynamics, in Thibault L (ed): *1982 Advances in Bioengineering.* ASME, 1982, pp 35–38.
53. Ewing CL, Thomas DJ: *Human head and neck response to impact acceleration, NAMRL Monograph 21* (USAARL 73–1). Pensacola, Naval Aerospace and Regional Medical Center, 1972.
54. Yang KH, Khalil T, Tzeng CR, et al.: Finite element model of a functional spinal unit, in Woo SL-Y, Mates RE (eds): *1983 Biomechanics Symposium,* 1983, pp 137–140.
55. Hakim NS, King AI: A three-dimensional finite element dynamic response

analysis of a vertebra with experimental verification. *J Biomech,* 1979; 12:277–292.

56. Eiband AM: Human tolerance to rapidly applied accelerations: A summary of the literature. NASA Memorandum No 5-19-59E, 1959.

CHAPTER 18

Joints:
Clinical and Experimental Aspects

Cyril B. Frank, David Amiel, Savio L-Y Woo, Wayne H. Akeson

INTRODUCTION

Joint injuries probably cause more concern and receive more attention from the anxious patient, physician, and therapist than any other form of musculoskeletal trauma. The disturbing possibility of producing painful, stiff, or functionless joints despite years of treatment, rehabilitation, and reconstruction makes the complete understanding of joint trauma physiology and joint mechanics, in particular, both relevant and important.

Our purpose in this chapter is to review joint injuries from both clinical and experimental points of view, to describe some of the general principles of their management, and supply supportive rationale for treatment decisions. We will begin with a background to understanding joint injuries by defining and classifying the joints. We will then describe their relevant anatomic and functional features along with the incidence, mechanisms, and patterns of injury to certain of their structures. We will then discuss the current diagnostic and treatment methods for joint injuries and introduce some of the practical possibilities for reducing joint trauma in the future. Finally, we will present the experimental rationale for various treatment options during joint recovery and mention some of the ways that a knowledge of mechanical principles may be helpful in salvaging or reconstructing damaged joints.

BACKGROUND

Definitions
A joint or articulation is a union of two or more bones.[1] It is any point of contact between the elements of an animal skeleton[2] designed to permit, or even facilitate, their controled movement in relation to one another. In analogous terms, "joints are the bearings of the body machine."[1]

Classification
Joints can be classified according to their structural composition and movability into: juncturae fibrosae (syntharthroses) or immovable joints, juncturae cartilaginae (amphiathroses) or slightly movable joints, and juncturae synoviales (diarthroses) or freely movable joints.[3] These major classifications can then be subdivided according to the details of their attachments, their locations, and the specific tissues involved.[4]

Immovable joints include three major types: syndesmoses, sutura, and gomphoses.[3,4] A syndesmosis is an articulation in which two bones are united by interosseous ligaments, as in the distal attachments of the tibia and fibula. Sutura are found only in the skull and are fitted articulations of irregularly shaped bones, separated only by thin layers of fibrous tissue. A gomphosis is an articulation of a conical process into a socket and is represented by the insertions of the teeth in the jaw. In general, these joints are nearly solid and are, therefore, highly resistant to injury except by direct force.

There are two major types of slightly movable joints: the synchondroses and the symphysis. Synchondroses (cartilage connections between bones) occur only in immature skeletons, as the cartilage is ultimately converted to bone when the skeleton matures. Symphyses are joints in which contiguous bony surfaces are connected by fibrocartilage, as in the intervertebral disks of the back and the pubic bones of the pelvis. These slightly movable joints are fairly resistant to injury and usually yield only to high-energy loads in various directions. Separations of these joints can occur with distraction or shear. Their compression can lead to injury to the intervening tissue (e.g., ruptured disk in the back) or occasionally to periarticular bony collapse.

The majority of joints in the body (and those most commonly injured) are freely movable diarthroses, having a distinct joint space covered by a synovial membrane. They are often, therefore, referred to as "synovial joints" and are thereafter subclassified according to the types and degree of motion that they permit, ranging from short arcs of single planar movement to nearly unrestricted motion. From a mechanical point of view, these motions are critical insofar as they define the planes of freedom for the joint and therefore also its planes of injury. There are uniaxal joints, modified uniaxial joints, multiaxial ball-and-socket joints, and gliding joints, each with its own characteristics and predispositions for injury.

There are, for example, two types of uniaxial joints. The first is the gynglymus or hinge joint, allowing motion only in one plane but over a considerable arc. Articular surfaces in hinge joints are held together and guided by ligaments—tough bands of fibrous tissue running longitudinally or diagonally across the joint. The elbow and finger joints are typical and relatively classic hinge joints. Hinge joints, in general, may be subject to injury from forces that occur in any plane other than their normal single plane of function. Lateral forces on the humeroulnar joint of the elbow, for example, cannot be absorbed by simple bending or accommodation of the joint and may lead to its serious injury. The second type of uniaxial joint is one that allows rotation and is called a "trochoid" or "pivot" joint. It is made up of bone and ligaments in a ringlike configuration, as in the radioulnar articulation of the elbow and the atlanto–axial joint in the neck. Forces in directions other than those planes of rotation or in excess of their limits of rotation will lead to injuries of some of their pivotal joint structures.

Condyloid and saddle joints are modified synovial articulations that have slightly greater amounts of freedom. In both types there is a complicated configuration of bones that allows multiaxial movement but prevents axial rotation. The wrist is a condyloid joint with ovoid surfaces fitting into an elliptical cavity. The thumb carpometacarpal joint is an example of a saddle joint with reciprocally concavoconvex surfaces. Both the wrist and the thumb can be injured by extremes of any of their normal functions (beyond the range of their surrounding soft tissues or articulating joint surfaces), by forces that they cannot absorb (axial rotation), or by excessive impact loading—forces directed onto their joint surfaces. Because of their peripheral locations, these particular joints are often damaged by combinations of these factors (e.g., during falls on outstretched arms).

Ball-and-socket joints (spheroidal joints) like the hip and shoulder allow even greater mobility. In each, the distal bone is capable of motion around many different axes with one common center of rotation. This axial rotation, in combination with movement over great arcs, is functional in that it allows placement of the extremity (hand or foot) over large areas in space. The cost of this freedom, however, is a relative lack of stability of these joints, particularly at extremes of motions, with subluxation (partial separation of joint surfaces) or complete dislocations being relatively common. When combined with axial loading (which is frequent) and strong muscular contractions across the joint, simultaneous fracturing of the ball and/or socket can also occur.

The most mobile joint, in the sense of having the least configurational stability, is the plane or gliding joint. Bony surfaces are almost flat and, therefore, rely almost exclusively on surrounding tissues for support and guidance through their gliding movement. Many of the small bones of the wrist and ankle articulate in this way. If the soft tissue supports of these joints get torn (by a sudden forceful loading or in some extreme motion), bony surfaces easily separate and dislocation occurs. It is unusual for fracturing of these bones to occur because of their size, shape, and material properties, but also because of their three-dimensional mobility and the ability of their supporting structures to absorb energy by allowing their displacement.

Certain joints are combinations of some or all of the above, therefore having various intrinsic combinations of their mechanical advantages and disadvantages. The knee, for example, is a modified hinge joint, with some spheroidal and some planar joint characteristics. In addition to simple flexion and extension of the tibia relative to the femur during knee movement, there are subtle combinations of rolling, gliding, sliding, and rotation occurring during its normal range of motion. Due to its unique placement at a large distance between a large moving mass (the torso) and the ground, the knee undergoes tremendous stresses. Since the knee serves as the best example of nearly all of the clinical and experimental principles that are important in a biomechanical discussion of the joints and joint injuries, we will concentrate on its characteristics in more detail in several sections that follow.

Normal Joint Loads
Despite the common distinction between synovial joints as being either load bearing or nonload bearing it is probably safe to say that some load is normally transmitted across virtually all joints. The sources of that load may be mainly gravitational (in the legs) or musculotendinous (in the hands). However, some force is clearly generated on the joint surfaces or intervening tissues in nearly all normal situations.

It is important in a discussion of injury mechanisms to characterize the nomenclature for describing the potential forces and rotational couples acting on a typical mobile joint (e.g., the knee) in both normal and abnormal situations. Goodfellow and O'Connor[5] classified the forces in three directions acting at right angles to each other, each with its own rotational couple (Fig. 1). This convention allows description of all normal movements of the joint as well as all forces acting on it. It has been estimated that during walking, for example, 2 to 6 times the body weight is transmitted as interpenetrating forces across both the hip[6] and the knee.[7] Shear forces on the knee during walking are much lower due to the highly efficient gliding surfaces, probably being less than the body weight. These forces are, of course, dependent on joint position, body masses, and the magnitudes of segmental momentum. The actual forces and stresses in each joint structure, however, have not been accurately measured and probably vary over a broad range as the joint moves.

Joint Tissues
Since the knee is among the most complicated joints from the point of view of its functional anatomy, some understanding of its components will demonstrate the principles of virtually all diarthrodial joints. The knee is made up of many different types of tissue (Fig. 2) that are organized into different structures. Each of these tissues is a composite material, having its own unique mechanical properties.[8]

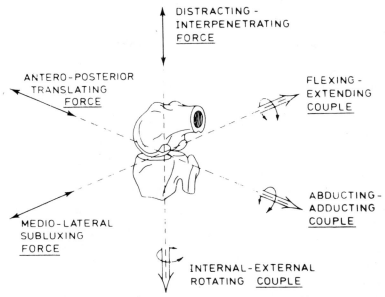

DISTRACTING -
INTERPENETRATING
FORCE

ANTERO-POSTERIOR
TRANSLATING
FORCE

FLEXING -
EXTENDING
COUPLE

ABDUCTING-
ADDUCTING
COUPLE

MEDIO-LATERAL
SUBLUXING
FORCE

INTERNAL-EXTERNAL
ROTATING COUPLE

Figure 1. Potential planes of motion and rotations of the knee joint. There are many complicated movements in this modified hinge joint. (*From Goodfellow and O'Connor.*[5])

The femur and the tibia are the bones that form the knee. Both are so-called long bones, having a shell of thick cortical bone covering spongy cancellous bone and marrow. The mechanical properties of cortical and cancellous bone are quite different,[8] but they act together to distribute stresses, resist compression, and form the major contours of the joint.

Between the two bones of the knee are two different types of cartilage. One is called "hyaline" or "glassy" cartilage and covers both joint surfaces. It forms the highly efficient gliding surfaces of the femur and tibia, being very resistant to both compressive and shearing stresses that it must endure. Hyaline cartilage probably does not serve much of a shock-absorbing role, however,[9] since it is so thin. The second type of knee cartilage is fibrocartilage, and it is much more able to absorb and distribute interpenetrating forces on the basis of its shape and its properties. These fibrocartilagenous structures called "menisci" are half-moon-shaped, central facing pads that form ideal sockets for the femoral condyles. Their rubbery nature (secondary to a mixture of collagen and cartilage) permits their deformation and movement through various joint positions; absorbing a significant amount of joint loads[10] while helping to lubricate and nourish the hyaline articular surfaces.

The ligaments are dense collagenous structures, highly oriented for the resistance of tensile stresses. Individual ligaments resist specific tensile stresses and work in concert to stabilize the joint in its full range of motion.[11] Cruciate ligaments (in the center of the knee joint) resist antero-

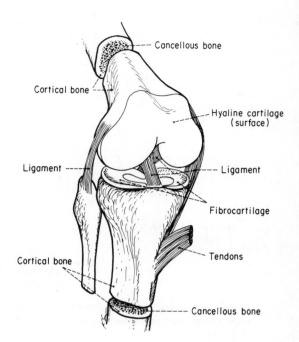

Figure 2. Structures in the normal knee. Each tissue has a specific function. Injury can compromise these functions by changes in their isolated (mechanical) or combined (structural) properties.

posterior translating forces, while collateral ligaments (on either side of the joint) resist medial lateral subluxations and abducting–adducting rotations.[12] These four ligaments and their adjacent capsular tissues also combine to resist tibial distraction and rotation during the complicated transfer of soft tissue tension during motion.[13,14]

Outside the joint, muscle–tendon units can be found inserting into the bone of the tibia or femur, mainly either anteriorly or posteriorly. These units power the joint but also protect it by making it more rigid when the muscles are contracted. Joint surfaces are forced together and become more congruous due to muscle forces, causing muscles to be called the "primary stabilizers" of the joint.

Although this brief overview of the structures at the knee does not accurately reflect their complex interactions, it will serve as a useful reference for the clinical material to follow.

Structure and Function
What then are the factors that allow controlled and stable joint motion? Normal joint stability clearly depends on what type of joint we are describing. In diarthroidial (synovial) joints, there are several important factors maintaining joint integrity and function. In the knee, these factors include bony contours, menisci, muscle function, ligamentous function, and even, to some extent, atmospheric pressure.[15] The ligaments, capsule, and menisci provide considerable stability under no-load conditions, while the bony con-

tours become increasingly important under compressive loading situations.[16] Increasing congruity of the distal femur with the tibial–meniscal surfaces occurs with increasing interpenetrating forces, providing an adaptive stabilization during times of functional need.

Normal joint motion depends on the integrity and strength of its movers (muscles and tendons) as well as the normal interaction of its surfaces and controls. In the knee, the normal rolling, gliding, and sliding of the tibia are often altered by loss of check rein (ligament) guidance or else by mechanical abnormalities of any number of intra-articular tissues. Tears of the menisci, fractures of articular surfaces, loose pieces of hyaline cartilage, proliferation of synovium, or even loose ends of ligament can get caught between the bones and block their interaction.

Since the joints are such finely tuned structures, virtually any interference with normal motion or stability usually leads to either signs or symptoms of joint pathology. Injury to almost any joint structure can lead to a mechanical dysfunction and joint degeneration (osteoarthritis). As we will also note in a later section, simply preventing the joint from performing its normal functions appears to injure it, having serious long-term complications.

For a more complete understanding of the anatomic and functional details of the various joint structures, the reader is referred to more specific references on those subjects.[15,16]

CLINICAL ASPECTS OF INJURY

Types of Injuries

Incidence of Joint Injury. According to a recent survey,[17] joint injuries are the second most common type of trauma (after contusions) encountered in clinical practice. Nearly 1 in 10 Americans will suffer a fracture, dislocation, sprain, or strain in each year.[18] Approximately 20 to 30 percent of all dislocations, sprains, and strains will require bed rest,[17] usually those involving the lower extremities. This group of joint injuries is second only to severe fractures for the average number of days of disability[17,18] with an average of 7 days of disability per sprain and an average of 25 days per dislocation.

Since over 50 percent of acute joint injuries seek medical advice,[17] the personal and health care costs of this spectrum of trauma is enormous.

Patterns of Joint Injury. Patterns of joint injury depend on what type of joint is involved, the location of that joint (central or peripheral), and its intrinsic anatomy or its functional capabilities.

Patterns of injury to specific joints can be grouped by their clinical appearances or more specifically by their tissue pathology. It is common, for example, to refer to a "joint sprain," where the patient presents with typical signs and symptoms of acute joint trauma (a red, hot, painful, swollen, functionless joint) but with an uncertain amount of articular tissue damage.

Sprains can also be used more specifically to describe specific ligament injuries, in synovial joints varying from mild (grade 1) with minimal tearing to severe (grade 3) with complete disruption of a specific ligament structure.

As indicated earlier, less mobile joints have greater intrinsic stability and are relatively resistant to external forces. On the other hand, they are less able to compensate for trauma (absorb energy) by moving or simply collapsing in the direction of the force. Since they are more rigid, less mobile joints usually suffer higher-energy types of injuries, with either fracturing or combined fractures and dislocations.

More peripheral joints are more predisposed to repeated types of trauma (e.g., running) and are those joints injured most often in common activities (e.g., falls, jumping). They most often suffer acute types of injuries, therefore, with various combinations of damage to bones and soft tissues. More central joints (e.g., the spine) are protected from more trivial trauma by the large central tissue mass, only being injured acutely by relatively large forces. On the other hand, these central joints must often bear the majority of the body weight and are often called upon to move to their limits, even during relatively simple activities. For that reason, these central joints usually suffer more chronic types of injuries and seem to be more prone to arthritic degeneration over time. Certain joints are, of course, more predisposed to this deterioration than others (e.g., the hip), probably as a result of combined genetic, biologic, and environmental (mechanical) factors.

Grossly, diarthroidal injury patterns can be classified as disruptions of stability or disruptions of motion. Disruptions of stability commonly involve combinations of injury to the stabilizing elements (as noted above) with sprains of various ligaments or capsule, bony injuries, or meniscal damage. The most common disruption of stability is an injury to the ligaments. The most common knee ligament injured is the medial collateral. When injured in association with the anterior cruciate ligament and the medial meniscus, this injury complex is often called the "terrible triad of O'Donoghue."[19] The medial collateral ligament is failed in tension by a valgus force. A continued valgus force can tear the cruciates and may compress the lateral joint structures (meniscus or bone). The medial meniscus can often be torn from its attachment to the capsular portion of the medial collateral ligament at the same time. Continued excessive force can lead to a subluxation of the joint, a partial loss of contact between the joint surfaces. The next step in increasing injury severity is, of course, total loss of joint integrity, known as a "dislocation." The worst pattern of synovial joint injury is a fracture-dislocation where bony and hyaline surfaces are disrupted in addition to being displaced.

Patterns of bony injury in joint trauma can be complex, with specific fracture patterns being associated with specific stresses[20] (Fig. 3). A combination of the material properties of bone (being weakest in shear) with the anatomic features of each joint lead to joint-specific fracture patterns. The

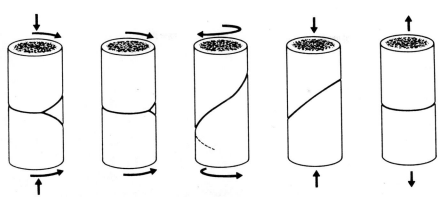

Figure 3. Schematic patterns of bony injury. From left to right, they are transverse failure with a large butterfly fragment from bending in compression, a smaller butterfly in the compression side of a transverse fracture secondary to pure bending, a spiral fracture from rotational failure, an oblique fracture in pure compression (since bone is weaker in shear), and a simple transverse fracture in tension. Higher energy increases comminution. Joint bone is less homogeneous than this model would suggest, and joint failures are, therefore, more complex. (*Adapted from Hayes.*[20])

knee, for example, often fractures with either a Y-shaped or T-shaped pattern of the distal femur[21] or a compression type of fracture of the tibial plateau. (Either the wedge-shaped tibial spines split the femoral condyles, or the femoral condyles depress the tibia.)

Disruptions of joint motion can occur with or without instability. The most common cause of interference with knee joint motion is a tear of one of the menisci, most commonly the medial meniscus. Meniscal injury patterns (e.g., bucket handle tears) have been extensively studied, with sites and types of injury depending to a large extent on internal meniscal architecture.[22] Mechanical interference of meniscal fragments with the motions of the knee cause the typical signs and symptoms of joint locking. Locking, as with any synovial joint, can also be due to loose bone, hyaline cartilage, ligament, synovium, or the production of other reactive materials in the joint (e.g., calcified soft tissues).

Mechanisms of Joint Injury. Mechanisms of joint injury can be demonstrated by discussing the interactions between forces (external, combined, or intrinsic) and tissues. The best demonstration of the effects of common external forces is to present a typical case history of an individual who had damage to numerous joint structures. This case will also be useful later, in discussing common diagnostic methods and treatments, as well as providing a valuable insight into the prognosis of joint trauma as compared with other musculoskeletal injuries.

Case History. J.C. was a 30-year-old male who was involved in a motor vehicle accident in which he sustained multisystem multiple injuries. Both

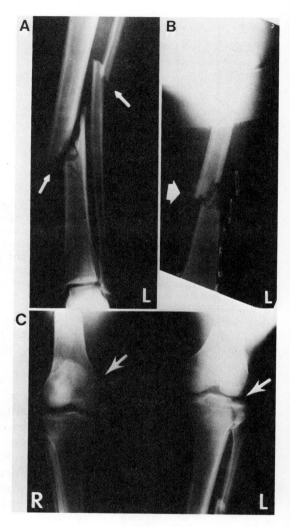

Figure 4. Acute x-rays of major injuries in patient J.C.: **A.** Left tibia–fibula fracture with undisplaced butterfly. Pattern suggests bending and rotation under axial load. **B.** Left femur fracture. Comminution suggests high-energy lesion. **C.** AP x-ray of both knees. Left knee has intra-articular fracture (arrow). Right knee shows small amount of air in joint (arrow), which could easily be missed. A congenital tripartite patella is also seen in the right knee.

lower extremities were injured by external forces applied near the level of the knees. He suffered pure extra-articular injuries to his left tibia and fibula (Fig. 4A) and his left femur (Fig. 4B), as well as suffering a fracture of his left lateral tibial plateau (Fig. 4C). Collectively, therefore, he had three relatively severe bony injuries to his left leg. It is worthy of note that in his original x-rays, there were no bony injuries to the right leg at all. However, there was a significant soft tissue injury to the lateral aspect of the right knee, with an open wound. The initial x-ray (Fig. 4C) also showed air in the right knee joint as a result of obvious communication with the atmosphere.

A basic mechanism of these injuries can be hypothesized (Fig. 5). A large lateral force, combined with probable axial loading on the left leg,

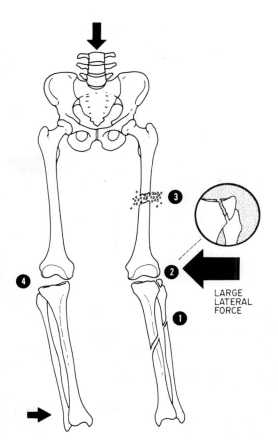

Figure 5: Schematic representation of major injuries in patient J.C. from large lateral force. Fractures of the left tibia and fibula **(1)**, the tibial plateau **(2)**, the femur **(3)**, and the soft tissue deficiency of right lateral knee **(4)** are shown. The tibial plateau fracture **(2)** is a clue to look for medial soft injury in that joint. Injury and inertial forces are shown by black arrows.

LARGE
LATERAL
FORCE

resulted in multiple high-energy bony lesions. Comminution of the femur, in particular, is a sign of considerable energy being absorbed to catastrophic failure of the cortical bone. The tibial lesion is typical of combined longitudinal loading, bending, and rotation[19] (refer back to Fig. 3) with relatively less comminution than the femur, probably indicating that some energy had been spent before this fracture occurred. The left lateral tibial plateau fracture is the result of compressive lateral forces,[23] with the lateral femoral condyle being driven downward onto the tibia. This injury mechanism implies a medial soft tissue injury to the left knee as well, since compressive failure loads of cortical bone (which were exceeded in the plateau fracture) are considerably higher than usual soft tissue limits. A continuation of this lateral force either directly or indirectly (through the upper body) to the right leg could then explain the failure in tension of the right lateral knee tissues (confirmed by examination). As shown schematically in Figure 5, the combined axial load, intertial mass of the upper body, and relative fixation of the right foot on the ground probably created a large varus bending moment at the right knee. This lateral bending moment was not in the usual

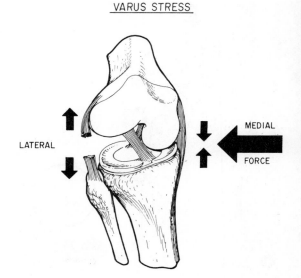

VARUS STRESS

LATERAL

MEDIAL

FORCE

Figure 6. Typical mechanism of injury of lateral joint structures. In patient J.C., a large medial force caused failure of lateral ligament supports and stretching of the anterior cruciate ligament. Compressive forces were not sufficient to fracture the bone in this knee, as they had been in the other leg.

plane of function of the knee joint, resulting in the failure of its ligamentous supports (the lateral collateral ligaments and anterior cruciate ligaments). The fact that there were relatively low-energy soft tissue injuries *only* to the right leg, further supports the source of energy as being from the left lateral side with progressive expenditure of energy from left to right (Fig. 6).

Other Examples. Another important mechanism of joint injury is through a combination of external and internal forces. A fracture of the olecranon, for example, may occur through the tip of the ulna but may be caused, in part, by the strong pull of the triceps muscle. A crack in the olecranon from a direct blow (fall on the elbow) may also displace secondary to this triceps muscle force. Overcoming these muscle forces is a necessary part of the treatment plan to restore and maintain the joint surface (Fig. 7).

Intrinsic forces (rapid acceleration or deceleration, loads out of the normal plane of the joint, exceeding the normal range of motion) can also damage joint surfaces or cause instability, leading to more chronic joint injuries.

Diagnostic Methods

Diagnostic methods of joint injuries include history and physical examination, noninvasive maneuvers (x-rays, tomograms, CT scans, radionuclide scans, stress x-rays), semi-invasive maneuvers (joint aspiration, arthrogram, CT arthrogram), and invasive maneuvers (arthroscopy and arthrotomy).[24] The minimum number of investigations yielding a complete diagnosis is preferable from many points of view, not the least of which is patient comfort, cost, and the optimal use of hospital facilities. Many diagnoses can be

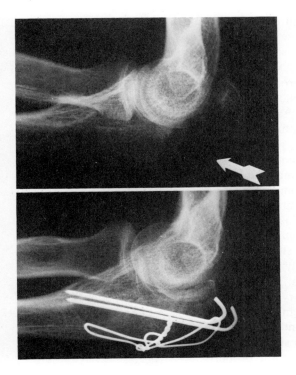

Figure 7. Typical intra-articular fracture of the elbow. (*Top*) Fragment of olecranon is displaced, leaving a large gap (arrow), and must be reduced and held by a technique known as tension band wiring (*bottom*) to overcome triceps pull.

made in the field through careful applications of the basic history and physical examination.

History. The history of joint injury is significant from a biomechanical point of view,[10] in that certain predictions of injury patterns can be derived from sources of energy, their magnitude, and directions of impact. In the previously mentioned case history, for example, the history of being hit from the left or from the right may have changed our expectations of where the most severe injuries may be found. In less forceful injuries, the history of the mechanism becomes even more important, since patients are often more aware of details, such as position of the joint at the time of impact (suggesting relative contributions of supporting structures), the point of impact (a clue to direction of force and relative compression or tension sides of the joint), and resulting deformity (which may have been corrected before examination). The history of specific joint pain or instability is often helpful as well in localizing the search for either intra-articular or extra-articular joint pathology.

Physical Examination. The physical examination of a joint that has been subjected to trauma follows the guidelines of inspection, palpation, and gentle manipulation to identify as atraumatically as possible the quantity, quality, and locations of injured structures.[10] Bleeding, bruising, and abrasions are signs of soft tissue injury and may overlie fractures. Obvious deformities are documented and corrected with careful observation of both nerve and vessel function. Neurovascular compromise can occur either acutely after a joint injury (from nerve or vessel impingement, laceration, or stretching) or chronically (following severe swelling or compartmentalization of pressure) and must be assessed serially. After careful palpation to localize tenderness, quantitate joint fluid, and examine joint surfaces, several specialized maneuvers of testing joint function are carried out.

As noted previously, the purpose of a joint is to provide painless, stable articulation through its normal range of motion. The specialized physical examination of an injured joint, therefore, should test for these factors, first with minimal applied forces and then with more provocative loads, if appropriate. Reproduction of the historical mechanism of injury, if possible, is the best way to reproduce the signs and symptoms of the injury and disclose any hidden instabilities. These tests are specific for the various stabilizing factors of each joint type and mainly test for the static components of stability. As long as the patient is awake, there is a significant dynamic contribution to joint stability (either voluntary or involuntary muscle contraction). This contribution may have to be removed by giving the patient a local or general anesthetic to reveal true joint pathology. This examination of joints under anesthesia is essential in any situation in which pain, swelling, and guarding make the joint difficult to examine properly.

Laboratory Tests. Noninvasive diagnostic tests include plain x-rays of the affected joints as well as the bones and joints above and below the suspected site of injury. They usually underestimate tissue pathology, however, and say nothing about the dynamic behavior of the injury, since they are taken without any stress on the joint. The previously mentioned diagnostic maneuvers (e.g., stressing the joint by hand) can be objectively documented and quantitated by measurement with x-rays. Two-dimensional and three-dimensional reconstructions of x-rays, tomograms, CT, or nuclear magnetic resonance (NMR) images can be used to add further diagnostic accuracy to soft tissue injury and/or bony displacement around joints, with or without stresses being applied.

Besides quantitation of joint instability by radiographic measurement, other objective measuring devices of joint function are also being developed. Measurement of displacement of the tibia relative to the femur, for example, is a way of determining deficiency of the cruciate ligaments.[25] By testing large populations of normal people, statistical comparisons with injured knees can allow statistical conclusions regarding the likelihood of cruciate pathology (Fig. 8). These methods are useful adjuncts to clinical examinations and offer some potential for objective diagnosis, quantitation, and comparison of joint treatment methods.

More invasive diagnostic methods are needed to more accurately assess

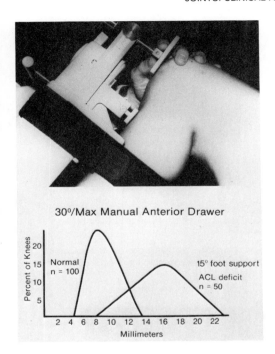

30°/Max Manual Anterior Drawer

Normal
n = 100

15° foot support

ACL deficit
n = 50

Percent of Knees

2 4 6 8 10 12 14 16 18 20 22
Millimeters

Figure 8. A. Instrumented testing of cruciate ligament function. Tibia is sagging posteriorly at 90° flexion in this example, suggesting possible posterior cruciate ligament deficiency. **B.** Measurements of maximum tibial displacement at various positions of the knee can be used to statistically diagnose cruciate pathology. Greater than 12 mm of maximum anterior tibial drawering with the knee tested at 30° of flexion is highly suggestive of an anterior cruciate ligament (ACL) deficit. (*Adapted from Daniel and Malcolm.*[25])

intra-articular pathology. Aspiration, for example, can reveal blood in the joint, and if it contains fat globules, an intra-articular fracture is suggested. Pure intra-articular blood is more suggestive of a capsular lesion or ligament tear. Removal of blood may be necessary to relieve intra-articular pressure (and pain) but also to remove its internal splinting effect on the joint and reveal a potential instability. Blood or fluid removal is also helpful in order to inject radiopaque contrast to assess the radiolucent joint structures, such as menisci in the knee (Fig. 9) or the hyaline articular cartilage of any of the synovial joints.

The most invasive diagnostic maneuvers involve actually looking at the joint structures and surfaces, either with magnification, through arthroscopic equipment in the larger joints (Fig. 10), or through an incision into the joint. More severe joint injuries demand more severe diagnostic methods in order to insure proper acute treatment, prescribe accurate rehabilitation, and make more accurate predictions of the prognosis for joint recovery.

Case History. The same patient who was described earlier is a good example of these diagnostic maneuvers and how they can help plan the treatment for a joint injury.

Referring back to Figure 4C, it can be recalled that the patient J.C. had a normal appearing right knee on plain x-ray with the exception of some air in the joint. Physical examination showed a contusion over the lateral side of the knee that communicated with the joint. It was not until the joint was

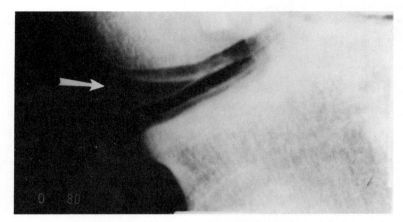

Figure 9. Arthrogram of knee. AP view shows dye in tear (arrow) that separates the medial meniscus from its normal peripheral attachment to the medial capsular ligament.

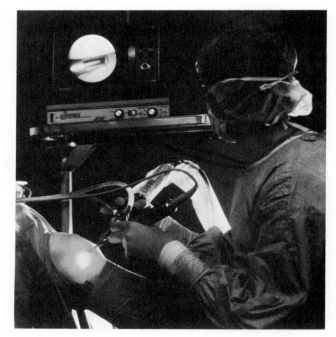

Figure 10. Operative arthroscopy. Image of joint surface is seen on television monitor in background. Fiberoptic light source with lavage system and cartilage shaving device are inserted into the knee.

stressed, however (Fig. 11), that the latent lateral instability of the joint was discovered. This was also documented on stress x-ray, with opening of the lateral joint compartment in full extension, suggesting serious lateral ligamentous deficiency. Similar stress testing in the anterior plane disclosed a serious anterior cruciate ligament deficiency in this patient as well. Both of these third-degree ligament ruptures required immediate surgical attention. J.C. had early arthrotomy with repair of the lateral structures as well as an anterior cruciate reconstruction of his right knee. His left leg was treated conservatively with traction for the femur fracture and casting for the tibial fracture (Fig. 4) after the tibial plateau was internally fixed.

Treatment

Treatment of joint injuries can be divided into nonsurgical and surgical treatments.

Nonsurgical Treatment. Nonsurgical management of joint injuries has often been advocated for the relative extremes of joint injury, either very mild or extremely severe joint damage. Mild injuries can be managed by symptomatic relief (ice, rest, elevation) for a short period of time, progressive remobilization of the joint with stretching, judicious passive motion, and an active exercise program to restore painless, stable, and full ranges of motion in as short a time as possible. External stabilization of the joint or external mechanical assistance may be necessary to protect, stretch, mobilize, or strengthen the respective stabilizers and movers of the injured joint during the rehabilitation period. Specific protocols of therapy are prescribed for individual needs, individual injuries, and individual joints. More aggressive methods are needed for larger weightbearing joints and in joints that are relatively resistant to remobilization (e.g., the elbow). A clear knowledge of the normal mechanical structure and functional demands of each joint is an important prerequisite to planning an appropriate nonsurgical rehabilitation program.

Severe joint injuries are sometimes best managed nonsurgically, as well, particularly in the cases of severe joint surface comminution.[26] Passive and active motion of those joints, instituted early after injury, may improve joint congruity and improve long-term results.[27] Complete joint dislocations with associated complications have also been managed successfully with conservative methods.[28] However, there is an increasing tendency toward acute surgical repairs of both joint bones and soft tissues in experienced hands.[13,29] As longer term results become available, there is an ever increasing appreciation for the subtle complexity of normal joint anatomic–functional interactions and the need for their restoration in optimizing joint recovery.

Surgical Treatment. Surgical treatment of joint injuries can be summarized by stating its idealized goal: Reconstructing the normal anatomy of the joint while maintaining its biologic properties will produce the best functional

386

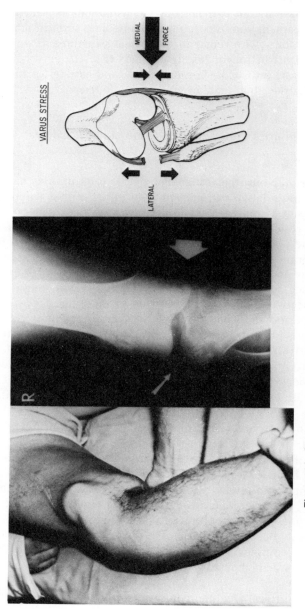

Figure 11. Clinical knee examination reproducing the mechanism of injury to the right knee (*left*). Under anesthesia the deformity was easier to demonstrate and document with AP stress x-rays (*middle*). The medial force (large white arrow) is used to open the lateral side of the joint (small white arrow). A schematic diagram (*right*) shows the medial force separating the lateral structures and compressing the medial side of the joint. This soft tissue injury required surgical repair and reconstruction.

results. The principles of internal fixation, as developed by the AO group,[29] were drafted mainly with this goal of joint function in mind. The principles of anatomic reduction (especially in articular fractures), preservation of blood supply through atraumatic operative technique, stable internal fixation with cognizance of biomechanical conditions, and early pain-free mobilization to prevent fracture disease are a clear statement of these desires. The actual reconstruction with either internal or external means is far too detailed and specific for this review. The methods can be extremely demanding and require considerable experience and expertise. In certain cases, these methods can make the difference between having a joint surface or none at all (Fig. 12). The reader is referred to the manuals on this subject for a comprehensive approach to the theories and techniques of bony fixation.[29]

Surgical treatment of soft tissue injuries around joints is somewhat more controversial than the goals expressed above. Although anatomic reconstruction of soft tissues (as with bones and joint surfaces) is the ideal, both technical and biological factors make the success of these procedures more variable[30] and their advisability, therefore, subject to debate. Acute repair of ligaments, for example, has long been advised.[22] However, more recent evidence suggests that the indications for these repairs or reconstructions may be more restrictive than initially thought.[31,32] The present state of knee repair, to be more specific, is dependent on which ligaments are torn, in what combinations, in what specific locations, and to what degree, not to

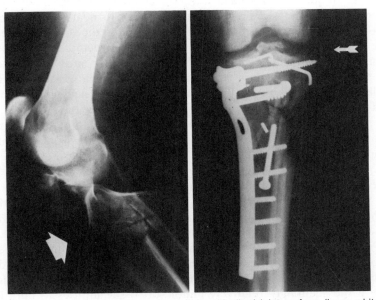

Figure 12. (*Left*) Destroyed proximal tibia with tilted joint surface (large white arrow) is shown being reconstructed (*right*) with multiple screws and a plate. Although not anatomically restored in this case due to severe comminution, the potential for a functional joint surface is recreated (small white arrow).

mention the more obvious surgical considerations of age, activity level, and occupation.

Basically, however, whether the bones or soft tissues are repaired or not, it is universally accepted that extensive therapy will be needed to restore normal joint function. Preservation of joint cartilage, healing of bones, and maturation of soft tissue scars are all known to be controlled to some degree by motion. On empirical grounds, therefore, prolonged joint immobilization is becoming more and more discouraged and joint motion more desirable in more highly selected cases.[32]

Prognosis

The prognosis of joint injuries depends on multiple factors including the number, type, and severity of joint structures injured, associated factors of severity, such as disruption of blood or nerve supply to the joint, the variable success of restoring joint anatomy and physiology, the unique characteristics of each individual joint and its loading demands, and a complicated combination of personal factors (age, weight, conditioning, tissue properties, healing capabilities, nutrition). To express it most simply, the joints are highly complex and highly evolved mechanical systems with equally sophisticated biologic properties. The more closely these normal characteristics are restored after injury (mechanics and biology), the more likely that the joint will have normal function.

Different joint tissues have different healing potentials. Bone has the distinction of having slow regenerative capacity to near normal form and function, depending on its anatomic restoration and the stresses placed on it during healing. The highly evolved soft tissue structures (ligaments and tendons) heal by variable degrees of scar formation, an inferior quality of tissue that may or may not hold up under functional demands. Hyaline articular cartilage has probably the least capacity to heal, often being replaced by mechanically inferior fibrocartilage on the joint surfaces. Anatomic restoration of joint surfaces and motion appear to be positive factors in optimizing the recovery of this gliding surface. However, a great deal of experimental work will be needed to better quantitate these effects.

Case History. Patient J.C. is an excellent example of many of these facts. While it appeared from initial x-rays that the long bone fractures in the left leg would be the most significant long-term problems in this case (Fig. 4), they healed very well with conservative (traction and immobilization) methods (Fig. 13, bottom right). The intra-articular fracture of the left knee was managed surgically, with anatomic restoration of the joint surface and good bony healing. The most debilitating injuries in this patient (at 3 year follow-up) were the seemingly relatively minor soft tissue injuries to both knees. Despite acute repairs of the right lateral collateral ligament and reconstruction of the right anterior cruciate ligament, the patient has had continued severe instability symptoms of that knee with signs of persisting ligamentous

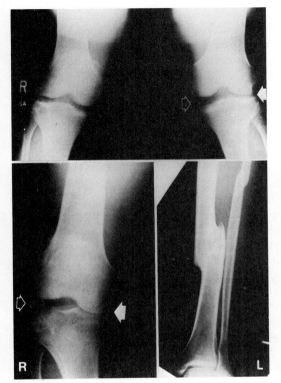

Figure 13. (*Bottom right*) Healed tibia and fibula in patient J.C. Although fractures had healed, there was some leg length discrepancy, aggravating soft tissue instabilities. Present soft tissue deficiencies of the patient are demonstrated by stress x-rays showing left knee medial ligamentous instability (*top*) with bilateral valgus stress (white arrow indicates force, black arrow shows medial widening) and continued right lateral knee instability (*bottom left*) despite acute repair (white arrow indicates force, black arrow shows lateral opening). Both of these instabilities are symptomatic.

insufficiency (Fig. 13, bottom left). He has also had increasing ligamentous insufficiency of the left knee (Fig. 13, top) combined with an ACL deficit, causing frequent collapsing of that side as well. This instability is not surprising in retrospect, based on the original mechanism of injury (Fig. 5), but it was extremely difficult to assess and treat acutely due to the floating knee (making it difficult to test joint stability until the bones were healed). He now has some early signs of arthritic deterioration of both joint surfaces due to these mechanical abnormalities, problems that will probably worsen over time.

In the common picture of the multiple trauma patient with life-threatening injuries, the lessons learned from this patient should be reinforced. If the patient survives the initial management crisis, the most likely causes of late complaints and disabilities will probably be orthopaedic and will almost certainly involve the joints.

Prevention

The prevention of joint injuries depends on recognizing every aspect of joint anatomy and function that we have described thus far, with particular emphasis on stabilizing factors and their limits.

The best protection for a joint is the optimum conditioning of its parts for the demands placed on them. Muscular strength, tone, and response are a critical stabilizing feature of the joints.[33] As will be discussed in the experimental aspects of joint injury, there is evidence to suggest that other joint tissues are similarly responsive to conditioning, with improvements in their mechanical and structural properties. Proper exercise is, therefore, probably the best means of minimizing joint injuries.

Protection against joint injuries also requires some knowledge of their limits of function (ranges, and so on) and the demands of activities placed on them. Equipment can be worn that can limit movements and may even passively protect joints from excessive stresses, as during sports activities.[34] These devices are commonly prescribed after injuries to the joint but, in some cases (in sports), are now being worn prophylactically. Protective devices can be defined as anything from a well-designed shoe to a complex joint brace, as long as they serve the basic function of preventing joint damage, either acutely or chronically.

Prevention of reinjury involves a combination of the above: avoiding potential injury stresses and situations, maximizing joint resistance through conditioning, and wearing all forms of proven supportive equipment. Artificial means of conditioning, such as through electrical stimulation of muscle tone or strength, may also prove effective for prevention of joint injuries. More careful documentation of its effects along with the growing body of data on other forms of conditioning will be needed to determine its indications and limits.

The increasing recognition of "prevention as the best means of treatment" applies to joints more so than probably any other area of trauma medicine.

EXPERIMENTAL ASPECTS OF JOINT INJURY

Treatment

The major treatment options of joint injuries involve either surgical or nonsurgical approaches. A considerable amount of experimental work has been carried out in both of these areas to document their relative pros and cons and indications for each technique. In general, there has been work done to determine the effects of trauma (and surgery) on the various joint tissues and a considerable amount of work done on the effects of different levels of stress or motion on both normal and healing joints.

Biomechanical evidence supports the restoration of normal joint architecture to ensure normal joint kinematics.[35–37] Abnormal kinematics clearly contribute to degenerative arthritis, as shown consistently in animal models.[38] Anatomical restoration of the joint surface produces the best potential for articular cartilage survival[39] and, therefore, minimizes the risk of surface breakdown and late osteoarthritis. Surgical restoration (or reconstruction) of joint surfaces where possible is indicated from both mechanical and biological points of view.

Conservative treatment methods, including rehabilitation, have received a great deal of attention. The three major conservative treatment options involve different levels of joint activity: immobilization, passive motion, or active exercise. Since it is often necessary to immobilize several joints (e.g., the wrist, hand, ankle, foot, spine) to immobilize *one* joint effectively (or to immobilize an adjacent long bone injury), the effects of immobilization on both normal and injured joint tissues have equal relevance to trauma rehabilitation.

There is considerable experimental evidence that immobilization causes synovial joint contracture and deterioration through injury-like processes in many of the joint tissues. Review articles have been written on the extensive histologic, chemical, and mechanical changes in the various tissues[40] as a result of being stress and motion deprived, suggesting that a Wolff's law of joints is just as appropriate as it is for the individual tissues.[41] More specifically, there are changes in both the structural and mechanical properties of immobilized joints, as demonstrated[42] in tensile tests of certain ligaments (Fig. 14). Several weeks of immobilization causes weakening of ligament insertions through bony resorption[43] as well as decreasing the stiffness of the ligament substance itself.[42] Analogous changes in other joint tissues as a result of stress deprivation have also been found,[44] rendering the immobilized joint a weakened and vulnerable structure.

Similar effects of immobilization have been found in healing synovial joint tissues. Stress and motion deprivation seems to inhibit scar formation and its maturation in ligaments,[45] tendons,[46] skin,[47] bone,[27] and articular cartilage.[48] The latter observations, combined with the noted degradative effects of immobilization on normal articular cartilage, have recently stimulated the use of passive joint motion[48] as a treatment for healing joints as well as its use as a prophylactic maneuver during limb immobilization for some other cause. The physiology and therapeutic values of passive joint motion have recently been reviewed,[49] with their combined mechanical and biologic implications.

Progressing upward on the scale of increasing joint motion, it has been found that active exercise further augments the structural properties of joint tissues.[50-52] These efforts are asymptotic, reaching a relative plateau at some point above normal properties.

In terms of both joint function and tissue properties, a relative scale of activity-related properties can, therefore, be proposed (Fig. 15), based on a percentage of ultimate values. On that scale, immobilization of a synovial joint causes relatively rapid declines in structural test parameters, while exercise results in very slow improvement in these parameters. Recovery of joints after immobilization is an area of current research interest. Different tissues apparently have different recovery rates and potentials.[53] The times, relative rates, and completeness of recovery of each joint tissue (both nor-

LATERAL COLLATERAL LIGAMENT

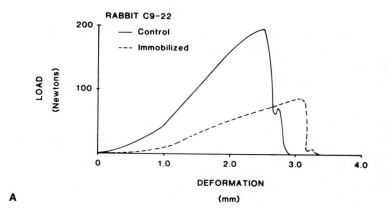

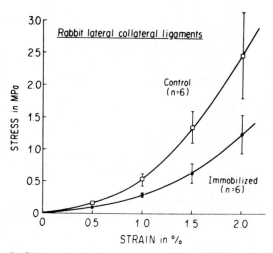

Figure 14. A. Change in structural properties of immobilized rabbit bone–lateral collateral ligament–bone complex as compared with cage activity controls. **B.** Mechanical properties of the LCL substance are also affected by several weeks of immobilization. (*Adapted from Woo et al.*[42])

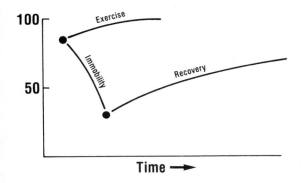

Figure 15. Hypothetical curves of joint and tissue responses to various degrees of stress and motion based on a scale of 100. The top point represents normal tissue properties and is estimated to be about 80 percent of optimum values. Response rates (slopes) and other points are speculative at this time but are probably joint and tissue specific.

mal and injured) are obviously of critical importance in designing optimum treatment protocols. In general, however, we might speculate that more highly ordered tissues with the lowest metabolic rates will be the first to deteriorate and the last to recover. These will be the limiting factors in determining how soon and how completely any joint can be mobilized and loaded after a severe joint injury.

Reconstruction
The failure to restore normal joint anatomy and function acutely leads to fairly certain deterioration of that joint over time. For that reason, as noted, more aggressive attempts at acute anatomical joint reconstitution are being recommended. If tissues are missing or if they are completely destroyed, however, some types of substitution or joint reconstruction become necessary. Replacement with either biological or nonbiological substances has been recommended to reconstruct joints. A complete review of these attempts is beyond the scope of this paper. It should be noted, however, that joint reconstruction and joint resurfacing with materials ranging from allografted animal materials[54] to stainless steel[55] have been extensively studied and are now an orthopaedic subspecialty in itself.

SUMMARY

As one of the most highly evolved and complex areas of the human body, the joint is perhaps the most susceptible and sensitive structure to trauma in the musculoskeletal system. It is impossible to accurately reflect the complexity and surprising specificity of each joint injury, just as it is difficult to generalize about their mechanisms of injury, diagnosis, and treatment.

In conclusion, we may say that we are only beginning to fully understand the subtleties of each joint. Through a better working knowledge of their normal mechanics and biology, better treatment of joint injuries will be planned. Anatomic reconstruction and early return to function are the

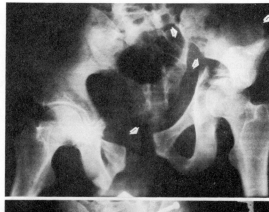
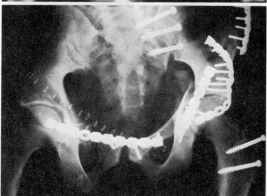

Figure 16. Multiple pelvic fractures and dislocations (*top*) repaired by extensive internal fixation techniques (*bottom*). The prognoses of such severe joint injuries are variable.

ideal goals of treatment. However, the lengths to which we must presently go to achieve these goals are sometimes nearly prohibitive (Fig. 16).

Further research into all areas of joint trauma is clearly needed.

ACKNOWLEDGMENTS

The authors gratefully acknowledge the financial support of the Alberta Heritage Foundation for Medical Research as well as the facilities of the University of California, San Diego, and the Veterans Administration Medical Center, San Diego. We would also like to thank Drs. D. Daniel and J. Moitoza for the use of their clinical material in this manuscript.

REFERENCES

1. Rosse C, Simkin PA: Joints, in Rosse C, Clawson DK (ed): *The Musculoskeletal System in Health and Disease.* Hagerstown, Harper & Row, 1980, pp 77–78.

2. Gove PB (ed): *Websters Third New International Dictionary.* Springfield, G & C Merriam, 1976, p 1219.
3. Goss CM (ed): *Gray's Anatomy of the Human Body,* ed 29. Philadelphia, Lea & Febiger, 1973, pp 287–369.
4. Walmsley R: Joints, in Romanes GJ (ed): *Cunningham's Textbook of Anatomy,* ed. 11. London, Oxford University Press, 1972, pp 207–214.
5. Goodfellow J, O'Connor J: The mechanics of the knee and prosthesis design. *J Bone Joint Surg* 1978; 60B:358.
6. Paul JB: Bioengineering studies of the forces transmitted by joints, in Kenedi RM (ed): *II. Biomechanics and Related Bioengineering Topics.* Oxford, Pergamon, 1965, p 369.
7. Morrison JB: The mechanics of the knee joint in relation to normal walking. *J Biomech* 1970; 3:51.
8. Frank C, Woo SL-Y: Biomechanics of sports injuries, in Nahum AM (ed): *Biomechanics of Trauma.* E. Norwalk, Conn., Appleton-Century-Crofts, 1984.
9. Fung YC: Bone and cartilage, in Fung YC (ed): *Biomechanics: Mechanical Properties of Living Tissues.* New York, Springer-Verlag, 1981, pp 383–400.
10. Radin EL, Paul IL: Does cartilage compliance reduce skeletal impact loads? *Arthritis Rheum* 1970; 13:139.
11. Shrive NG, O'Connor JJ, Goodfellow JW: Load-bearing in the knee joint. *Clin Orthop* 1978; 131:279.
12. Noyes FR, Grood ES, Butler DL, et al.: Clinical biomechanics of the knee– Ligament restraints and functional stability, in Funk FJ (ed): *The American Academy of Orthopaedic Surgeons: Symposium on the Athletes' Knee; Surgical Repair and Reconstruction.* St. Louis, Mosby, 1980, pp. 1–35.
13. Hughston JC, Andrews JR, Cross MJ, et al.: Classification of knee ligament instabilities. I. The medial compartment and cruciate ligaments. *J Bone Joint Surg* 1976; 58A:159.
14. Mueller W: *The Knee Form, Function, and Ligament Reconstruction.* Berlin, Springer-Verlag, 1983.
15. Johnson RJ, Pope MH: Knee joint stability without reference to ligamentous function, in *American Academy of Orthopaedic Surgeons: Symposium on Reconstructive Surgery of the Knee.* St. Louis, Mosby, 1978, pp 14–25.
16. Hsiech HH, Walker PS: Stabilizing mechanisms of the loaded and unloaded knee joint. *J Bon Joint Surg* 1976, 58A:87.
17. Wilder CS: Types of injuries: incidence and associated disability. *Vital Health Stat* 1965; 57(10):1.
18. Kelsey JL, White AA, Pastides H, et al.: *Musculo-Skeletal Disorders: Their Frequency of Occurrence and their Impact on the Population of the United States.* New York, Prodist, 1978, pp 51–59.
19. O'Donoghue DH: Surgical treatment of fresh injuries to the major ligaments of the knee. *J Bone Joint Surg* 1950; 32A:721.
20. Hayes WC: *Response of Bone Tissue to Loading.* Orthopaedic Biomechanics Course, White Plains, New York, June 1982.
21. Hohl M, Larson RL: Fractures and dislocations of the knee, in Rockwood CA, Green DP (eds): *Fractures.* Philadelphia, Lippincott, 1975, pp 1131–1181.
22. Wagner HJ: Die Kollagen faserarchilektur der Menisken des menschlichen Kniegelenkes. *Z Mikrosk Anat Forsch* 1976; 90:302.
23. Martin AF: The pathomechanics of the knee joint. I. The medial collateral ligament and lateral tibial plateau fractures. *J Bone Joint Surg* 1960; 42A:13.

24. Noyes FR, Bassett RW, Grood WS, et al.: Arthroscopy in acute traumatic hemarthrosis of the knee. Incidence of anterior cruciate tears and other injuries. *J Bone Joint Surg* 1980; 62A:687.
25. Daniel D, Malcolm L: Unpublished data, 1983.
26. Apley AG: Fracture of the lateral tibial condyle treated by skeletal traction and early mobilization. *J Bone Joint Surg* 1956; 38B:699.
27. Salter RB, Ogilvie-Harris DJ: Healing of intra-articular fractures with continuous passive motion. *AAOS Instruct Course Lect* 1979; 28:102.
28. Kennedy JC: Complete dislocation of the knee joint. *J Bone Joint Surg* 1963; 45A:889.
29. Muller ME, Allgower M, Schneider R: *Manual of Internal Fixation. Technique Recommended by the AO Group.* Berlin, Springer-Verlag, 1979, pp 3–14.
30. Frank C, Beaver P, Radenmaker F, et al.: A computerized study of knee ligament injuries: Repair versus removal of the torn anterior cruciate ligament. *Can J Study* 1982; 25:454.
31. Noyes FR, Matthews DS, Mooar PA, et al.: The symptomatic anterior cruciate-deficient knee. Part I. The long-term functional disability in athletically active individuals. *J Bone Joint Surg* 1983; 65A:154.
32. Noyes FR, Matthews DS, Mooar PA, et al.: The symptomatic anterior cruciate deficient knee. Part II. The results of rehabilitation, activity modification and counseling on functional disability. *J Bone Joint Surg* 1983; 65A:163.
33. Pope MN, Johnson RJ, Brown DW, et al.: The role of the musculature in injuries to the medial collateral ligament. *J Bone Joint Surg* 1979; 61A:398.
34. Brody DM: Running injuries. *Clin Symp* 1980; 32:2.
35. Yablon IG, Heller FG, Shouse L: The key role of the lateral malleolus in displaced fractures of the ankle. *J Bone Joint Surg* 1977; 59A:169.
36. Radin EL: Relevant biomechanics in the treatment of musculoskeletal injuries and disorders (editorial). *Clin Orthop* 1979; 144:107.
37. Radin EL, Paul IL, Rose RM: Current concepts of the etiology of idiopathic osteoarthritis. *Bull Hosp Joint Dis* 1977; 38:117.
38. McDevitt C, Gilbertson E, Muir H: An experimental model of osteoarthritis: early morphological and biomechanical changes. *J Bone Joint Surg* 1977; 59B:24.
39. Mitchell N, Shepard N: Healing of articular cartilage in intra-articular fractures in rabbits. *J Bone Joint Surg* 1980; 62A:628.
40. Akeson WH, Amiel D, Woo S L-Y: Immobility effects on synovial joints. The pathomechanics of joint contracture. *Biorheology* 1980; 17:95.
41. Pauwels F: *Biomechanics of the Locomotor Apparatus.* Berlin, Springer-Verlag, 1980, pp 375–407.
42. Woo S L-Y, Gomez MA, Woo YK, et al.: Mechanical properties of tendons and ligaments. *Biorheology* 1982; 19:397.
43. Laros GS, Tipton CM, Cooper RR: Influence of physical activity on ligament insertions in the knees of dogs. *J Bone Joint Surg* 1971; 53A:275.
44. Michelsson JE, Videman T, Langenskiöld A: Changes in bone formation during immobilization and development of experimental osteoarthritis. *Acta Orthop Scand* 1977; 48:443.
45. Long ML, Frank C, Schachar NS, at al.: The effects of motion on normal and healing ligaments. (Abstract) *Trans Orthop Res Soc* 1982; 7:43.
46. Gelberman RH, Woo S L-Y, Lothringer K, et al.: Effects of early intermittent passive mobilization on healing canine flexor tendons. *J Hand Surg* 1982; 7:170.

47. Arem AJ, Madden JW: Effects of stress on healing wounds: I. Intermittent noncyclical tension. *J Surg Res* 1976; 20:93.
48. Salter RB, Simmonds DF, Malcolm BW, et al.: The biological effect of continuous passive motion in the healing of full thickness defects in articular cartilage. *J Bone Joint Surg* 1980; 62A:1232.
49. Frank C, Akeson WH, Woo S L-Y, et al.: Physiology and therapeutic values of passive joint motion. *Clin Orthop* 1984; 185:113.
50. Tipton CM, Matthes RD, Maynard JA, et al.: The influence of physical activity on ligaments and tendons. *Med Sci Sports* 1975; 7:165.
51. Woo S L-Y, Gomez MA, Amiel D, et al.: The effects of exercise on the biomechanical and biochemical properties of severed digital flexor tendons. *J Biomed Eng* 1981; 103:51.
52. Woo S L-Y, Kuei SC, Ameil D, et al.: The effect of prolonged physical training on the properties of long bone: A study of Wolff's law. *J Bone Joint Surg* 1981; 63A:780.
53. Woo S L-Y, Ameil D, Akeson WH: Unpublished data, 1983.
54. Salama R, Weissman SL: The clinical use of combined xenografts of bone and autologous red marow. A preliminary report. *J Bone Joint Surg* 1978; 60B:111.
55. Elloy MA, Wright JTM, Cavendish ME: The basic requirements and design criteria for total joint prosthesis. *Acta Orthop Scand* 1976; 47:193.

CHAPTER 19

Biomechanics of Chest Injury

Richard M. Peters

Injuries to the chest can be classified into two major groups, blunt trauma[1] or penetrating trauma of the chest,[2] and each of these can be divided into subgroups.[3] The blunt trauma can result in enough disruption of the chest to seriously alter the structural mechanics of the chest which limits motion of the chest only as the result of pain and muscle spasm. Penetrating injury can result in injury to the internal organs in the chest, causing bleeding from the heart and great vessels or disruption of the lung, with only a small wound of entry and/or exit, or it can result in a large defect in the chest wall that adds the additional problem of disrupting the structural mechanics of the chest cage by creating a sucking chest wound.

THE CHEST CAGE

It is difficult, if not impossible, to define for the chest as a whole what forces are necessary to result in any particular type of injury.[4] The chest is not uniform in its physical structure. The upper portion of the chest is heavily protected by the scapula, posteriorly by the spine, by the pectoral muscles and clavicles in the anterior superior portion of the chest, and by the humerus in the lateral portion of the chest. The lower anterolateral portion of the chest is least protected, the ribs being in some areas almost subcutaneous, in others covered only by a thin layer of muscle. It is this area that is most vulnerable to direct injury. Since the contents of the chest are suspended within the bony chest cage, just as the brain is in the skull, sudden deceleration of the chest can stop the chest wall while the contents continue to move, thus resulting in tearing of the contents. The seriousness of penetrating wounds again is largely affected by the location. Lateral wounds are less likely to injure major blood vessels or the heart and so are of less consequence than wounds over the central and superior portions of the chest, where the heart and great vessels are very likely to be injured.

In addition to the anatomic nonuniformity of the chest, there are also marked differences as an individual progresses from infancy to the late decades of life. In infancy the chest cage is very flexible and does not afford a great deal of body protection to the chest contents. During growth the chest gains increasing stiffness but still remains quite flexible so that a young child or early teenager can be run over by an automobile without any fracture of ribs, though he may sustain serious internal injury. In the elderly, the flexible anterior costal cartilages and the supple and movable posterior joints may become calcified and the ribs lose calcium and become more brittle. In these patients, compressive forces are almost certain to result in fractures. An example of this difference is that the very flexible chest of a young person withstands compression of the sternum for cardiopulmonary resuscitation without significant injury. In an elderly patient, unless CPR is done carefully, the whole sternum may be ripped free from its costochondral attachments to the ribs, producing a significant chest injury.

The chest cage and its muscles must function as a pump to draw air into and expel air from the lungs. The respiratory pump is a suction pump in contrast to the heart, which is a positive ejection pump. Air is drawn by creating a subatmospheric pressure in the chest to overcome the elastic recoil and airway resistance of the lungs. When the inspiratory muscles are used to expand the lungs and chest cage, the elastic energy stored by the stretching provides force for expiration so that in quiet breathing expiration is passive.

Effective and efficient action of any muscle requires adequate but not excessive stretch, or preload, on the muscle fibers (Fig. 1). For the heart and the lungs, this muscle contraction must periodically occur to maintain life. If the preload or stretch on the muscles is inadequate or the force required is excessive, the respiratory muscles will fatigue and ventilatory failure will occur. One of the major problems in the treatment of thoracic injury is to understand the mechanism of contraction and prevent the development of respiratory muscle fatigue by providing mechanical ventilatory support if needed.

The chest cage acts principally as a suction pump, enlarging the volume of the chest to create a subatmospheric pressure that stretches the lung and sucks air in through the trachea and bronchi to ventilate the lungs. The force across the lung must be adequate to overcome the lung's inherent airway and elastic resistance. The elastic resistance in the lung results from both the elastic elements and the geometric arrangement of the components of the

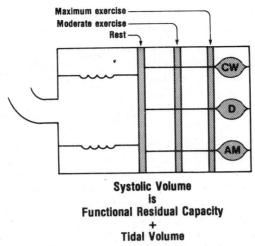

**Systolic Volume
is
Functional Residual Capacity
+
Tidal Volume**

Figure 1. The relationships among lung mechanics, functional residual capacity, and work of ventilation. The residual volume is determined by the elastic recoil of the lungs and the chest wall. The inspiratory muscles pull air into the lung. In the normal lung, the diastolic volume is the functional residual volume. The inspiratory muscles are stretched and have a normal preload. The injection fraction may be considered tidal volume over functional residual volume. As ventilation demands of exercise increase, tidal volume increases and all the inspiratory muscles of the chest wall and the diaphragm contract. CW, chest wall; D, diaphragm; AM, accessory muscles. (*From Peters.*[5])

lung. When the muscles of inspiration relax, removing the stretching force on the lung, the elastic energy stored in the lungs and the chest wall during inspiration expels the air from the lung to prepare the patient for the next inhalation. If not all of the air is expelled, the stretch on the muscles will be diminished and the next contraction will be less efficient. On the other hand, if the volume of the lung is too small because the lung is stiff or there is something compressing the chest, the residual volume, the volume at end expiration, will be small. The residual volume of the lung is determined by the elastic recoil of the lungs, the chest cage, and resistance to air flow through the airways. Any abnormality that causes increased fluid in the lungs, whether due to direct injury and bleeding, leak from capillaries, or inflammatory reactions, makes the lungs stiffer. Stiffer lungs require increased force to stretch the lungs during inspiration and increase the elastic recoil and decrease the end-expiratory lung volume. When the end-expiratory lung volume is decreased, the partially filled alveoli and alveoli that are poorly ventilated tend to collapse completely, further increasing the proportion of lung that is unventilated. Therefore, the volume at end-expiration of the lung is critical. If it is too small, the lungs tend to collapse, and if it is too large, the stretch on the inspiratory muscles, or preload, is inadequate.

ACUTE RESPIRATORY DISTRESS SYNDROME

Unventilated lung creates an intrapulmonary shunt, so that blood can reach the arterial circulation without gas exchange, and severe hypoxemia results.[5] This is characteristic of the condition called "acute respiratory distress syndrome" (ARDS). As presently defined, the acute respiratory distress syndrome is caused by the leakage of fluid from the alveolar capillaries due to a capillary injury. This capillary injury may have many causes. It is thought now to be principally the result of some kind of immune reaction, activation of complement, and so is often the result of multiple transfusions. When the capillaries leak, a greater amount of fluid crosses the capillary wall and accumulates in the interstitial space, causing ARDS. When such a capillary leak occurs, it usually takes a number of days to clear up. Another etiology of increased lung water and acute alveolar collapse in the postinjured patient is not related directly to the injury but to the resuscitation with excessive volumes of fluid, causing pulmonary edema by simply fluid overloading the patient.

Collapse of alveoli and accumulation of blood in the lungs results from direct injury to the lung, pulmonary contusion, with bursting of the capillaries and small vessels and bleeding into the lung tissue. A pulmonary contusion is a direct, extensive injury to the lung and takes many days to clear up, while aspirated blood on x-ray can have the same appearance but may clear very promptly. Good tracheal toilet, the patient being able to cough and breathe deeply, will remove the blood but will not affect ARDS or contusion. The mistake of labeling pulmonary contusion or direct blast injury to the lung as aspirated blood from a laceration has more than theoretical importance in the understanding of biomechanics and the treatment of these patients. The injury to the lung that causes a major pulmonary contusion seriously disrupts the structure and function of the lung and repair takes days to weeks. In contrast, bleeding within the trachobronchial tree, while it may interrupt the function of the lung, can be relatively easily treated by improving the patient's cough and removing the retained blood or secretions by bronchoscopy and other forms of endotracheal aspiration.

Alveolar collapse due to leaky capillaries, ARDS, from a biomechanical or immunologic injury, on the other hand, depends on removal of the source of that agent, which is most frequently the result of infection or multiple transfusion. In the patient who develops the syndrome of ARDS, the source of infection must be sought and effectively treated. If this source is removed, early recovery is possible. If not, the lung's alveolar network is destroyed by progressive fibrosis.

Regardless of the source of increased fluid in the lung, whether from blood retained in the alveoli, secretions that are not coughed up, capillary leak into the alveoli, or direct hemorrhage, the biomechanical deficit is increased stiffness. The stiff lung lowers end-expiratory volume and invites collapse of poorly ventilated alveoli. The biomechanical method of preventing alveolar collapse due to stiff lungs is to provide a greater transpulmonary expanding force. One of the most effective ways of increasing the transpulmonary expanding force is to get the patients out of the supine posture, sit them up, and turn them from one side to the other to use the mechanical weight of the mediastinum or the abdominal contents to expand the lung. This moving also tends to move secretions about so they are more easily coughed up. The patient's mobility is primarily dependent on the control of the pain, and the control of the pain is best achieved by some kind of a regional block. If such simple measures fail, the end-expiratory airway pressure can be elevated to counter the elastic recoil force. This is called positive end-expiratory pressure (PEEP).

The stiffness of the lungs and the increased force required for breathing may exceed the muscular strength of the injured individual, and muscle fatigue may result (Fig. 2). If pain is not controlled and part of the chest cage is splinted and therefore not doing its share of the work, early muscle fatigue and respiratory failure can result. If there is severe disruption of the chest cage with multiple fractured ribs, the patient may be able to maintain ventilation if the lungs are relatively normal, but if the lungs become stiff, the increased pressure required to ventilate them, as noted below, results in increased paradoxical motion of the chest cage, and fatigue will result. An essential part of the care of the patient with chest injury is to assess the mechanical effects of the injury on the lung and the chest cage and to alleviate the deterioration of function of the lung to minimize the work of the chest cage. It is also essential to recognize when the mechanical properties of the lung and the deficiencies in the function of the chest cage will exceed the muscle strength of the individual and result in respiratory fatigue and respiratory arrest.

PNEUMOTHORAX AND HEMOTHORAX

The coordination between the mechanical pump and its struts, the rib cage, and the gas exchange apparatus, the lungs, determines in large part the physiologic consequences of injury to the chest.[1,4] There are elastic elements in both the lung and the chest cage. The visceral pleura of the lungs is in intimate contact with the encasing membrane lining of the chest, the parietal pleura. Each pleural cavity in the normal lung is a closed space, but the two pleural cavities are separated only by the very mobile mediastinum, which can move into one or the other chest to equalize the pressure difference between the two chest cages. At resting lung volume, the elastic forces of the chest cage oppose those of the lung. The resting volume of the chest cage would be larger without the lungs in it, and the resting volume of the lungs would be smaller without the chest around them. As a result, during resting ventilation the intrapleural pressure is subatmospheric throughout the total respiratory cycle. On inspiration it drops from 2 or 3 cm of water

High Permeability Pulmonary Edema
A. R. D. S.

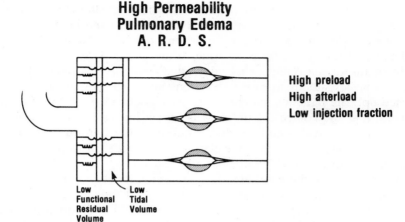

High preload
High afterload
Low injection fraction

Low Functional Residual Volume

Low Tidal Volume

Figure 2. This is similar to Figure 1, showing the effects of high permeability edema on the lung. The edema causes the lung to be stiff, represented by the added springs, and lowers the functional residual volume, increasing the preload, but the stiff lung raises the afterload, making the lung hard to ventilate with a low injection fraction. (*From Peters RM: Postoperative respiratory failure, in Proceedings First Boehringer Ingelheim Chest Symposium, Manila. Hong Kong, Excerpta Medica Asia, Ltd, 1983*)

down to 8 to 10 and on end-expiration rises back to 3 cm of water subatmospheric. In the upright posture, the weight of the abdominal contents pulls down the diaphragms, increasing the end-expiratory lung volume, making the pleural pressure more negative at end-expiration. In the supine posture, the abdominal contents push up the abdomen, raising the diaphrams and making the end-expiratory intrapleural pressure less subatmospheric.

When a rib is broken or with a penetrating injury of the chest, the parietal pleura, the visceral pleura, and the lung may also be lacerated. If the visceral pleura and lung are lacerated, on inspiration the drop in intrapleural pressure sucks air into the pleural cavity through the leak in the lung. During expiration the laceration of the lung tends to fall together, blocking the escape of the air that is in the pleural cavity and creating a pneumothorax of increasing size as each subsequent breath allows air to leak through the lung into the pleural cavity and prevents its escape. In this way a significant pneumothorax develops, which acts to collapse the lung. If there is bleeding from a laceration of the lung or a laceration of one of the vessels of the chest cage, blood can also accumulate in the chest, resulting in hemothorax and further loss of lung volume.

Pneumothorax and hemothorax are treated by insertion of a catheter through an interspace into the pleural cavity. The catheter is connected to a tube that leads down to a bottle containing a small amount of fluid to cover the lower centimeter of the tube. This bottle is vented through another tube so that air can escape when the pleural pressure rises above atmospheric.

Air cannot enter the pleural cavity through the catheter because with the bottle on the floor, a higher negative pressure than the patient could generate, 100 cm of H_2O, would be necessary to pull air back into the chest. Because the pleural pressure is subatmospheric during the entire cycle of ventilation, if all air is to be evacuated, the air pressure in the bottle must be subatmospheric also. To achieve this, a controlled suction is applied to the bottle to drop the pressure in the bottle to 10 to 30 cm below atmospheric pressure. A similar method can be used to drain blood if the tube is placed posteriorly in the chest where blood tends to accumulate.

THE VENTILATORY PUMP

The subatmospheric pressure in the chest is created by lowering the diaphragms and lifting the rib cage.[5] In a healthy person this coordinated movement in quiet, upright breathing is performed by both the chest cage and the abdomen. In the supine posture where the abdominal contents push the diaphragms up, abdominal respiration is more prominent. For a subatmospheric pressure to be created in the chest, there must be an intact bony framework to prevent the chest from being collapsed by the pressure differential across it. This is provided by the ribs and contracture of the intercostal muscles to provide a stiffness between them. The diaphragms act as piston pumps when the abdominal muscles are relaxed, pushing the abdominal contents down as they expand the chest. The chest wall is expanded when the ribs are pulled up and out laterally by the respiratory muscles to expand the volume of the chest. If the abdominal muscles are contracted at the same time that the diaphragms contract, the diaphragms cannot descend because the volume of the abdomen is limited. Under these circumstances, the diaphragms act as a lever to raise the lower rib cage. This mechanism of breathing can be important in the patient who suffers injury to the cervical spinal cord and only has the diaphragms available for breathing. By compressing the abdomen or when muscle spasm develops late after injury, these patients may be better able to breathe because the diaphragm can act to raise the chest cage as well as acting just as a piston pump.

Injury to the chest cage inhibits the actions of this pump in two ways. All injuries of the chest result in inhibition of the motion of the chest cage, called splinting, the result of the pain due to the injury. This splinting associated with the injury results in a decrease in the tidal volume, or volume of each breath, speeding of the respiratory rate, and inhibition of motion of the hemithorax that is involved. Since coordinated relaxation and contraction of the diaphragms and the abdominal muscles are necessary to efficient and effective ventilation, injuries to the upper abdomen or lower chest cage that result in spasm and splinting of the abdominal muscles can inhibit appropriate motion of the chest cage, just as can splinting due to injury of the ribs. Increased rate and the inhibition of motion of the involved lung greatly decrease the efficiency of the ventilatory pump and the gas exchange system of the lungs. To prevent the failure to ventilate the lung due to the destructive splinting, a critical part of the treatment of chest injury is the early provision of pain relief. This is best done by some type of nerve block, either direct injection of the intercostal nerves or, much more satisfactory, the introduction of a catheter into the epidural space to provide more complete regional anesthesia to the chest wall region. This control of pain must be achieved early after injury to prevent the deterioration of lung function that occurs rapidly if splinting ventilation continues more than 2 or 3 hours.[6]

FLAIL CHEST

If an injuring force results in the fracture of many ribs, particularly in the unprotected anterolateral portion of the chest cage, the bony structure of the chest is disrupted. On inspiration, the disrupted region of the chest wall is sucked in, thereby reducing the expansion of the chest. During expiration, the area moves out. The paradoxical motion of the flail segment of the chest wastes a significant amount of ventilatory effort. The amount wasted depends on three factors: the extent of the injury, how well the pain is controlled, and the mechanical properties of the lung.

These effects can better be understood by considering the mechanism in a patient with an open wound, a hole in the chest cage. Through this hole air enters the pleural cavity during inspiration when the intrapleural pressure is subatmospheric and will escape during expiration if the intrapleural pressure rises above atmospheric pressure. While the two chest cavities are separated by the structures of the mediastinum, the heart, great vessels, trachea, and esophagus, the mediastinum is totally mobile so that the pressures in the two hemithoraces are almost equal. If there is a hole in the left chest, on inspiration as air rushes into the left chest the mediastinum will move over toward the right chest so that the ventilation of both lungs is significantly affected. The amount of air that enters the lung through the trachea, as compared to what will enter through the hole in the chest, determines the required increase in chest cage volume and thus the amount of effort that is necessary to achieve an adequate gas exchange. If the hole in the chest is small, the lungs are not stiff, and airway resistance is normal, the patient is likely to be able to maintain adequate ventilation to the lungs without assistance. If the hole is large and/or the lungs are stiff and the airway resistance is high, even a small open pneumothorax may not be tolerated by the patient.

The mechanism is exactly the same with multiple rib fractures. While air does not actually rush into the chest, the chest wall moves in during inspiration instead of out, called a "flail chest." If a large number of ribs are injured, the inward movement of the chest cage during inspiration will be greater and more effort will be required to maintain a normal tidal volume. If the lungs become stiff due to edema or retained secretions, greater subatmospheric pressures will be required to ventilate the lungs. The inward movement of the chest during inspiration will be greater. These effects of a stiff lung vs a normal lung during ventilation are illustrated in Figure 3.

DIAPHRAGMATIC RUPTURE

Disruption of the diaphragm can occur if a blow to the abdomen acutely raises the intra-abdominal pressure and results in a rupture of the dome of the diaphragm.[2] These tears occur most frequently on the left side, which is not protected by the liver, but rupture can occur on the right side. With the more common use of seat belts, more ruptures of the right diaphragm are being seen as a result of high-speed collisions. As the patient is arrested by the seat belt across the abdomen, it raises the intra-abdominal pressure. If the diaphragm is disrupted in this manner, the abdominal contents move into the chest and can significantly compress the lungs, diminishing ventilatory reserve. Frequently when the abdominal contents are in the chest, gas is retained within the stomach and upper intestine. It is important under these circumstances to introduce a tube through the esophagus to decompress the stomach and prevent accumulation of air in the gastrointestinal tract, which would further compress the abdominal contents. Ruptured diaphragm requires urgent surgical repair, and in the interim ventilatory assistance may be needed.

INJURIES OF THE CHEST CONTENTS

Since the chest is a hollow, partially bony and partially muscular box containing the delicate organs, the lungs and the heart and major blood vessels, nonpenetrating forces acting on the chest cage can injure the contents as well as the chest cage itself. When the chest cage is suddenly arrested, as might occur in a head-on crash, the contents of the chest may continue to move after the cessation of motion of the chest cage. This can result in a number of specific types of injury. One of the more common is a disruption of the aorta. The posterior aorta is attached along the spine to the chest cage, whereas the transverse arch is freely mobile. When the patient is suddenly arrested, a tear can occur in the intima and muscularis of the aorta at the junction between the mobile transverse arch and the arrested posterior aorta. Resulting bleeding into the aortic wall is only contained by a thin adventitia. If the force is great enough, the total wall of the aorta may be disrupted, resulting in a fatal hemorrhage. Similar mechanisms can occur in the ascending aorta where tears can extend into the sinus of the aortic valve. Tears can also occur where the vessels to the upper extremities and head

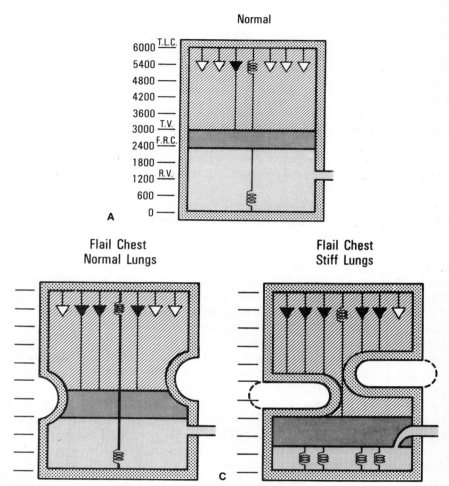

Figure 3. A. This figure depicts the normal chest cage and lung. The light stippled area at the bottom represents the end-expiratory lung volume, the darker stippled area above it the tidal volume, and the diagonal-lined area the inspiratory reserve. The coils at top and bottom represent the balancing elastic forces of lung and chest cage, and the triangles at the top the available units of muscle force. In normal quiet breathing, only one is needed, the black triangle. TLC, total lung capacity; TV, tidal volume; FRC, functional residual capacity; RV, residual volume. (*From Peters RM: Respiratory care of the postoperative patient, in Nyhus LM (ed): Mastery of Surgery, Boston, Little, Brown, in press.*) **B.** With flail chest and normal lungs, more work is required. Three black triangles are needed because the chest wall moves in. This is a level of work the patient can tolerate. **C.** Flail chest with stiff lungs. The stiff lungs require more force for expansion so pleural pressure must be subatmospheric. As a result, the paradoxical motion of the chest is greater. The patient is working at near maximal capacity, five black triangles. The patient will fatigue because both chest cage and lung mechanics are disrupted.

branch off the aortic arch. Unless promptly recognized, these injuries are fatal, but if promptly recognized, they can be repaired before the adventitia ruptures.[7]

Since the heart lies just behind the sternum, a blow to the anterior chest can forcibly compress the heart against the sternum and cause a serious myocardial contusion. A contusion has an effect similar to a myocardial infarction. Likewise, sudden compression of the heart can result in ruptures of the aortic or mitral valves, causing acute insufficiency that often requires emergency repair.

Compression injuries to the upper chest may cause rupture of the bronchi.[8] The mechanism thought to produce these injuries is a compressing force on the upper chest, such as being run over by a car. If escape of air is blocked by a closed glottis or a force closing the trachea, there can be an acute rise in the intrabronchial pressure. The acute rise in airway pressure causes a blowout rupture in the large bronchi and the trachea because the wall is greatest in airways with the largest radius. Depending on the pressure distribution, the blowout will occur in the large bronchi or the trachea. Another mechanism of rupture of the trachea is acute hyperextension of the head, which elongates the trachea beyond its elastic limit. The result is a transverse tear transection of the trachea, usually in the upper thoracic portion. These injuries can lead to acute respiratory distress and may require immediate endotracheal intubation and repair.

Penetrating injuries can cause lethal injuries due to blood loss if the heart or great vessels are entered. In addition to direct injury to vessels, high-velocity missiles may cause a blast injury, the result a massive contusion of the lung with uncontrollable intrabronchial bleeding. These patients may require pulmonary resection. There is considerable argument about the mechanisms of these injuries and how frequently they occur. They are probably quite rare and are often confused with lacerations of vessels that communicate directly with the bronchi, causing flooding of the bronchi with blood.

PREVENTION AND TREATMENT

While we know very little about the actual forces that are responsible for the various injuries to the chest, we know something about the nature of the forces and can make some points about their prevention. Since periodic contraction of ventilatory muscles is not automatic, fracture of the cervical spine, which interrupts central control, has the most destructive effect on the biomechanics. This injury may be almost completely prevented by passive restraints, seat belt, and, even better, the air bag.

Many of the injuries to the anterior chest will be controlled by the proper use of simple restraints, such as the shoulder and lap belt. These prevent a particular portion of the vehicle hitting the victim's chest and causing a point injury that has enough force to disrupt the chest cage.

There has been some increase in rupture of the right diaphragm with lap belts inappropriately positioned above the iliac chest. However, one may state with confidence that without the lap belt, there would have been a much more serious injury to the patient's cervical spine or chest cage. The increase in treatable diaphragmatic rupture may be the price one pays for the seat belt prevention of more serious injuries. The air bag replacing the seat belt should prevent both types of injury.

For prevention of penetrating injuries of the chest, particularly bullet wounds, the bulletproof vest is an effective form of protection but not a very practical one for the average citizen. Restriction on handguns would seem to be a more likely public health measure here.

Most patients with blunt chest trauma and gunshot wounds have multiple system injuries. The effective treatment of chest injuries depends on a recognition of the major cause of immediate death, which is blood loss. Prompt resuscitation in the field with volume replacement is critical, followed by quick evacuation to a regional trauma center where complex treatment of those who are multiply injured can be undertaken. The priorities in treatment are (1) control of blood loss and prompt blood volume restoration, (2) operative repair if there is excessive bleeding and/or intrathoracic injury, such as ruptured bronchus or aorta, and (3) total definitive treatment of the patient's other injuries, particularly internal fixation of long bone fractures so that maximum mobilization is possible. For chest injuries and upper abdominal injuries, mobilization depends on the provision of adequate local pain relief so the patient can cough, breathe deeply, and be allowed out of bed.

With this type of early therapy, the retention of clotted blood in the chest, the development of the acute respiratory distress syndrome, and stiff lungs all can be markedly diminished. If injury to the lung leads to severe contusion, intrabronchial bleeding, high permeability pulmonary edema, or ARDS, early recognition and the provision of ventilator support of a patient with increased positive end-expiratory airway pressures (PEEP) to keep the lung from collapsing are essential. These are complex therapies that require careful monitoring of cardiac output, blood pressure, and respiratory mechanics. They are dependent on adequate resources in both facilities and medical care skill and demand that treatment for the high percentage of success should be done in a regional trauma center.

REFERENCES

1. Christophi C: Diagnosis of traumatic diaphragmatic hernia: Analysis of 63 cases. *World J Surg* 1983; 7:277–280.
2. Grover FL, Ellestad C, Arom KV, et al.: Diagnosis and management of major tracheobronchial injuries. *Ann Thorac Surg* 1979; 28:384–391.
3. Kirsh MM, Sloan H: *Blunt Chest Trauma, General Principles of Management.* Boston, Little, Brown, 1977.

4. Peters RM: Trauma to the chest wall, pleura, and thoracic viscera, in Shields TW (ed): *General Thoracic Surgery,* ed 2. Philadelphia, Lea & Febiger, 1983, chap 31.
5. Peters RM: The lung, in Peters RM, Peacock EE Jr, Benfield JR (eds): *The Scientific Management of Surgical Patients.* Boston, Little, Brown, 1983, chap 14.
6. Shackford SR, Virgilio RW, Peters RM: Selective use of ventilator therapy in flail chest injury. *J Thorac Cardiovasc Surg* 1981; 81:194–201.
7. Shackford SR, Virgilio RW, Smith DE, et al.: The significance of chest wall injury in the diagnosis of traumatic aneuryms of the thoracic aorta. *J Trauma* 1978; 18:493–497.
8. Peters RM, Loring WE, Spring WH: Traumatic rupture of the bronchus—A clinical and experimental study. *Ann Surg* 1958; 148:871–884.

CHAPTER 20

Clinical Aspects of Extremity Fractures

David H. Gershuni

INTRODUCTION

The extremities are frequently injured in domestic and work-related accidents, but much more serious injuries tend to occur in motor vehicle accidents. Even then, extremity injuries are rarely dangerous to life as compared to cranial, thoracic, and abdominal injuries.[1] The upper extremity is injured in 7 to 10 percent and the lower extremity in 10 to 13 percent of motor vehicle accident victims.[1,2] Car occupants were found to sustain 50 percent of their upper extremity and 60 percent of their lower extremity injuries on the instrument panel, and an additional 20 percent of the lower limb injuries were produced from the floor area.[2] Lower extremity trauma in pedestrians is frequently caused by direct vehicle impact with the car bumper and other front end structures and by collision of the pedestrian with the ground after being struck by a car.[3]

The main problem concerning limb injuries is that they often require long-term hospitalization and frequently cause permanent disability. Injuries to the upper limb may have critical effects on prehensile functioning, fine movements, and important sensory requirements. Many of the functions of the lower limb are performed subconsciously, and sensory finesse is at a lower level. However, limb length inequality following pelvic and lower extremity trauma may be much more debilitating than in the upper limb.

Whether an extremity injury is sustained by a child, adult, or old person has a distinct bearing on individual management. Children normally have bone of excellent quality, which heals rapidly and remodels to a surprising degree. If internal fixation is ever required, it usually poses no problem from the point of view of implant anchorage. Children also tolerate immobilization of their limbs extremely well, so that postinjury rehabilitation is a lesser problem than in the adult or aged. A relatively less severe extremity injury to an old person may have life-threatening consequences, especially if it must be treated with confinement to bed. Enforced bedrest in such traumatized patients may provoke venous thrombosis, pulmonary embolus, urinary stasis and infection, decubitus skin ulceration, demineralization of the skeleton, and psychologic upsets related to transfer from home to a strange hospital environment. Methods of cast immobilization are poorly tolerated, slower healing usually occurs, and poor bone quality may defy internal fixation techniques. Prosthetic replacement with cement fixation on one or both sides of a joint may, therefore, be required following some juxta-articular fractures where such inferior bone is present. The mature patient will usually possess bone of normal structure and healing potential, but cast immobilization is also not well tolerated. Economic considerations related to hospitalization and enforced absence from gainful employment become preeminent. These patients have many years to develop degenerative arthritis so that precise reduction of intra-articular fractures and prevention of angular and rotational deformities of the limb bones is mandatory.

This chapter will further discuss extremity injuries involving bone. While it is not intended to be comprehensive, attention will be drawn to overall aims and principles of diagnosis and management, complications that may occur, and their prevention. The impact of biomechanical factors on the management of extremity injuries is emphasized rather than details of specific techniques.

DIAGNOSIS

Together with a history and physical examination, routine two-plane radiography supplemented with oblique and special views is the basis of diagnosis of bony extremity trauma. In some instances further evaluation and elucidation may be necessary, in which case one or more of the following techniques may be utilized.

Conventional tomography can be helpful in delineating suspected femoral neck and osteochondral fractures, defining the fragments and their positions in tibia plateau fractures,[4] and confirming clinical suspicion of nonunion.

Computerized tomography (CT) provides a further multidimensional appreciation of the problem trauma case. It has become invaluable in the assessment and treatment planning of pelvic injuries. Routine anteroposterior 45° inlet and outlet views of the pelvis supplemented with iliac and obturator oblique views for the acetabulum provide basic information.[5] Suspected sacral fractures, sacroiliac disruptions, and acetabular fractures can then be more clearly defined on CT scanning (Fig. 1). Decisions as to pelvic stability, need for hip arthrotomy to remove osteochondral fragments, and the best surgical approaches to reduce and internally fix acetabular fractures are more easily made after evaluating plane radiographs followed by appropriate CT scans. Thus, with careful usage of CT techniques, unique diagnostic information can be obtained that will give the correct diagnosis, demonstrate the extent and severity of the trauma, and facilitate therapeutic planning.

Bone scanning by radionuclide techniques may sometimes prove useful in diagnosing extremity trauma when conventional radiography results are negative or equivocal in the presence of positive clinical findings. Thus, patients presenting with exercise-related pain in the extremities can be evaluated between 24 and 72 hours following injury by bone-seeking radioisotopes.[6] These studies have a high degree of sensitivity.[7,8] The ability to make a specific diagnosis of stress fracture in the femoral neck by such a scanning technique allows the surgeon to take measures to avoid subsequent fracture displacement, usually by internal fixation.[9] Bone scanning of the wrist for suspected scaphoid fractures may also help avoid the traditional 2 to 3 weeks casting while awaiting the possible appearance of conventional radiographic signs.[10]

TREATMENT

The management of significant bony injury to the extremities is by either nonoperative or operative means. Either of these approaches in any particular circumstance may be the conservative approach. This is because the technical ability of the treating physician, the facilities and instrumentation available to him, the safety and speed of the employed method, and the attained end result all have a significant influence on whether the approach is judged to be moderate and cautious or radical. The open anatomic reduction and stable internal fixation of a comminuted acetabular fracture by an experienced surgeon, allowing the patient to be walking with crutches the next day and producing a painfree mobile joint, is indeed a conservative approach compared to a treatment regime of 3 months in bed with skeletal traction with little chance for either an anatomic reduction or a functional result.[5,11] The complications of inexpert anesthesia, failure of fracture reduction and stable fixation, and onset of infection may seem to be dramatic but not more so than deep venous thrombosis, pulmonary embolism, and decubital ulcers following prolonged immobilization in bed.

Fractures with little displacement and no comminution will usually heal well because the surrounding soft tissues are little disturbed and will provide stability and a good blood supply for the healing process; nonoperative methods are clearly indicated in such cases. Compression loading, relative immobility, and good oxygen supply encourage osteogenesis, whereas ten-

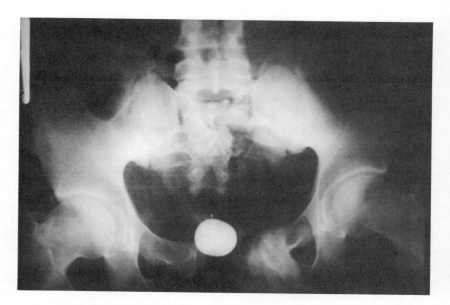

A

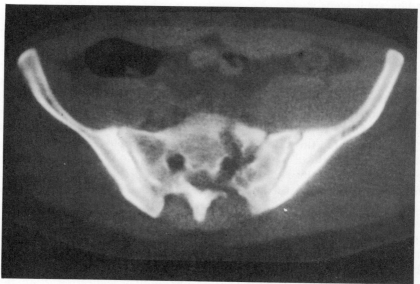

B

Figure 1. A. Anteroposterior radiograph of pelvis shows a symphysis pubis separation with pubic fracture and probable left-sided sacral fracture. **B.** Computerized tomography scan of pelvis defines a displaced comminuted sacral fracture with intact sacroiliac joints.

sile loading, excessive movement, and poor oxygen supply stimulate cartilage and fibrous tissue production. Motion at the fracture site appears to be a key factor in determining fracture healing. In the case of a rib fracture, excellent and rapid healing by callus formation occurs in the presence of continual motion, whereas a femoral neck fracture demands rigid immobilization for primary bone healing. These are the ends of the spectrum of immobilization requirements of extremity fractures. The art and science of fracture management imply a good appreciation of the basic mechanisms of fracture healing and an understanding of the personality of the individual fracture. The implication, then, is that there rarely is a perfect treatment for every fracture in all patients. Informed and intelligent judgment is required to produce the best result for a particular patient in a given setting.

Nonoperative Techniques

Recently emphasis has been placed on maintaining reduction of extremity fractures by casting and bracing techniques that facilitate early functional treatment of the muscles and joints of the part and allow weightbearing in the lower limb. The old orthopaedic adage that a fracture should be immobilized in a cast from the joint above to the joint below is not soundly based. Fixing the joints above and below the fracture may be necessary for rotational control, but preventing joint movement may actually encourage musculotendinous units crossing the joint to act on the fracture fragment distal to the articulation.[12] The lack of a requirement to immobilize the joints contiguous to a shaft fracture is well supported by the treatment of humeral diaphyseal fractures with a brace extending from the axilla to above the elbow. This method permits early shoulder and elbow motion while maintaining fracture alignment and has produced excellent results.[13] Similarly, functional below-knee bracing can be used for tibia fractures after a short period of long leg casting.[14] It cannot, however, be accepted that the most careful molding of a so-called patellar–tendon-bearing cast can relieve a tibial fracture of longitudinal stresses when the patient bears weight.[15] Nevertheless, weight relief does not seem to be necessary to achieve fracture union. Orthoses that are custom-made or have a construction allowing ongoing adjustment to maintain perfect limb conformity have recently gained favor, particularly for tibial fracture treatment. The effect of casting and bracing is to provide a rigid external shell that helps to control displacement by a hydraulic effect on the soft tissue envelope around the bone. In situations where the soft tissues are lacking, such as in certain severe open fractures, the hydraulic principle cannot be as effectively utilized. Similarly, where extensive fasciotomy has been performed in the treatment of compartment syndromes, the hydraulic effect is lost, and casting techniques lead to loss of fracture fragment position. In such cases, internal fixation performed at the time of the fasciotomy procedure is recommended to obtain a satisfactory result.[16]

Methods of fracture management in the extremities utilizing traction still have their place at times. More often, skin or skeletal traction is a temporizing approach prior to definitive treatment by cast bracing or internal fixation. However, in a situation such as a comminuted fracture of the upper end of the tibia, with poor quality of overlying skin, the technique described by Apley,[17] employing a Steinman pin through the upper tibia and traction bow followed by early joint motion, can lead to satisfactory results.

Operative Techniques
Sir Arbuthnot Lane,[18] from Guys Hospital in London, was probably the first to use internal fixation initially by wires, screws, and later with plates. Corrosion, plate breakages, and inflammatory reactions were common with the original methods used. However, significant advances in metallurgy along with improved surgical techniques and appreciation of biomechanical principles have changed the whole approach to internal fixation.

In recent years, there has been a significant move in many trauma centers in the USA to use open reduction and internal fixation more often as the primary method of treatment of extremity fractures; this approach will therefore be emphasized in this chapter. There are several reasons for this trend to operative treatment. (1) Significant advances have taken place in the development of instruments and implants of high quality and innovative design. Concomitantly, a better understanding has been gained of the principles of using the instrumentation and implants largely formulated by Swiss orthopaedic surgeons.[19] The emphasis has been on gentle care of the soft tissues, anatomical reduction, stable fixation, and functional aftertreatment of the fractured limb. Complications and poor results can therefore be kept to a minimum. (2) The distinct lowering of morbidity and mortality has been demonstrated in the multiply traumatized patient when treatment by early internal stabilization of the fractures has been performed.[20,21] This operative approach facilitates maximal respiratory function by allowing verticalization of the chest, decreases risks of fat embolism, permits earlier overall mobilization, and, hence, decreases the risk of venous thrombosis, skin breakdown, urinary stasis, and skeletal demineralization. (3) Patient demands and expectations have increased so that limb shortening and poor joint and muscle function are less easily accepted than in the past. Finally, economic considerations have become more dominant, and earlier discharge from hospital and return to gainful employment and weightbearing (the latter usually possible after intramedullary nailing procedures of the femur and tibia) are expected following operative treatment.

However, operative treatment has the potential for complications, such as infection, implant loosening, and failure. Later need for implant removal with possible subsequent refracture through screw holes or incompletely remodeled bone are further known problems.

Internal Fixation

Once a decision is made to perform open reduction and internal fixation of a fracture, the operation is usually done as expeditiously as possible. Although there is evidence that healing is improved by a 10 to 14 day delay prior to operating in some long bone fractures,[22-24] other factors make such a delay unwise. Thus, the early healing process interferes with defining and reducing fracture fragments. A greater amount of soft tissue dissection is required, and even then some structures will need to be more forcibly stretched than in the early postinjury phase to obtain fracture reduction. Nevertheless, delay for 1 to 2 days may be advantageous to allow severely swollen parts or traumatized skin to recover.

In the operative treatment of fresh fractures, several important principles of biomechanics and biology must be employed to attain a satisfactory result. The aim is to obtain a perfect anatomic reduction and reduce the gap between the fragments to a minimum. This situation is maintained with suitable implants that are inserted whenever possible under tension so that the fracture surfaces are compressed. There is no evidence that anatomic reduction and interfragmentary compression per se accelerate fracture healing, although it has been suggested that a piezoelectrically induced effect stimulates osteogenesis.[25,26] However, compression does provide an optimum environment for healing[27,28] and can prevent delayed union and nonunion.

Compression may be of a static or dynamic nature. Static compression is attained by screws, inserted in a lag fashion, or by a compression plate and screws. To obtain maximum stability, optimum placement of the lag screw must be performed. This requires an appreciation of the major forces (axial compression, torsional, or bending) to which the fractured bone will be exposed. To resist axial compression, the lag screw should be inserted perpendicular to the long bone axis.[29] Thus, when an axial load is applied, oblique fracture fragments will tend to override, which increases the transverse diameter of the bone. This leads to increased tension in the screw and increased compression at the fracture surfaces. A screw perpendicular to the fracture line will be subject to decreased tension following axial compression; the screw toggles in the gliding hole and the fragments are allowed to glide upon one another. However, a lag screw inserted perpendicular to the fracture line provides the best compressive force to the fragments and, hence, maximal resistance to torsion.

A bending movement applied to a fracture parallel to the plane of a lag screw is well resisted by any of the various screw positions. However, a bending movement applied perpendicular to the screw causes the bone surfaces in an oblique fracture to slide on one another. A lag screw inserted perpendicular to the fracture applies the greatest interfragmentary compression and is most effective in countering such shear.

In fixing a butterfly fragment to the shaft of a long bone, insertion of lag screws has been recommended in such a way that they bisect the angle

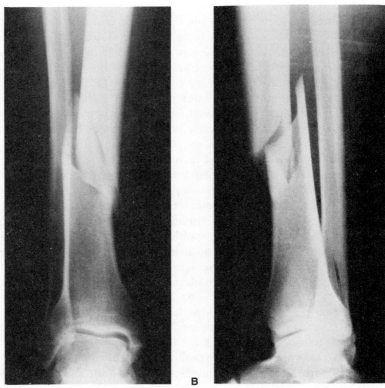

Figure 2. A and **B.** Anteroposterior and lateral radiographs of spiral fracture of the lower third of the tibia. (*cont.*)

subtended by the perpendiculars to the fracture plane and the long axis of the bone.[19] In other fracture patterns, the main distorting forces should be analyzed as precisely as possible, and based upon the above principles, screws are inserted in one of the three modes (perpendicular to the long bone axis, perpendicular to the fracture line, or a compromise between the two perpendiculars), or individual screws may be placed in different modes (Fig. 2). Once interfragmentary lag screw fixation is obtained, it is often necessary to add a countoured neutralization plate to protect the lag screw fixation from excessive torsional, bending, and shearing forces (Fig. 3).

Compression through a plate may be performed with the use of a compression device that is fixed to the bone distal to the plate to which the latter is hooked. By a turnbuckle effect the plate, attached to the bone by screws at its opposite end, is drawn toward the device, hence compressing the fracture fragments (Fig. 4). Alternatively, using a plate with oval holes and asymmetric insertion of screws through the holes, a lateralization motion of the plate along the bone is similarly obtained, which also produces interfrag-

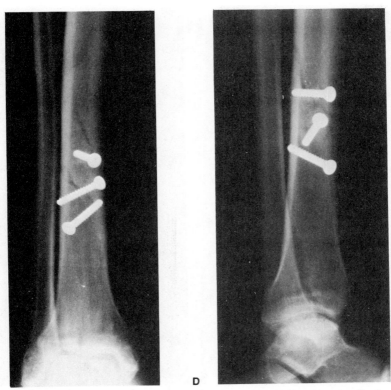

C   D

Figure 2. (*cont.*) **C** and **D.** Anteroposterior and lateral radiographs after fixation by three lag screws, two inserted perpendicular to the fracture line and one perpendicular to the long axis of the bone.

mentary compression.[30] Providing the compression applied exceeds the potential distraction or shearing forces exerted across the fracture during active limb motion, the compression will maintain a stable reduction by frictional resistance between the fragments. The addition, whenever possible, of a lag screw passing obliquely across the fracture further enhances the stability of the reduction. Prebending the plate so that there is a slight elevation of the plate away from the bone at the fracture site is an important technique. This maneuver greatly increases flexoral rigidity, results in nearly uniform compressive contact stresses across the fracture site, and prevents the fracture from gapping opposite the plate during functional loading.[31,32]

Decay of compression applied via a plate is slow. It is explained by viscoelastic stress relaxation in the first 2 weeks and internal remodeling of the haversian systems, whereby prestressed bone is replaced by nonprestressed bone in the later weeks.[33]

Dynamic compression implies the use of predetermined forces that will pass across the fracture site after fixation on commencement of active usage

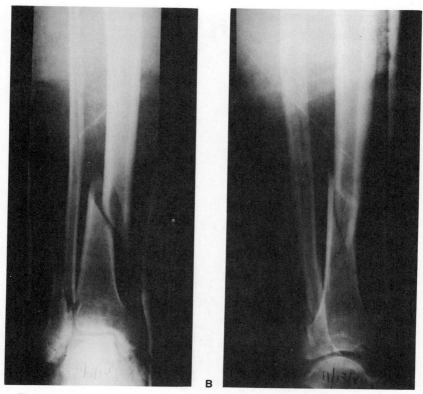

A   B

Figure 3. A and **B.** Anteroposterior and lateral radiographs of a spiral fracture, with an undisplaced butterfly fragment, of the lower third of tibia. (*cont.*)

of the limb. Such compression may then be applied in two situations. The first possibility is in fresh fractures, e.g., of the patella. In such a case it is clear that on active flexion of the knee a transverse fracture would be distracted. However, if a suitable, strong wire is passed over the anterior surface of the patella under the quadriceps tendon and patellar ligament and then tightened, it will absorb all tensile forces and allow further interfragmentary compresion of the bone on dynamic loading coincidental with knee flexion (Fig. 5). The anterior patellar surface is thus called the tension band side of the bone. A relatively small wire implant, which however has great strength in tension, is thus most suitably inserted to allow functional aftercare of the limb.[19] In some fresh fractures, when the tension band side is known, dynamic compression plating can be applied. Thus, in the femur the lateral side of the shaft is clearly the tension band side because the mechanical axis of the limb passes medial to the shaft. Another situation in which the dynamic compression principle can be employed is in intramedullary nailing of a long bone fracture. Active use of the limb will then encourage

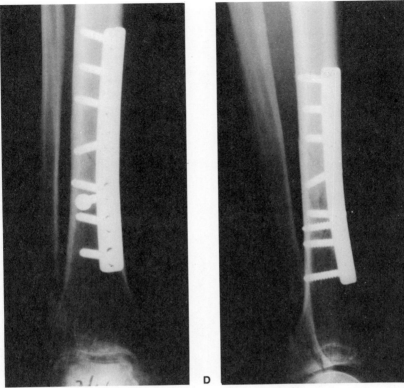

C   D

Figure 3. (*cont.*) **C** and **D.** Anteroposterior and lateral radiographs of tibia after lag screw fixation and neutralization plate.

the longitudinally arranged muscles to force the fragments together in a dynamic fashion.

Dynamic compression can also be utilized in the fixation of a nonunion. When the radiographs of, for example, a tibial shaft nonunion are analyzed, the convex side of the bone can be defined; this is the tension band side (Fig. 4A). The application of a plate, under tension, on that side (Fig. 4B) will provide compression across the nonunion site, which will be enhanced by any subsequent varus-provoking force. Following such internal fixation, fracture healing is usually very rapid and can be guaranteed in over 90 percent of cases.[34,35] Additionally, the freedom from cast immobilization allows functional joint and muscle activity. In the treatment of fracture nonunion by bioelectrical techniques 70 to 80 percent success is usually claimed,[36] and 6 or more months of limb and joint immobilization, usually with nonweightbearing[37] is required.

When performing internal fixation an important aim should be to use a technique, such as a lag screw or plate interfragmentary compression, that

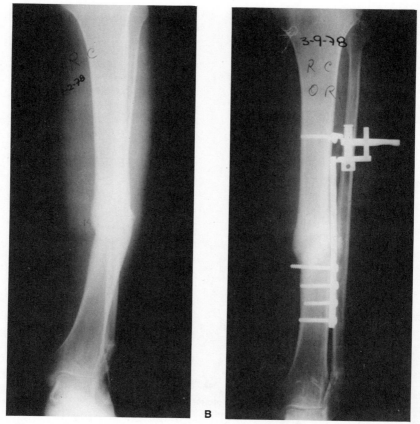

Figure 4. A. Anteroposterior radiograph of tibia with a diaphyseal nonunion and varus deformity. **B.** Anteroposterior radiograph taken during surgery. The plate was fixed to the lateral convex side of the diaphysis, and the compression device has corrected the deformity and compressed the nonunion site. Correction was obtained without any prior interference with the nonunion site.

allows the bone to transmit the maximum load. This approach should minimize the effects of stress deprivation on the bone, which may lead to a localized increase in cortical porosity.[38] The remodeling response of the bone under the plate may also be due to interference with the blood supply.[39,40] The development of less rigid plates, such as the carbon fiber reinforced plate of Tayton et al.,[41,42] allows more transmission of stress to the bone and has been claimed to permit periosteal callus formation and prevent localized bone loss under the plate. Failure of the bone to transmit load may have other ill effects, such as metal fatigue. This occurs due to instability of the internal fixation where there is lack of bone support, osteoporotic bone, massive overload by the patient, or technical errors. The

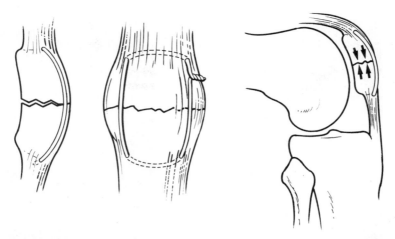

Figure 5. Transverse patellar fracture. The anterior tension band wire converts distraction forces into dynamic compression forces on knee flexion. The drawing on the left shows slight opening of the fracture posteriorly during full knee extension. The drawing on the right shows the fracture to be compressed on knee flexion.

instability leads to bending and torsional stresses in the implant and final failure.

Properly performed, stable internal fixation leads to primary bone healing by osteons growing across the bone ends without the formation of external callus.[43,44] Thus, if all the advantages of plate and screw fixation are required in the treatment of a specific fracture, particularly one in the diaphysis, the efficiency of periosteal callus formation must usually be dispensed with.[45] In fact, the appearance of periosteal callus after a stable internal fixation with plates and screws is an ominous sign, suggesting failure of fixation. The cuff of callus forming around a fracture treated by a plaster cast increases the bone's diameter and hence the rigidity efficiency by the fourth power of the diameter. Thus, periosteal callus is more efficient than intercortical new bone, which in turn is more efficient than endosteal callus.[46] If however, intramedullary nailing is performed, periosteal callus again assumes importance because endosteal bridging callus is largely inhibited.

Fractures of the femoral and tibial shafts often, and the humeral shaft occasionally, lend themselves to internal fixation by intramedullary nailing techniques that have significant advantages over other methods of internal fixation and over nonoperative treatment modalities.[47-51] Open nailing results in minimal angular deformity, good maintenance of bone length, early ambulation, good joint motion, short hospitalization, and high rate of union.[47,49,52] Closed nailing techniques have the same advantages as open nailing, with a lower infection rate and a smaller scar, but the required image intensifier implies more irradiation of the surgeon and patient and sometimes a less anatomic reduction than obtained during an open proce-

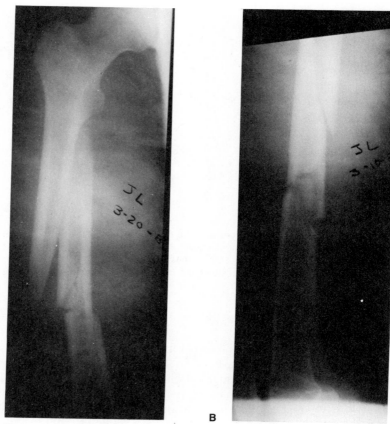

Figure 6. A and **B.** Anteroposterior and lateral radiographs of femur with comminuted diaphyseal fracture. *(cont.)*

dure. The medullary canal of the femur and tibia is narrow in its middle third. By reaming, a larger nail can be used and a large area of contact between nail and internal cortex can be obtained. This helps to make the nail a load-sharing device and to increase fracture stability, especially against rotational forces.[50,53] The intramedullary vasculature is only transiently disturbed by such reaming and nailing procedures, and a highly vascularized endosteal membrane soon forms around the nail tract.[54]

Implant failure is much less common with intramedullary nails than with plate and screw fixation because the latter are eccentrically placed and exposed to more bending stresses than is the case of a nail lying within the medullary canal. The potential problem of refracture following removal of plates and screws is largely avoided following intramedullary nail extraction, as the nail insertion site provides no stress riser effect in a critical position.

A comminuted or spiral fracture of the diaphysis can be treated with an intramedullary nail supplemented with lag screws around the nail (Fig. 6).

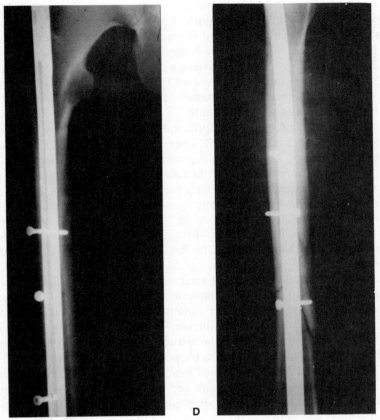

C   D

Figure 6. (*cont.*) **C** and **D.** Anteroposterior and lateral radiographs after open intramedullary nailing and insertion of three lag screws to stabilize the comminuted segment.

However, the development of interlocking nails has broadened the indications for closed intramedullary fixation from the transverse fracture, which has always been considered the primary indication, to include spiral fractures, more comminuted shaft fractures, and segmental fractures.[55] The interlocking nail can be inserted with a closed technique, then the proximal and distal interlocking screws are inserted without exposing the fracture site itself. The interlocking nail has two to four transverse holes through which transverse screws can be inserted from one cortex to the other. In a fracture of the femur or tibia in which longitudinal stabilization would not be attained by a nail alone, the use of interlocking nail osteosynthesis should be considered. The locations of fractures suitable for interlocking nails can also be much more proximal or distal in the shaft than normally accepted for intramedullary nailing.

Another form of intramedullary fixation is by Ender rods. These are relatively thin (4 mm diameter), flexible rods that may be inserted in stacks from the supracondylar region of the femur under image-intensifier control. The rods are passed into the femoral head and when necessary into the greater trochanter.[56] The Ender rod is placed in a biomechanically advantageous position in the middle of the femoral canal closer to the mechanical axis of the limb. It is thus much less exposed to bending forces than a laterally placed plate, and hence implant failure is an uncommon event.

Ender rods have been recommended for treatment of femoral fractures from the base of neck to lower shaft.[56-58] However, a relatively high incidence of complications, such as supracondylar fractures (when both medial and lateral insertion sites are used), knee pain from rods backing out with fracture settling, and femoral head penetration, suggests that the device should be used in a more limited fashion. Thus, the minimally displaced intertrochanteric fracture in the elderly patient with a wide medullary canal is a good indication. The treatment of an intertrochanteric fracture then rests between the use of Ender rods or, more commonly, a hip screw designed to allow controlled collapse of the fracture as the screw slides into a barrel attached to the side plate. The latter is fixed by screws to the upper femoral shaft.

Nonunions of tibial or femoral fractures also often lend themselves well to intramedullary nailing, as do incipient or actual pathologic fractures. In the latter case, methyl-methacrylate cement supplementation may aid stability when cortical destruction is significant.[59] When the limiting factor in any osteosynthesis procedure is the bone tissue, its quality, and/or its presence or absence, the overall strength of a proposed implant becomes less important. Thus, if the bone in a femoral head is very osteoporotic, a fixation screw can easily cut its way out, and the intrinsic strength of the metal is never tested. In such instances, supplementation of the bone stock by methyl-methacrylate cement, into which the hip screw could gain purchase, should also be considered.

External Skeletal Fixation
External skeletal fixation is an important modality of treatment to be considered, especially in open or closed pelvic and lower radial fractures and in other open fractures of the limbs. The technique involves inserting full or half-pins into the bone and connecting the protruding pin ends to external rods to construct full frames or clamps (Fig. 7). The full pins are threaded in their middle section, and the Schanz type half-pins are threaded just proximal to the pointed tip. The threads facilitate firm fixation in the bone over long periods of time and by their stability help to prevent soft tissue irritation, secretions, and infection.

Probably the greatest utility of external skeletal fixation is in grades II and III open tibial fractures. The tibia has one surface that is subcutaneous, and so the minimal soft tissue envelope at this point is easily and frequently broached. Grade I injuries are perforations of the skin by bone spicules. They are treated by a limited debridement procedure and immobilization

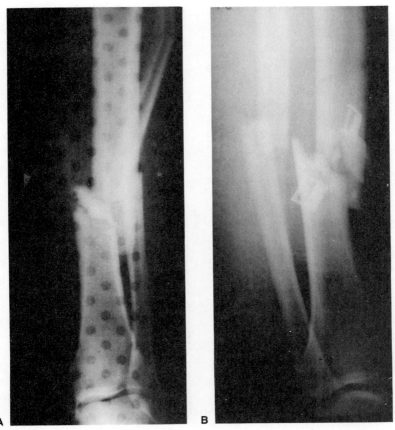

Figure 7. A and **B.** Anteroposterior and lateral radiographs of a grade II open, comminuted, tibial diaphyseal fracture. (*cont.*)

techniques applicable to closed fractures. Grade II open fractures imply a skin wound of at least 3 cm in one direction, with contusion and damage to the surrounding musculature. A grade II injury is due to high-energy trauma resulting from a road traffic, industrial, or agricultural accident or a gunshot wound. It involves the skin, entire muscle groups, vessels, and nerves and is usually associated with bone comminution.

In grades II and III open tibial fractures, following radical wound debridement and fracture reduction, an external fixator can be applied, preferably using sets of half-pins inserted via the subcutaneous surface to avoid muscle and other significant soft tissue penetration. Two or three strategically placed lag screws may be inserted through the open wound across fracture fragments (Fig. 8). The screws may assist considerably in obtaining a stable reduction, in conjunction with the external fixator, especially if the fixator then can be applied in a compression mode.[60] In that case, the fixator

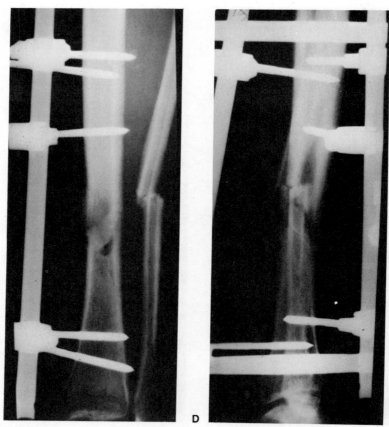

C   D

Figure 7. (*cont.*) **C** and **D.** Anteroposterior and lateral radiographs after debridement of wound and application of a double half-frame fixator.

functions as a plate neutralizing bending, shearing, and torsional forces that pass across the fracture segment.

Following stable fixation with the fixator, care of the wound (which is always initially left open) and rehabilitation of the limb are greatly facilitated. This is in comparison to casting techniques, which obstruct access to the wound and limit joint and muscle activity. The fixator is maintained in place until at least soft tissue healing is obtained and function regained but then can be sequentially dismantled and substituted by a below-the-knee cast. This allows the bone to be gradually more stressed and encourages secondary bone healing. If the external fixator is maintained in situ, time to final bone healing is slower than with the previously described techniques, but the tendency to malunion is decreased.[60] Where fracture fragment reduction is not well achieved or bone loss is significant, cancellous bone grafting, as early as 2 to 3 weeks after injury, is well advised to speed up a predictable delayed union or nonunion. Grafting can be performed through the open granulating wound by the Papineau technique[61] or via a separate incision. The soft tissue wound often heals by secondary intention or, failing that, may require a skin or composite graft for closure.

A second major use for external skeletal fixation is in the treatment of pelvic trauma. In this situation, half-pins of the Schanz type are inserted into the ilium superior and/or inferior to the anterior superior iliac spines. Anterior frames of various types[62,63] may then be constructed. Disruption of the symphysis pubis, which books open, in association with anterior sacroiliac ligament tearing only, lends itself well to a simple anterior frame.

More complex pelvic disruptions frequently occur. The Malgaigne type is a very unstable injury. It consists of a posterior fracture or sacroiliac disruption together with anterior, double pubic arch fractures or symphyseal dislocation. The hemipelvis then tends to migrate cranially. A trapezoidal frame theoretically can maintain reduction of posterior sacral and ilial fractures or sacroiliac joint disruptions so that mobilization of the patient into a chair or

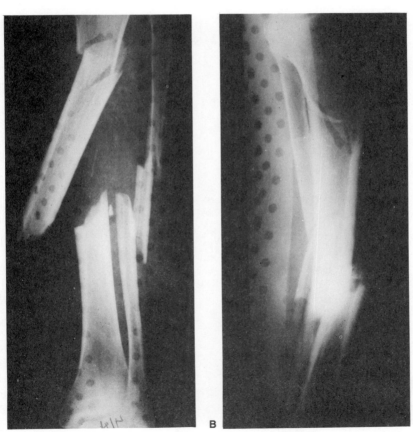

A   B

Figure 8. A and **B.** Anteroposterior and lateral radiographs of a grade III open, segmental tibial fracture. (*cont.*)

even to allow weightbearing has been recommended.[62] However, in practice the stability obtained with the trapezoidal frame is insufficient for weightbearing until at least 6 weeks have passed when healing of the tissues is well advanced. A better approach to stabilizing these injuries is to internally fix the posterior lesion and treat the now simple anterior problem with a compression frame (Fig. 9) or by double plating across the symphysis pubis.

There are many advantages to this more aggressive approach to treating pelvic disruptions. Thus, continual bleeding from very vascular cancellous bone surfaces is greatly diminished by fracture reduction and immobilization. Pain is markedly relieved by stabilizing the injury; the patient then may become more mobile, and nursing is made considerably easier. Marked deformities, often with limb length inequality, can be eliminated, and hospitalization is usually abbreviated by avoiding long periods of time in traction.

There are some disadvantages to external skeletal fixation in the pelvis.

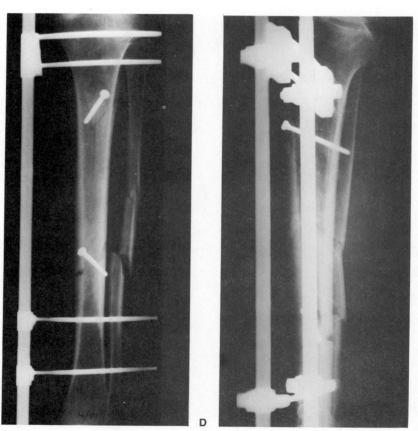

C   D

Figure 8. (*cont.*) **C** and **D.** Anteroposterior and lateral radiographs of fracture stabilized with two lag screws and a frame fixator placed with compression.

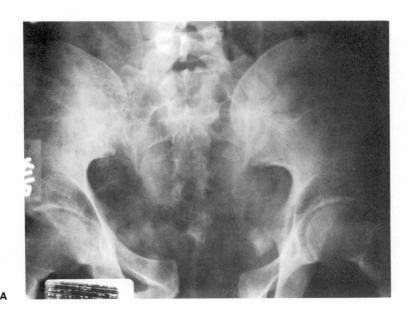

A

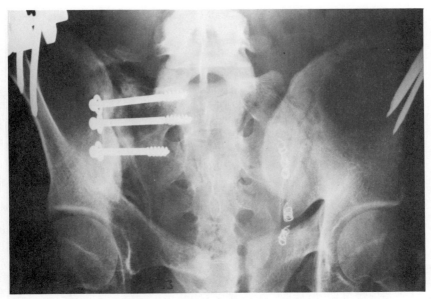

B

Figure 9. A. Anteroposterior radiograph of pelvis with a right sacroiliac disruption and a symphysis pubis diastasis. **B.** Anteroposterior radiograph after reduction, fixation, and fusion of the sacroiliac joint and application of an anterior compression external fixator.

There is the potential for intrapelvic pin insertion. This can be avoided by exposing the bone by a separate incision and inserting the pins under direct vision. Because the pins often pass through significant thicknesses of soft tissue, movement of the tissues around the pins occurs; this encourages secretions and infections. Finally, the frames are rather bulky for the patient.

A recommended technique for comminuted fractures of the lower end of the radius, which may or may not be intra-articular, is a combination of pins and plaster.[64] The pins are inserted into the metacarpus and radius or ulna, and following reduction, the hand and forearm including the pins are incorporated into a circular cast. While this method may maintain the fracture reduced as regards the radial length, it cannot maintain the normal, approximately 15°, volar tilt of the distal articular radial surface. Any movement of the cast obviously toggles the incorporated pins and may cause pin tract reactions. Several other complications, such as pin tract infection, osteomyelitis, pin loosening, and precipitation of carpal tunnel syndromes, have also been described with this technique.[65] A special light-weight model of the external fixator maintains all the attributes of the pins and plaster method, while allowing volar tilt of the distal radius and dispensing with the bulky forearm cast, thus facilitating functional use of the hand (Fig. 10).

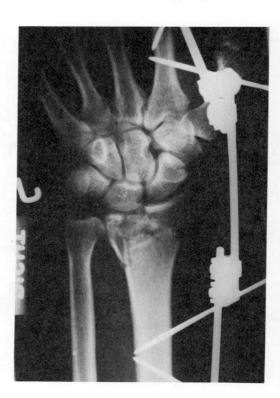

Figure 10A. Anteroposterior radiograph of a comminuted intra-articular fracture of the lower end of radius. The external fixator holds the fracture out to length and maintains an acceptable reduction of the multiple fragments. (*cont.*)

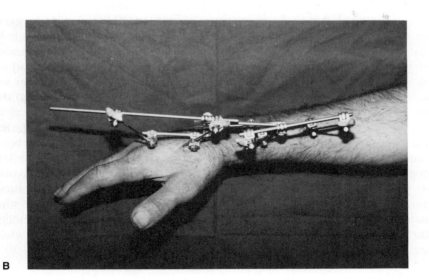

B

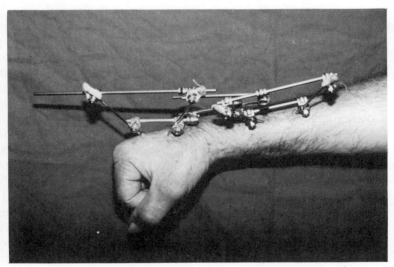

C

Figure 10. (*cont.*) **B** and **C.** Ulnar view of hand, wrist, and forearm with external fixator in place. The fixator maintains the wrist in some degree of flexion and permits finger function.

REHABILITATION OF THE EXTREMITY

Whichever method of fracture fixation is used, it is most important to commence functional care of the extremity as soon as possible following surgery. The development of continuous passive motion machines appears to hold great promise for maintaining joint range of motion, promoting venous drainage, and decreasing edema fluid collection. However, it is important not to neglect the aspect of possible loss of muscle power if such devices are solely relied upon. Active use of the limb should be complementary to the continuous passive motion machine.

The necessity for supplementation of internal fixation by external means implies that the original objective of the operation has not been attained or that the overall concept of internal fixation has not been understood. The original trauma to the limb is the presenting and thus unavoidable problem. Surgery must add further tissue damage with even the most careful technique, but to then immobilize the muscles and joints by a nonfunctional cast is an avoidable and unnecessary insult. Fracture management clearly has as its main objective healing of bone, but, to obtain union while disregarding the ultimate functions of the extremity is not acceptable.

COMPLICATIONS OF INTERNAL FIXATION

Infection

Any operative procedure may be complicated by infection, but in the presence of metallic implants subsequent treatment of the infection becomes more difficult. The metal per se has no causal or promotional responsibility for infection.[66] However, by harboring organisms away from the vasculature and preventing the supply of antibody, cellular, and antibiotic defenses by the circulation, the infection becomes difficult or impossible to eradicate. This does not mean that an implant should automatically be removed in the presence of an infection at the time of the required debridement and lavage of the infected wound. Such implant removal may convert a *stable* infected fracture into an *unstable* infected fracture—a much more difficult problem. Implants providing no stability to the fracture act as foreign bodies and thus serve no useful purpose; therefore, they should be removed. Nevertheless, some stability must then be restored with either an external fixator, cast, or traction. The infected wound should be left open and whirlpool therapy provided. The appropriate antibiotic should be administered by the intravenous route. To establish the true pathogen, as distinct from contaminants, culture of the bone and not the purulent discharge is necessary.

When the fracture heals, the metal implant may be removed and an attempt at that time should be made to eradicate the chronic infection using radical debridement and antibiotic therapy. If the fracture shows no signs of healing in the presence of stable fixation, cancellous bone grafting should be performed by the open Papineau technique.[61] In a *controlled* infected environment around a nonstabilized nonunion, plate fixation may be indicated. Intramedullary nailing, which can disseminate infected material widely in the medullary canal, is unwise. The treatment of infected nonunion by bone grafting via the Harmon approach followed by further long periods of casting is not likely to produce a good functional or cosmetically acceptable result.[67] Such bone grafting, however, when supplemental to stable bone fixation, can be critical to final bone union.

Delayed Union and Nonunion

It is impossible to give precise definitions of delayed union and nonunion. Delayed union is present when, according to general experience, the period of time is exceeded that the fracture in question would normally be expected to heal. The site and degree of comminution of the fracture, the soft tissue injury (which will be greater in an open than in a closed fracture), the age, and the overall nutritional and health status of the patient are some of the factors modifying the expected healing time. Thus, delayed healing may be declared present at times between 2 and 8 months.

Nonunion is present when the treating surgeon considers that healing will not take place without further intervention. Once again, a large element of subjectivity enters into the definition so that a fracture may be declared a nonunion at 6 to 12 months following injury.[68] Nonunion can further be described as infected or noninfected, fibrous or true synovial pseudoarthrosis, atrophic or hypertrophic. The latter descriptions respectively imply a nonreactive (absence of callus), poorly vascularized or a reactive (elephant foot), well-vascularized response.

When a delayed union is noted, the treatment plan is usually not altered until such a time as a definitive nonunion is diagnosed whether the initial treatment has been operative or nonoperative. An alternative philosophy, which the author recommends, is to be more active at the delayed union stage. Most cases of potential nonunion or implant failure can then be avoided by proceeding with the solutions to be described, or established nonunion. Much loss of time and suffering can thus be avoided.

Sir Astley Cooper (1798–1841) advised that to heal a nonunion, one must "ensure all the mechanical conditions which are essential for the consolidation of callus, comprising perfect rest and immobility and contact and pressure of the broken surfaces against each other." These principles, together with the addition at times of cancellous bone graft and the maintenance of limb function, formulate the modern approach to management of fracture nonunions. The overall aims of treatment should be to obtain bone union and a pain-free, cosmetically acceptable, functional limb in a reasonable time-frame.

The surgical approach to nonunions related to their location and type is set out in two algorithms (Figs. 11 and 12). Noninfected diaphyseal nonunions should be stabilized where possible with intramedullary nails and a

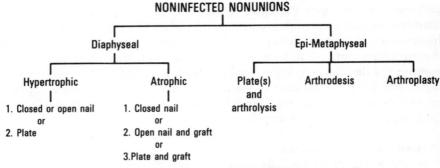

Figure 11. Algorithm for treating noninfected fracture nonunions.

closed technique (Fig. 13). If this is impossible because of severe deformity, an open procedure is required. In that case the periosteum around the fracture site is removed with flakes of attached bone, which later will fall back as living autografts. Further decortication of the bone, which stimulates osteogenesis, is then done around the nonunion site. The deformity is corrected, preferably without taking down the fibrous tissue between the bone ends, and fixation with a plate or nail is performed.[68] When plate fixation is chosen, it is placed on the tension band side of the nonunion and is compressed[34] as previously described. Cancellous bone grafting is added in atrophic nonunions and when open reduction, which damages the periosteal circulation, is combined with intramedullary nailing, which in turn damages the medullary circulation.

Noninfected epiphyseal and/or metaphyseal nonunions are similarly treated, but frequently medial and lateral plates are required, especially if the epiphyseal fragment is small. Usually vascularity near the joints is sufficient so that bone grafting, which might also subsequently interfere with joint motion, is rarely necessary. In this location joint motion is frequently poor by the time a nonunion is diagnosed so that arthrotomy, arthrolysis, and later treatment by continuous passive motion are indicated. Occasion-

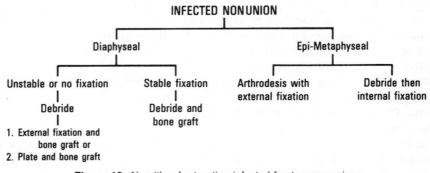

Figure 12. Algorithm for treating infected fracture nonunions.

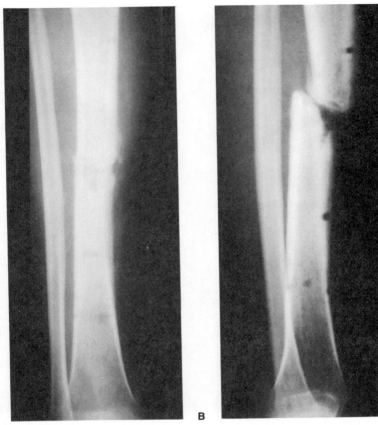

Figure 13. A and **B.** Anteroposterior and lateral radiographs of an atrophic, noninfected, tibial diaphyseal nonunion after original fracture treatment with an external skeletal fixator. (*cont.*)

ally, joint function is irreparable, and arthrodesis, at the time of treating a nonunion, or arthroplasty is the salvage procedure.

Infected diaphyseal nonunions must first have the infection brought under control by debridement of necrotic and infected material. The fracture is then stabilized with external skeletal fixation or sometimes by a compression plate.[69] Cancellous bone grafting, usually by the Papineau technique, is then required. Debridement and external fixation can normally be performed in one session, but debridement, internal fixation, and bone grafting usually should be staged. Appropriate antibiotic therapy in the perioperative period is probably advantageous. In the unusual event of a stable infected nonunion, staged debridement and then cancellous bone grafting are indicated.

It is very difficult to salvage an infected epiphyseal or metaphyseal

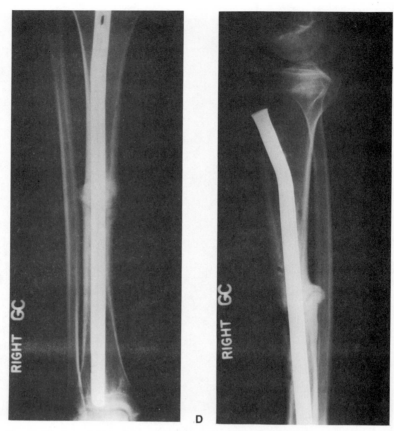

C   D

Figure 13. (*cont.*) **C** and **D.** Anteroposterior and lateral radiographs after closed intramedullary nailing with no additional bone grafting. Fracture union is well advanced 3 months after surgery.

nonunion so that debridement followed by external skeletal fixation, spanning the contiguous joint and fracture with the aim of obtaining fracture union and arthrodesis, is usually the only treatment (Fig. 14). In all classes of extremity fracture nonunion, as soon as stable external or internal fixation is obtained, physical therapy to the limb must be commenced to finally achieve as functional a limb as possible.

REFRACTURE

The incidence of refracture after fracture union has been reported as approximately 1 to 3 percent[68,70,71]; this includes operatively and nonoperatively treated patients, with the latter sometimes being in the majority.[70] Refrac-

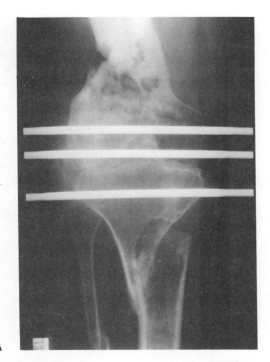

A

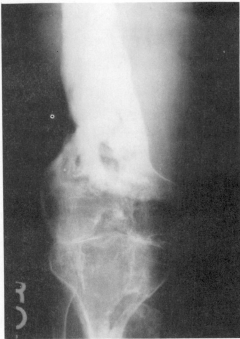

B

Figure 14. A. Anteroposterior radiograph of an infected lower femoral metaphyseal nonunion, infected knee joint, and upper tibial metaphyseal nonunion. The external fixator spans both metaphyses and the knee. (The upper femoral and lower tibial transverse pins are not seen in the illustration.) **B.** Anteroposterior radiograph after fracture healing and knee fusion have been obtained.

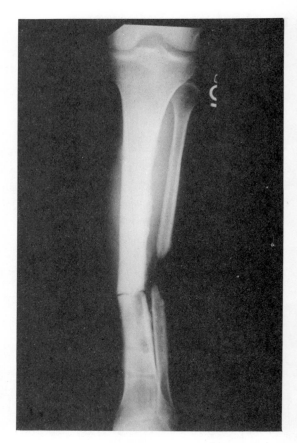

Figure 15. Anteroposterior radiograph. The refracture is seen traversing the original fracture site and not through the pin tracts of the external fixator originally applied to treat a grade III open fracture.

tures occur through the old fracture line (Fig. 15) or a screw hole defect. Any defect in the bone can act as a concentration point of stress and precipitate a refracture on application of less trauma than that which caused the original fracture. Abnormalities in the material properties of the bone may also lead to refracture.

Following fracture union, patients should be advised to refrain from strenuous activity for a period of time until all stress raisers have been removed and the bone has again become a hollow tube with a closed cross-section[72]; this may take more than a year. Following removal of a screw or plate and screws in the lower limb, protected weightbearing for 4 to 8 weeks appears to be indicated.[73] This initial period should be followed by abstaining from sporting activities for a year or more, as previously described.

REFERENCES

1. Braunstein PW: Medical aspects of automotive crash injury research. *JAMA* 1957; 163:244.
2. Nahum AM, Siegel AW, Hight PV, et al.: *Lower Extremity Injuries on Front Seat Occupants.* Paper No. 680483. New York, Society of Automotive Engineers, 1968.
3. Huelke D: Extremity injuries produced in motor vehicle collisions. *J Trauma* 1970; 10:189.
4. Elstrom J, Bamkovitch AM, Sassoon H, Rodriguez J: The use of tomography in the assessment of fractures of the tibial plateau. *J Bone Joint Surg* 1976; 58A:551.
5. Letournel E, Judet R: *Fractures of the Acetabulum.* Berlin, Heidelberg, New York, Springer-Verlag, 1981.
6. Matin P: The appearance of bone scans following fractures including immediate and long-term studies. *J Nucl Med* 1979; 20:1227.
7. Wilcox JR, Moniot AL, and Green JP: Bone scanning in evaluation of exercise related stress injuries. *Radiology* 1977; 123:699.
8. Gelsien GE, Thrall JH, Espinosa, JL, Older RA: Early detection of stress fractures using ^{99m}Tc-polyphosphate. *Radiology* 1976; 121:683.
9. El-Khoury GY, Wehbe MA, Bonfiglio M, et al.: Stress fractures of the femoral neck: A scintigraphic sign for early diagnosis. *Skelet Radiol* 1981; 6:271.
10. Ganel A, Engle J, Oster Z, Farine I: Bone scanning in the assessment of scaphoid fractures. *J Hand Surg* 1979; 4:540.
11. Gershuni DH: The treatment of acetabular fractures. *West J Med* 1982; 137:410.
12. Hicks JH: External splintage as a cause of movement of fractures. *Lancet* 1960; 1:667.
13. Sarmiento, A, Kinman PB, Galvin EG, et al.: Functional bracing for fractues of the shaft of the humerus. *J Bone Joint Surg* 1977; 59A:596.
14. Sarmiento A: A functional below-the-knee brace for tibial fractures. *J Bone Joint Surg* 1970; 52A:295.
15. Svend-Hansen H, Bremerskov V, Ostri P: Fracture suspending effect of the patellar-tendon bearing cast. *Acta Orthop Scand* 1979; 50:237.
16. Gershuni DH, Mubarak SJ, Yaru NC, et al.: Fracture of the tibia complicated by acute compartment syndrome. Submitted for publication, 1984.
17. Apley AG: Fractures of the lateral tibial condyle treated by skeletal traction and early immobilization. *J Bone Joint Surg* 1956; 38B:699.
18. Lane WA: The direct fixation of fractures. *Trans Clin Soc London* 1893; 27:167.
19. Müller ME, Allgöwer M, Schneider R, Willenegger H: *Manual of Internal Fixation.* Berlin, Heidelberg, New York, Springer-Verlag, 1979.
20. Border JR, La Duca J, Sabel R: Practices in the management of patients with polytrauma. *Progr Surg* 1975; 14:84.
21. Riska EB, Von Bonsdorff H, Hakkmen S, et al.: Prevention of fat embolism by early internal fixation of fracture in patients with multiple injuries. *Injury* 1976; 8:110.
22. Charnley J, Guindy A: Delayed operation in the open reduction of fractures of long bones. *J Bone Joint Surg* 1961; 43B:664.
23. Piekarski K, Wiley AM, Bartels JE: The effect of delayed internal fixation on fracture healing. *Acta Orthop Scand* 1970; 40:543.

24. Smith JEM: Results of early and delayed internal fixation for tibial shaft fractures. *J Bone Joint Surg* 1974; 56B:469.
25. Lavine LS, Lustrin I, Shamos MH, et al.: The influence of electric current on bone regeneration in vivo. *Acta Orthop Scand* 1971; 42:305.
26. Martin RB: Theoretical analysis of the piezoelectric effect in bone. *J Biochem* 1979; 12:55.
27. Perren SM, Huggler A, Russenberger M: The reaction of cortical bone to compression. *Acta Orthop Scand* 1969; 125(suppl):1.
28. Ruedi T, Webb JK, Allgöwer M: Experience with the dynamic compression place (DCP) in 418 recent fractures of the tibial shaft. *Injury* 1976; 7:252.
29. Arzimanoglou A, Skiadaressis G: Study of internal fixation by screws of oblique fractures in long bones. *J Bone Joint Surg* 1952; 34A:219.
30. Allgöwer M, Matter P, Perren SM, et al.: *The Dynamic Compression Plate.* Berlin, Heidelberg, New York, Springer-Verlag, 1978.
31. Perren SM, Hayes WC: Biomechanik der Plattenosteosynthese. *Med Orthop Techn* 1974; 2:55.
32. Nunamaker DM, Perren SM: Radiological and histological analyses of fracture healing using prebending of compression plates. *Clin Orthop Rel Res* 1979; 138:167.
33. Perren SM, Rahn BA: Biomechanics of fracture healing, in Uhthoff HK (ed): *Current Concepts of Internal Fixation of Fractures.* Berlin, Heidelberg, New York, Springer-Verlag, 1980, p 15.
34. Müller ME: Treatment of non-union by compression. *Clin Orthop Rel Res* 1965; 43:83.
35. Rosen H: Compression treatment of long bones pseudarthroses. *Clin Orthop Rel Res* 1979; 138:154.
36. Brighton CT, Black J, Friedenberg ZB, et al.: A multicenter study of treatment of non-union with constant direct current. *J Bone Joint Surg* 1981; 63A:2.
37. Bassett CAL, Pilla AA, Pawluk RJ: A non-operative salvage of surgically resistant pseudarthroses and non-union lay by pulsing electromagnetic fields. *Clin Orthop Rel Res* 1977; 124:128.
38. Akeson WH, Woo SL-Y, Coutts RD, et al.: Quantitative histological evaluation of early fracture healing of cortical bones immobilized by stainless steel and composite plates. *Calcif Tissue Res* 1975; 19:27.
39. Rhineland FW: Circulation in bone, in Bourne G (ed): *The Biochemistry and Physiology of Bone.* New York and London, Academic Press, 1972, Vol 2.
40. Gunst MA: Interference with blood supply through plating of intact bone, in Uhthoff HK (ed): *Current Concepts of Internal Fixation of Fractures.* Berlin, Heidelberg, New York, Springer-Verlag, 1980, p 268.
41. Tayton K, Johnson-Nurse C, McKibbin B, et al.: The use of semi-rigid carbon fiber and reinforced plastic plates for fixation of human fractures. *J Bone Joint Surg* 1982; 64B:105.
42. Tayton K, Bradley J: How stiff should semi-rigid fixation of the human tibia be? A clue to the answer. *J Bone Joint Surg* 1983; 65B:312.
43. Schenk R, Willenegger H: Morphological findings in primary fracture healing. *Symp Biol Hung* 1967; 7:75.
44. Rahn BA, Gallinaro P, Baltensperger A, et al.: Primary bone healing. *J Bone Joint Surg* 1971; 53A:783.

45. McKibbin B: The biology of fracture healing in long bones. *J Bone Joint Surg* 1978; 60B:150.
46. Perren SM: Physical and biological aspects of fracture healing with special reference to internal fixation. *Clin Orthop Rel Res* 1979; 138:175.
47. Küntscher G: Intramedullary surgical technique and its place in orthopaedic surgery. My present concept. *J Bone Joint Surg* 1965; 47A:809.
48. Böhler J: Closed intramedullary nailing of the femur. *Clin Orthop Rel Res* 1968; 60:51.
49. Rokkanen P, Slätis P, Vankka E: Closed or open intramedullary nailing of femoral shaft fractures? A comparison with conservatively treated cases. *J Bone Joint Surg* 1969; 51B:313.
50. Laurence M, Freeman MAR, Swanson SAV: Engineering considerations in the internal fixation of fractures of the tibial shaft. *J Bone Joint Surg* 1969; 51B:754.
51. Clawson DK, Smith FR, Hansen ST: Closed intramedullary nailing of the femur. *J Bone Joint Surg* 1971; 53A:681.
52. Hansen ST, Winquist RA: Closed intramedullary nailing of the femur. Küntscher technique with reaming. *Clin Orthop Rel Res* 1979; 138:56.
53. Anderson LD, Gilmar WS, Tooms RE: Experimental fractures treated with loose and tight fitting medullary nails. *Surg Forum* 1962; 13:445.
54. Rhinelander FW: Effects of medullary nailing on the normal blood supply of diaphyseal cortex. *AAOS Instructional Course Lectures*. St. Louis, Mosby, 1973, vol 22, p. 161.
55. Hempel DR, Fischer S: *Intramedullary Nailing*. Stuttgart, New York, George Thieme Verlag, 1982.
56. Küderna H, Böhler N, Collon DJ: Treatment of intertrochanteric and sub-trochanteric fractures of the hip by the Ender method. *J Bone Joint Surg* 1976; 58A:604.
57. Pankovitch AM, Goldflies ML, Pearson RL: Closed Ender nailing of femoral shaft fractures. *J Bone Joint Surg* 1979; 61A:222.
58. Goldie I, Althoff B, Hallström E: Ender's nailing of femoral shaft fractures. *Injury* 1983; 14:312.
59. Harrington KD, Sin FH, Enis JE, et al.: Methylmethacrylate as an adjunct in the internal fixation of pathological fractures. *J Bone Joint Surg* 1976; 58A:1047.
60. Gershuni DH, Halma G: The AO external fixator in the treatment of severe tibia fractures. *J Trauma* 1983; 23:986.
61. Papineau LJ, Abfugeme A, Dalcourt JP, et al.: Osteomyelite chronique: Excision et greffe de spongieux à l'air libre aprés mises a plat extensives. *Int Orthop* 1979; 3:165.
62. Slätis P, Karaharju EO: External fixation of the pelvic girdle with a trapezoid compression frame. *Injury* 1975; 7:53.
63. Mears D, Fu FN: Modern concepts of external skeletal fixation of the pelvis. *Clin Orthop Rel Res* 1980; 151:65.
64. Green DP: Pins and plaster treatment of comminuted fractures of the distal end of the radius. *J Bone Joint Surg* 1975; 57A:304.
65. Chapman DR, Bennett JB, Bryan WJ, et al.: Complications of distal radial fractures. Pins and plaster treatment. *J Hand Surg* 1982; 7:509.
66. Gristina AG, Rovere GD: An in-vitro study of the effects of metals in internal

fixation on bacterial growth and dissemination. *J Bone Joint Surg* 1963; 45A:1104.

67. Gershuni DH, Pinsker R: Bone grafting for non-union of fractures of the tibia: A critical review. *J Trauma* 1982; 22:43.

68. Müller ME, Thomas RJ: Treatment of non-union in fractures of long bone. *Clin Orthop Rel Res* 1979; 138:141.

69. Meyer S, Weiland AJ, Willenegger H: The treatment of infected non-union of fractures of long bones. *J Bone Joint Surg* 1975; 57A:836.

70. Chrisman OD, Snook GA: The problem of refracture of the tibia. *Clin Orthop Rel Res* 1968; 60:217.

71. Olerud S, Karlström G: Tibial fracture treated by AO compression osteosynthesis. Experience from a five year material. *Acta Orthop Scand* 1972; Suppl 140:1.

72. Frankel VJ, Burnstein AH: The biomechanics of refracture of bone. *Clin Orthop Rel Res* 1968; 60:221.

73. Burstein AH, Currey J, Frankel VH, et al.: Bone strength: The effect of screw holes. *J Bone Joint Surg* 1972; 52A:1143.

CHAPTER 21

Extremities:
Experimental Aspects

John W. Melvin, F. Gaynor Evans

INTRODUCTION

Biomechanical studies of trauma to the extremities can be grouped into those dealing with the bones of the extremities, those dealing with the joints and ligamentous structures of the extremities, and those dealing with the soft tissue structures (fascia, muscles, and blood vesels) of the extremities. This chapter will be primarily limited to a discussion of trauma to the bones of the extremities because there is a paucity of experimental biomechanical studies in the other two areas. By far, the greatest body of literature on experimental bone fracture deals with the bones of the lower extremities. This is primarily due to (1) the significant mobility impairments associated with lower extremity trauma and (2) the significant involvement of the lower extremities in serious injuries and permanent impairments resulting from automobile accidents, both for occupants and pedestrians.

This chapter will review and discuss the significant findings of research studies related to experimental determination of the biomechanical factors associated with fracture of whole bones of the extremities.

STATIC FAILURE OF BONES

The most comprehensive, and one of the earliest studies of the failure characteristics of whole bones of the extremities was the work of Messerer in 1880.[1] He tested the mechanical characteristics of bones from many regions of the body. His study involved 500 bones or bone combinations from 90 cadavers. In many cases, the bones he tested and the manner in which they were loaded still provide data of utility to present-day researchers.

Messerer's work on whole bones was preceded, however, by that of Weber in 1856.[2] Weber determined the loads required to fracture entire bones by three-point bending transverse to the long axis of the bone. The extremity bones were mounted in a test fixture while still articulated with the entire body. The load measurement indicated the applied force in 245 N increments. The tests included 509 bones from four men and five women. The distance between the two support points was identical for the same type of bone from each subject, and the load was applied at the center. Table 1 lists the average load at fracture for the five bone types tested and their associated support span. The clavicles were tested in the body without support. The maximum bending moments were calculated by multiplying the average load times the average support distance divided by four. Messerer[1] criticized Weber's work on the basis that he did not account for the effect of variation of bone dimensions on the failure loads.

Messerer's experiments on the bending and compression of long bones were conducted with a hydraulic testing machine that today would be considered a quasi-static universal testing machine. Load measurement was performed by an accurate beam balance system with a resolution of load differences on the order of 10 to 50 N. For very weak bones, dead weight loads were applied up to about 1174 N. The torsional strength of long bones was determined with a specially developed torsion tester.

The results of lateral bending tests with center loading and a support span of two-thirds the length of each bone are compiled in Table 2. The average load and a range are given for each bone type for male and female test subjects. Bones were obtained from six males with ages ranging from 24 to 78 years and six females with ages ranging from 20 to 82 years.

Messerer also noted that the direction in which the load is applied relative to the cross-sectional shape has a great influence upon the failure load. Since most bones have triangular or elliptical cross-sectional shapes, rather than circular ones, the ultimate load-carrying ability in bending depends on whether the central load is applied at the narrow or wide side of

TABLE 1. FRACTURE LOADS DUE TO BENDING

	Femur	Tibia	Humerus	Ulna	Clavicle
Male, kN	4.87	3.06	3.55	2.21	1.27
Female, kN	3.98	2.33	2.26	1.19	1.06
Support distance, CM	18.3	21.6	13.0	11.8	—
Maximum moment, male N·m	233	165	115	65	—
Maximum moment, female, N·m	18	125	73	35	—

After Weber.[2]

TABLE 2. FRACTURE LOADS DUE TO BENDING

	Clavicle	Humerus	Radius	Ulna	Femur	Tibia	Fibula
Male, kN	0.98 (0.78–1.18)	2.71 (2.35–2.94)	1.20 (0.98–1.77)	1.23 (0.98–2.16)	3.92 (3.43–4.66)	3.36 (2.30–4.90)	0.44 (0.35–0.54)
Average support length, cm	12	22.4	16	16	31.7	24.7	24.7
Average maximum moment, N·m	30	151	48	49	310	207	27
Female, kN	0.60 (0.49–0.69)	1.71 (1.18–2.35)	0.67 (0.54–0.88)	0.81 (0.69–0.98)	2.58 (2.26–3.33)	2.24 (1.86–2.65)	0.30 (0.21–0.39)
Average support length, cm	11.5	20	14	14	28	22.2	22.3
Average maximum moment, N·m	17	85	23	28	180	124	17

After Messerer.[1]

the bone. He gives an example of the tibia, for which the average failure load with contact on the crest was 3.19 kN and with contact on the inner surface was 2.21 kN. In several cases, a wedge-shaped piece of bone broke out at the site of load application, with the base of the wedge on the compression side of the bone. Messerer thought that this was a strange form of fracture, although it had been noted in steel and other uniformly grained materials. He concluded that this form of fracture is normal in bending failures of bones, and that this fracture indicated that bone is a uniform, homogeneous material, despite its obviously fibrous nature. This conclusion, of course, is not true. He was confusing the anisotropic nature of bone material and its influence on crack propagation with the net result of a fracture form attributed, at that time, to uniform, homogenous materials.

Comparison of the maximum bending moment values from Table 1 and Table 2 shows that, although Weber's fracture load values were sometimes greater than those found by Messerer, the generally longer support distances used by Messerer resulted in higher maximum bending moment values at failure for many of Messerer's tests. The only significant exception to this was for the ulna. Weber's method of loading the bones in situ might account for this, since the influence of the radius, even though it was not directly loaded, would be to share some of the load applied to the arm.

Messerer determined the cross-sectional geometry of each bone near the point of loading and from this calculated the tensile strength of the bone material, assuming a linear-stress–strain relationship. He found that the resulting strength varied from 102 MPa to 194 MPa. The average value for the combined male and female data, which had similar mean values, was 153.5 MPa. Messerer noted that the strength values for different bones of an individual were in fairly good agreement, with the exception of one 78-year-old male. He also noted that the strength values tended to increase with age up to middle age and then to decrease again. The modulus of elasticity was also calculated for tests with the humerus and femur bones from one 32-year-old male. The values ranged from 14.71 GPa to 17.65 GPa, with a mean value of 15.59 GPa.

The torsion tests conducted by Messerer were the first of their kind. He tested the same types of bones as were in his bending tests. Bones from four males (ages 27 to 56 years) and seven females (ages 19 to 81 years) were tested. Table 3 gives the average bone fracture torques and associated

TABLE 3. FAILURE TORQUES DUE TO TORSION ABOUT THE BONE AXIS

	Clavicle	Humerus	Radius	Ulna	Femur	Tibia	Fibula
Male, N·m	15	70	22	14	175	89	9
	(12–17)	(55–78)	(16–27)	(8–21)	(141–222)	(63–110)	(6–12)
Female, N·m	10	55	17	11	136	56	10
	(8–11)	(39–80)	(13–23)	(9–13)	(78–207)	(47–63)	(8–16)

After Messerer.[1]

ranges for males and females. The torsional fractures typically were initiated at regions of the bones where the cross-sections were the smallest. Particularly weak sites were noted in the upper and lower third of the humerus, femur, and fibula, in the upper third of the radius, in the lower fourth of the ulna and tibia, and in the center of the clavicle. The calculation of the torsional strength and the shear modulus of elasticity of the bone material was performed for the tests of the femurs of a 29-year-old male. The mean torsional strength was found to 56.4 MPa, and the average shear modulus was 4.9 GPa. All of the bones fractured with a spiral path at an angle of about 45 degrees.

Messerer conducted a number of different types of compression tests on long bones. The first of these tests was a direct transverse compression of the shaft of the bones. The technique basically crushed the bone cross-section. Only a few experiments were conducted: one set with the bones of a 31-year-old male and one with the bones of a 24-year-old female. The value at which an initial drop in the load occurred was taken as the indication of compressive failure. These values for the various bones of the two test subjects are shown in Table 4.

Messerer himself recognized the rarity of the clinical occurrence of injuries related to the mode of loading represented by transverse compression, and, in his second type of compression tests, he loaded whole bones in axial compression. The ends of the bones were padded with felt to prevent local failures at the bone ends. Even so, two forms of failure occurred in these tests: fracture of the shaft and/or crushing of the ends. In the case of the humerus, only one shaft failure occurred. This was at the lower third and was produced by a load of 5.87 kN in the bone of a 30-year-old female. All other tests of the humerus produced crushing failures. In the other bones, both forms of fractures were present in varying amounts. Table 5 presents the average axial compressive failure loads and associated ranges for the various bones of males and females. All failure loads are for shaft failure unless otherwise noted. Because the lengths of many of the bone types were great compared to the cross-sectional dimensions, the bones exhibited significant lateral deflections and failed in what should be termed a column buckling failure when the fracture occurred in the shaft. Such behavior was very pronounced in the fibula tests. The natural curvature of

TABLE 4. TRANSVERSE CRUSHING LOADS FOR DIRECT COMPRESSION OF THE SHAFTS OF BONES

	Humerus	Radius	Ulna	Femur	Tibia	Fibula
31-year-old male, kN	8.33	5.15	5.39	12.74	5.88	2.95
24-year-old female, kN	5.88	3.83	3.04	10.78	6.37	3.04

After Messerer.[1]

TABLE 5. FAILURE LOADS FOR COMPRESSION ALONG THE BONE AXIS

	Clavicle	Humerus (End Failures)	Radius	Ulna
Male, kN	1.89	4.98	3.28	2.21
	(1.22–2.64)	(2.15–7.83)	(2.35–4.21)	(1.76–2.84)
Female, kN	1.24	3.61	2.16	12.93
	(0.88–2.06)	(2.45–5.09)	(1.03–3.18)	(0.88–1.71)
	Femur (Shaft Failures)	Femur (Neck Failures)	Tibia	Fibula
Male, kN ·	7.72	27.99	10.36	0.60
	(6.85–8.56)	(6.85–10.52)	(7.05–16.39)	(0.24–0.88)
Female, kN	7.11	4.96	7.49	0.48
	(5.63–8.56)	(3.91–5.81)	(4.89–10.37)	(0.20–0.83)

After Messerer.[1]

many bones, such as the femur and radius, was noted by Messerer to contribute to buckling failures.

Two special types of compression tests were performed by Messerer for the neck of the femur and the patella. The neck of the femur was loaded, in a few cases, by a compressive load aligned with the axis of the femoral neck. It was found to be difficult to mount the head and neck of the femur securely, and only a few tests were successful. In one case, the neck penetrated the head as the head crushed. In another singular occurrence, a fracture was produced between the head and the neck of the femur of a 50-year-old male at a load of 5.38 kN. In three other femurs, the fracture was located between the shaft and the neck of the bone. The loads were 8.32 kN (a 32-year-old female and a 78-year-old male) and 4.40 kN (an 82-year-old female). The male was noted to have had massive bones.

To test the patella, Messerer found he had to support the posterior surface on felt and apply loads against the anterior surface. He performed six tests on bones from male subjects and seven tests on bones from female subjects. The average compressive load for males was 5.87 kN (range 5.24 to 7.58 kN) and for females was 4.11 kN (range 2.20 to 5.87 kN). All of the patellas failed by a longitudinal fracture that divided the patella in half.

A much more recent study of the static strength of long bones of the body was conducted by Motoshima and summarized by Yamada.[3] Motoshima determined the bending properties of the major long bones of 35 persons. The bones were tested wet in lateral bending, with a central load applied in the anteroposterior direction. The average breaking loads for five age groups are shown in Table 6. Most of the average breaking loads were lower than the corresponding values given by Messerer (Table 2). This may be due to anthropometric differences between the two populations. Motoshima found that the tibia, rather than the femur, had the highest breaking load of any bone, in contrast to the findings of Weber and Messerer.

TABLE 6. FRACTURE LOADS DUE TO BENDING (kN)

Bone	Age Groups					Adult Average
	20–39 Yrs	40–49 Yr	50–59 Yr	60–69 Yr	70–89 Yr	
Femur	2.72±0.11	2.47±0.05	2.35±0.09	2.33±0.06	2.14±0.11	2.45
Tibia	2.90±0.11	2.52±0.11	2.43±0.05	2.39±0.09	2.29±0.09	2.60
Fibula	0.44±0.02	0.40±0.04	0.39±0.03	0.37±0.02	0.33±0.02	0.39
Humerus	1.48±0.12	1.39±0.10	1.28±0.10	1.23±0.09	1.13±0.08	1.33
Radius	0.59±0.07	0.53±0.04	0.52±0.08	0.48±0.04	0.43±0.03	0.52
Ulna	0.71±0.05	0.63±0.08	0.61±0.06	0.59±0.04	0.55±0.04	0.63

After Motoshima, in Yamada.[3]

DYNAMIC FAILURE OF BONES

The studies of Weber and Messerer involved, of necessity, the static measurement of failure loads. Virtually all forms of skeletal trauma, with the exception of some crushing injuries and sports injuries, such as those in wrestling, involve rapid or dynamic loading of the bones. It is only in the last 30 years that methods to produce dynamic loads experimentally have been developed and accurate measurements have.been made. Interest in this area has largely been generated by the automotive safety problem of protecting vehicle occupants in crashes. Bone material, like many biological materials, exhibits viscoelastic or rate-sensitive behavior. Thus, it is expected that under dynamic loading, a particular bone would require a greater load and a smaller deflection to produce a fracture when compared to the static loading situation.

Mather[4] demonstrated this effect using 32 pairs of human femurs. One member of each of the paired sets was tested statically in a materials testing machine, and the other matching member was tested in a drop-weight testing apparatus with an impact velocity of 9.75 m/sec. The impact load was not measured directly, but the test apparatus was capable of indicating the impact energy absorbed by the bone. These data were compared to the static energies absorbed by the companion bones as calculated from the areas under the load-deflection curves for the static tests. The mean value of the ratio of dynamic energy to static energy was 1.66, and wide variations among the ratios were evident (standard deviation 0.77). The mean static energy to failure was 28.7 joules, while the mean impact energy to failure was 42.5 joules. This represents a 48 percent increase due to impact loading.

Studies of dynamic loading of the upper extremities are rare. Recent interest in lateral impact in automobiles has led to impact testing studies of cadavers with localized loading to the upper arm and the supporting thorax. One such study by Cesari et al.[5] produced a single example of a fractured humerus with a peak load of 416 kg, which is approximately 50 percent greater than the mean value for males reported by Messerer (Table 2). Such a finding is consistent with Mather's results on the effect of dynamic loading.

Torsional loading of the long bones of the lower extremities has become of interest in relation to skiing injuries. Martens et al.[6] reported on a study involving femoral and tibial bone samples obtained from 65 autopsy subjects ranging from 27 to 92 years of age. The ends of the bones were embedded in gripping blocks of plastic and torsionally loaded by an impact torsional loading machine. The time of loading to failure was less than 100msec. The femoral tests produced a mean value of failure torque for males of 204 N·m (range 122–291). For the tibia the mean failure torque values were 111N·m (range 70 to 179) for males and 71.4 N·m (range 61 to 159 for females. Comparison of these values with those of Messerer (Table 3) shows that, for the male data, the mean dynamic failure torques are 16 percent to 24 percent greater than the comparable static values. The female data for the tibia are in agreement with the male data, with a 27 percent increase. The female femur dynamic value, however, is actually lower than the static value. This is most likely due to bone dimensional effects, with Messerer's smaller sample (7 versus 13 tests) being influenced by two large-boned subjects. The average energy to failure for dynamic torsional loading were, for the femur, 37.5 joules (male) and 29.8 joules (female) and, for the tibia, 27.3 joules (male) and 18.4 joules (female). These values are comparable in magnitude to the bending failure energies of Mather.[4]

The bones of the lower extremities can be subjected to a variety of types of dynamic loads in automobile crashes. This is true for both restrained and unrestrained vehicle occupants. The first research on the impact tolerance of the lower extremities with respect to the automobile occupant was the work of Patrick et al.[7] Ten unrestrained, seated, embalmed cadavers were impacted into instrumented chest, head, and knee targets to simulate a vehicle interior. The knee targets were covered with 3.7 cm of padding in most cases. Fractures of the femur were produced at loads as low as 6.67 kN, while loads as high as 17.13 kN were sustained with no fracture of the femur but with a fractured patella and pelvis. The majority of the femoral fractures were found to occur at the distal end of the bone. The authors concluded that failure of the femur occured at slightly lower load levels than those of either the patella or pelvis. They suggested a conservative overall injury threshold load level of 6.23 kN.

In a later paper,[8] they obtained loads of 6.54, 7.61, 8.68, and 8.76 kN on two embalmed cadavers without fractures and suggested raising the threshold load level for injury to 8.68 kN. Two features of this work should be noted. The first is that the material used was embalmed, not fresh, as in all the work presented above. Second, multiple high-load-level tests were run on the test subjects prior to obtaining fractures. This technique introduces the uncertainty of possible progressive damage to the skeletal structure, particularly the pelvis, prior to the test that produced the fracture. Comparison of these suggested threshold values with those static values of Messerer (Table 5) for axial femur compression indicates that they are within or near the range of static strengths noted by Messerer.

Subsequent knee impact tests have employed stationary, seated test subjects that were struck on each knee by a moving impactor mass. Powell et al.[9] struck the flexed knees of embalmed cadavers with a 15.6 kg striker mass with a rigid, flat striking surface. The test subjects were seated with back support. A total of six tests on four cadavers were reported. All but one fracture involved the femoral condyles or patella. The average failure load was 10.49 kN, with a range of 8.73 to 12.51 kN. The authors indicated that bending effects in the femoral shaft play a significant role in femur response to longitudinal impacts, but these effects were not measured in the study.

The first seated knee impact study to use unembalmed fresh cadavers as test subjects was reported by Melvin et al.[10,11] A total of 26 cadavers (15 males and 11 females) with an age range of 45 to 90 years was tested. A 20.9 kg impact-piston/load-cell striker system was used to impact each flexed knee, one at a time, at velocities ranging from 6.3 to 23.2 m/s. The impactor surface was either rigid with no padding, padded lightly with 2.54 cm of padding, or padded thickly with 2.54 cm of padding and 5.08 cm of aluminum honeycomb material. The different impactor surface conditions were used to vary the force-time waveform of the impact. Only one test was conducted on each knee in most cases. For rigid surface impacts, the average peak axial fracture load was 19.6 kN, with a range of 17.1 to 23.7 kN. Fractures of the patella only were produced in two tests, fractures of the patella and the femoral condyles were produced in one test, and condylar fractures were produced in the remaining test. In the lightly padded tests, the average peak axial fracture load was 18.4 kN, with a range of 13.3 to 28.5 kN. Patellar and condylar fractures were produced in those tests in which fracture occurred. Only one of the thickly padded tests in which the load was recorded produced a fracture. A fracture of the distal third of the shaft was produced with a peak load of 19.7 kN. In none of the tests was a fracture produced in which the peak input load was less than 13.35 kN. The authors thought that, in addition to a threshold force level for failure, an associated impact energy level of 542 joules was necessary to produce distal failures of the femur and patella. These values were found to be significantly lower if osteoporosis or other bone degeneration was present.

The force-time histories of the tests ranged in duration from 2.6 to 22 m/s. In order to extend the impact time duration and to study the effect of loading of both knees simultaneously, a series of six cadaver sled tests was conducted by Melvin and Nusholtz.[12] A load cell arrangement was used to allow the test subject to strike its knees against the impact surface at velocities of 8.3 to 10.1 m/s. The ages of the test subjects (four males and two females) ranged from 49 to 79 years. These tests produced proximal skeletal damage as well as the distal damage typical of the impactor tests. In the sled tests that produced a combination of distal and proximal upper leg injuries, a particular load-time waveform was noted to occur. This waveform, which was trapezoidal in general form, occurred whenever a femoral neck or pelvic fracture was produced in combination with femoral shaft or condylar fractures.

The trapezoidal nature of the waveform suggests the occurrence of two major fracture events. The first event would initiate an abrupt break in the slope of the loading trace to form the trapezoidal shape. The second major event would cause an abrupt drop-off in load that ends the trapezoidal portion of the load-time history. Consideration of the probable effects of the various types of skeletal fractures on the load-carrying ability of the upper leg led the authors to hypothesize that the first fracture event was most likely associated with femoral neck or pelvic fractures, while the second or terminal event was the fracture of the femoral shaft or condylar region. A femoral neck fracture can give rise to an instantaneous change in load-carrying ability but would be followed closely by a resumption in loading at a lower rate or by a gradual decrease in load. The ligamentous structures around the hip joint and the proximal end of the femur are thought to carry load around such fractures and thereby continue the loading of the body through the upper leg. Fractures in other regions of the femur can subsequently occur. In contrast, when a fracture of the shaft of the femur or an extensive fracture of the condyles occurs, there would be an abrupt loss in load-carrying ability, with no immediate alternative load path. This would result in a severe drop-off in load.

Viano and Stalnaker[13] analyzed the data from a series of axial knee impact tests on a total of six seated cadavers. A 10.1 kg impactor was used in conjunction with varying degrees of padding. The flesh was removed from the shaft of the femur to allow high-speed photography of the femur as it was loaded. This procedure was used in an effort to measure the time of fracture initiation, based on analysis of the high-speed movies.

All six of the tests with rigid impactors produced both patellar and femoral shaft fractures and also produced either a condylar or femoral neck fracture. The peak force ranged from 13.4 to 28.5 kN, with an average of 18.3 kN. The two tests conducted with lightly padded impactors both produced bilateral condylar fractures. The peak loads were 16.0 and 15.4 kN. Only two of the five tests with thickly padded impactors produced fractures (one condylar and one femoral shaft), and both of these involved cadavers that were considered by the authors to have bones in "abnormal" condition. The three normal specimens produced peak loads of 5.3, 13.8, and 14.0 kN and presented no fractures. Many of these specimens had multiple fractures, and the authors concluded that femoral shaft and condylar fractures occurred after peak load. They associated an average load of 10.6 kN with those fractures. In four of the rigid impacts, neck fractures were produced as well as shaft fractures. The femoral neck was not observable in tests, and thus observation of the occurrence of the neck fractures with respect to load level was not possible. All four of the tests produced load-time waveforms with characteristics similar to those discussed in the sled test study by Melvin and Nusholtz.[12]

The cancellous bone center of the patella makes it vulnerable to concentrated loadings. This phenomenon was studied by Melvin et al.,[14] employing three different impactor sizes, all rigid. Two of the impactors were flat-surfaced circular areas with diameters of 15.5 mm and 10.9 mm, while the third impactor was ring-shaped with an outer diameter of 12.7 mm and an inner diameter of 6.4 mm. Minimum failure loads ranged from 2.49 to 3.11 kN, and average failure loads ranged from 4.58 to 5.87 kN. To determine the effect of velocity, these tests were performed statically and at impact speeds of 4.5 m/s and 9 m/s. The patellar damage pattern varied dramatically with speed. The impactors caused a clean punch-through of the patella during the static tests but multiple fractures or near total destruction of the patella occurred during the 4.5 m/s and 9 m/s impacts.

In all the preceding knee impact studies, the loading of the femur was intentionally through the patella and femoral condyles and resulted primarily in patellar, femoral, and/or pelvic fractures. If the knee is not flexed greater than 90 degrees or the loading is applied below or across the knee joint, damage to the knee ligaments and/or fractures of the tibia and fibula may result. Viano et al.[15] impacted seated cadavers on the anterior portion of the tibia, just below the knee joint. Knee ligament tearing and/or tibia–fibula fractures were produced by impactor forces ranging from 3.28 to 6.89 kN, with an average of 5.09 kN. In two tests, no damage was observed for peak loads of 4.87 and 5.74 kN. For eight impacts that spanned the knee joint (involving both the patella and tibia), knee joint damage was produced for impactor forces ranging from 5.91 to 8.36 kN, with an average of 7.02 kN. The predominant injury mode was avulsion of the posterior cruciate ligament from the tibial plateau.

Sled tests were conducted by Viano and Culver[16] in which shoulder-belted cadavers (without lap belts) impacted knee bolsters. The two below-the-knee leg impacts produced significant ligament tears at peak bolster loads of 3.5 kN and 4.2 kN per leg.

Low-speed, ligament tolerance tests were also conducted on five isolated knee joints mounted in a universal testing machine. In these tests, the knee joint angle was maintained at 90 degrees, while the tibia was displaced rearward relative to the femur until complete joint failure occurred. Loads corresponding to the initiation of joint failure ranged from 1.43 to 2.56 kN, with an average of 2.02 kN. The corresponding displacement of the tibia relative to the femur at the initiation of joint failure ranged from 9.5 mm to 30.0 mm, with an average of 14.4 mm. The load corresponding to complete joint failure ranged from 1.67 kN to 3.0 kN, with an average of 2.48 Kn.

The only extensive dynamic loading study of the lower leg was conducted by Kramer et al.[17] for the purpose of investigating pedestrian impact trauma. In this type of impact, transverse loading of the bones is predominant. Kramer conducted 209 transverse impacts against the lower legs of cadavers with cylindrical impactors. The axes of the cylinders were horizontal to simulate vehicle bumpers striking a standing pedestrian's leg. A 145 mm diameter impact cylinder produced a 50 percent frequency of fracture at a force of 4.31 kN, and a 216 mm diameter cylinder produced a corresponding force of 3.29 kN. These values are comparable to those recorded by Messerer and Weber (Tables 1 and 2), since the entire lower leg was impacted, not just the tibia.

SUMMARY

Early studies of the static mechanical failure of whole bones of the extremities still represent some of the most useful sources of information on extremity trauma, particularly for the upper extremities. Although dynamic loading studies produce failure information for a more realistic loading circumstance, such studies have not been conducted for a full range of bones of interest. An understanding of the effect of loading rate on bone properties can allow one to adjust the results of static tests to the dynamic situation, but only for similarly simple loading configurations. However, failure mechanisms of whole bones under complex dynamic loading conditions, such as the femur subjected to axial loading from knee impact, will require further study before comprehensive injury criteria for bones can be established.

REFERENCES

1. Messerer O: *Uber Elasticitat und Festigkeit der Menschlichen Knochen.* Stuttgart, J.G. Cotta, 1880.
2. Weber CO: *Chirurgische Erfahrungen and Untersuchungen.* Berlin, 1859 (cited by Messerer, 1880).
3. Yamada H: In Evans FG (ed): *Strength of Biological Materials.* Baltimore, Williams & Wilkins, 1970.
4. Mather BS: Observations of the effects of static and impact loading on the human femur. *J Biomechan* 1968; 1(4):331-335.
5. Cesari D, Ramet M, Bloch J: Influence of arm position on thoracic injuries in side impact, in *Proceedings of the Twenty-fifth Stapp Car Crash Conference.* Warrendale, Pa, Society of Automotive Engineers, 1981, pp 271–297.
6. Martens M, van Audekercke R, de Meester P, Mulier JC: The mechanical characteristics of the long bones of the lower extremity in torsional loading. - *J Biomechan* 1980; 13(8):667-676.
7. Patrick LM, Kroell CK, Mertz HJ Jr: Forces on the human body in simulated crashes, in *Proceedings of the Ninth Stapp Car Crash Conference.* Minneapolis, University of Minnesota, 1966, pp 237–259.
8. Patrick LM, Mertz HJ Jr, Kroell CK: Cadaver knee, chest and head impact loads, in *Proceedings of the Eleventh Stapp Car Crash Conference.* Society of Automotive Engineers, 1967, pp 106–117.
9. Powell WR, Ojala SJ, Advani SH, Martin RB: Cadaver femur responses to longitudinal impacts, in *Proceedings of the Nineteenth Stapp Car Crash Conference.* Warrendale, Pa, Society of Automotive Engineers, 1975, pp 561–579.

10. Melvin JW, Stalnaker RL, Alem NM, et al.: Impact response and tolerance of the lower extremities, in *Proceedings of the Nineteenth Stapp Car Crash Conference*. Warrendale, Pa, Society of Automotive Engineers, 1975, pp 543–559.
11. Melvin JW, Stalnaker RL: *Tolerance and Response of the Knee-Femur-Pelvis Complex to Axial Impact*. Final Report No. UM-HSRI-76-33. Ann Arbor, The University of Michigan, Highway Safety Research Institute, 1976.
12. Melvin JW, Nusholtz GS: *Tolerance and Response of the Knee-Femur-Pelvis Complex to Axial Impacts—Impact Sled Tests*. Final Report No. UM-HSRI-80-27. Ann Arbor, The University of Michigan, Highway Safety Research Institute, 1980.
13. Viano DC, Stalnaker RL: Mechanisms of femoral fracture. *J Biomech* 1980; 13(8):707–715.
14. Melvin JW, Fuller PM, Daniel RP, Pavliscak GM: *Human Head and Knee Tolerance to Localized Impacts*. SAE Paper No. 690477. New York, Society of Automotive Engineers, 1969.
15. Viano DC, Culver CC, Haut RC, et al.: Bolster impacts to the knee and tibia of human cadavers and an anthropometric dummy, in *Proceedings of the Twenty-second Stapp Car Crash Conference*. Warrendale, Pa, Society of Automotive Engineers, 1978, pp 401–428.
16. Viano DC, Culver CC: Performance of a shoulder belt and knee restraint in barrier crash simulations, in *Proceedings of the Twenty-third Stapp Car Crash Conference*. Warrendale, Pa, Society of Automotive Engineers, 1979, pp 105–131.
17. Kramer M, Burow K, Heger A: Fracture mechanism of lower legs under impact load, in *Proceedings of the Seventeenth Stapp Car Crash Conference*. Warrendale, Pa, Society of Automotive Engineers, 1973, pp 81–100.

Index

Page numbers followed by f indicate figures; followed by t indicate tables.